Edited by
Vikas Mittal

Thermoset Nanocomposites

Polymer Nano-, Micro- & Macrocomposites Series

Mittal, V. (ed.)

Surface Modification of Nanotube Fillers

Series: Polymer Nano-, Micro- & Macrocomposites (Volume 1)

2011
ISBN: 978-3-527-32878-9

Mittal, V. (ed.)

In-situ Synthesis of Polymer Nanocomposites

Series: Polymer Nano-, Micro- & Macrocomposites (Volume 2)

2012
ISBN: 978-3-527-32879-6

Mittal, V. (ed.)

Characterization Techniques for Polymer Nanocomposites

Series: Polymer Nano-, Micro- & Macrocomposites (Volume 3)

2012
ISBN: 978-3-527-33148-2

Mittal, V. (ed.)

Modeling and Prediction of Polymer Nanocomposite Properties

Series: Polymer Nano-, Micro- & Macrocomposites (Volume 4)

2013
ISBN: 978-3-527-33150-5

Related Titles

Mittal, V. (ed.)

Polymer Nanotubes Nanocomposites

Synthesis, Properties and Applications

2010
ISBN: 978-0-470-62592-7

Mittal, V. (ed.)

Miniemulsion Polymerization Technology

2010
ISBN: 978-0-470-62596-5

Mittal, V. (ed.)

Optimization of Polymer Nanocomposite Properties

2010
ISBN: 978-3-527-32521-4

Galimberti, M. (ed.)

Rubber-Clay Nanocomposites

Science, Technology, and Applications

2011
ISBN: 978-0-470-56210-9

Fink, J.

A Concise Introduction to Additives for Thermoplastic Polymers

2010
ISBN: 978-0-470-60955-2

Xanthos, M. (ed.)

Functional Fillers for Plastics

Second Edition

2010
ISBN: 978-3-527-32361-6

Thomas, S., Joseph, K., Malhotra, S.K., Goda, K., Sreekala, M.S. (ed.)

Polymer Composites

Volume 1

2012
ISBN: 978-3-527-32624-2

Edited by Vikas Mittal

Thermoset Nanocomposites

WILEY-VCH

WILEY-VCH Verlag GmbH & Co. KGaA

The Editor

Dr. Vikas Mittal
The Petroleum Institute
Chemical Engineering Department
Bu Hasa Building, Room 2204
Abu Dhabi
UAE

■ All books published by **Wiley-VCH Verlag GmbH** are carefully produced. Nevertheless, authors, editors, and publisher do not warrant the information contained in these books, including this book, to be free of errors. Readers are advised to keep in mind that statements, data, illustrations, procedural details or other items may inadvertently be inaccurate.

Library of Congress Card No.: applied for

British Library Cataloguing-in-Publication Data
A catalogue record for this book is available from the British Library.

Bibliographic information published by the Deutsche Nationalbibliothek
The Deutsche Nationalbibliothek lists this publication in the Deutsche Nationalbibliografie; detailed bibliographic data are available on the Internet at <http://dnb.d-nb.de>.

© 2013 Wiley-VCH Verlag GmbH & Co. KGaA, Boschstr. 12, 69469 Weinheim, Germany

All rights reserved (including those of translation into other languages). No part of this book may be reproduced in any form – by photoprinting, microfilm, or any other means – nor transmitted or translated into a machine language without written permission from the publishers. Registered names, trademarks, etc. used in this book, even when not specifically marked as such, are not to be considered unprotected by law.

Print ISBN: 978-3-527-33301-1
ePDF ISBN: 978-3-527-65967-8
ePub ISBN: 978-3-527-65966-1
Mobi ISBN: 978-3-527-65965-4
oBook ISBN: 978-3-527-65964-7
ISSN: 2191-0421

Cover Design Grafik-Design Schulz, Fußgönheim, Germany
Typesetting Toppan Best-set Premedia Limited, Hong Kong
Printing and Binding Markono Print Media Pte Ltd, Singapore

Printed on acid-free paper

Contents

Preface *IX*
List of Contributors *XI*

1 Epoxy–Vermiculite Nanocomposites *1*
Vikas Mittal
1.1 Introduction *1*
1.2 Experimental *2*
1.2.1 Materials *2*
1.2.2 H_2O_2 Treatment *3*
1.2.3 Milling and De-agglomeration *3*
1.2.4 Delamination *3*
1.2.5 Cation Exchange Capacity *4*
1.2.6 Ion Exchange and Nanocomposite Preparation *4*
1.2.7 Characterization of the Fillers and Composite Films *5*
1.3 Results and Discussion *6*
1.4 Conclusions *14*
Acknowledgment *14*
References *15*

2 Polymer Nanocomposites with UV-Cured Epoxies *17*
Roberta Bongiovanni, Muhammad Atif, and Marco Sangermano
2.1 Introduction *17*
2.2 Photopolymerization of Epoxides *20*
2.3 Limits in Curing Epoxy Composites by UV Irradiation *24*
2.4 Top-Down UV-Cured Epoxy Nanocomposites *25*
2.5 Bottom-Up UV-Cured Epoxy Nanocomposites *32*
2.6 Conclusions *35*
References *36*

3 Influence of Organic Modification and Polyurethane Structure on Clay Dispersion in Polyurethane–Clay Nanocomposites *39*
Shroff R. Mallikarjuna and Swaminathan Sivaram
3.1 Polymer Nanocomposites: An Introduction *39*

3.2 Polyurethane–Clay Nanocomposites 41
3.2.1 Polyurethane Products and Chemistry 41
3.2.2 Polyurethane Clay Composites 41
3.2.2.1 Thermoplastic Polyurethane Elastomers and their Composites 42
3.2.2.2 Thermoset Polyurethanes and Their Composites 42
3.2.2.3 Polyurethane Foams and Their Composites 43
3.2.2.4 Polyurethanes Dispersions and Their Composites 43
3.3 Influence of Organic Modification of Clay and Structure of PU on PU/Clay Nanocomposites Structure 43
3.3.1 Reactive and Nonreactive Modifiers 43
3.3.2 Effect of Nature of Modifier on Clay Dispersion in PU 44
3.3.3 Effect of Nature of Modifier on Properties of PU Nanocomposites 56
3.4 Conclusions 58
Glossary 59
References 59

4 Thermal Properties of Formaldehyde-Based Thermoset Nanocomposites 69
Byung-Dae Park
4.1 Introduction 69
4.2 Theoretical Background of Thermal Kinetics 72
4.2.1 Conventional Kinetics of Thermal Cure and Degradation 72
4.2.2 Theory of Temperature-Modulated DSC (TMDSC) 74
4.2.3 Theory of Temperature-Modulated TG (MTG) 75
4.3 Thermal Properties of Nanocomposites 76
4.3.1 Cure Kinetics of MF Resin/Clay/Cellulose Nanocomposites 76
4.3.2 Cure Kinetics of PF Resin/Clay/Cellulose Nanocomposites 80
4.3.2.1 Cure Kinetics by the Ozawa and Kissinger Methods 80
4.3.2.2 Cure Kinetics by the Isoconversional Method 83
4.3.2.3 Cure Kinetics by TMDSC 84
4.3.3 Thermal Degradation Kinetics of PF Resin/MWCNT/Cellulose Nanocomposites 88
4.3.3.1 FT-IR Analysis of Surface-Modified MWCNTs 88
4.3.3.2 Thermal Degradation Kinetics Analysis of Conventional Thermogravimetry 90
4.3.3.3 Isoconversional Analysis of CTG 94
4.3.3.4 Thermal Kinetic Analysis of MTG 95
4.3.4 Dynamic Mechanical Analysis of PF Resin/MWCNT/Cellulose Nanocomposites 98
4.4 Mechanical Properties of the Nanocomposites 100
4.5 Summary 103
Acknowledgment 105
References 105

5	**Mechanical Performance of Thermoset Clay Nanocomposites** *109*
	Nourredine Aït Hocine, Said Seghar, Hanaya Hassan, and Saïd Azem
5.1	Introduction *109*
5.2	Viscoelasticity Analysis: Dynamical Mechanical Thermal Analysis (DMTA) *111*
5.3	Rigidity–Young's Modulus *114*
5.4	Strain at Break *118*
5.5	Stress at Break–Fracture Toughness *120*
5.6	Conclusion *124*
	References *124*

6	**Unsaturated Polyester Resin Clay Hybrid Nanocomposites** *129*
	Kanniyan Dinakaran, Subramani Deveraju, and Muthukaruppan Alagar
6.1	Introduction *129*
6.2	Reinforced Unsaturated Polyester Composites *130*
6.3	Clay Minerals *131*
6.3.1	Layered Structure *131*
6.3.2	General Characteristics *132*
6.3.3	Surface Modification *133*
6.3.4	Processing and Characterization *134*
6.4	Mechanical and Thermal Properties of Clay–UP Nanocopomposites *135*
6.5	Flame Retardance *141*
6.6	Bio-Derived Unsaturated Polyester–Clay Nanocomposites *144*
	References *145*

7	**Hyperbranched Polymers as Clay Surface Modifications for Nanocomposites** *147*
	Teresa Corrales, Fernando Catalina, Iñigo Larraza, Gema Marcelo, and Concepción Abrusci
7.1	Introduction *147*
7.2	Hyperbranched Polymers for Antimicrobial Surface *149*
7.3	Hyperbranched Polymers on Adsorbents for Cr(VI) Water Treatment *157*
	Acknowledgment *162*
	References *162*

8	**New Methods for the Preparation of Metal and Clay Thermoset Nanocomposites** *165*
	Kübra Doğan Demir, Manolya Kukut, Mehmet Atilla Tasdelen, and Yusuf Yagci
8.1	Introduction *165*
8.2	Thermoset Nanocomposites Based on Nanoclays *166*
8.3	Thermoset Nanocomposites Based on Metal Nanoparticles *176*

8.3.1	Polymer–Silver Nanocomposites	177
8.3.2	Polymer–Gold Nanocomposites	181
8.4	Concluding Remarks	184
	References	185

9 Bio-Based Epoxy Resin/Clay Nanocomposites 189
Mitsuhiro Shibata

9.1	Introduction	189
9.2	Bio-Based Epoxy Resins and Hardeners	191
9.2.1	Bio-Based Epoxy Resins	191
9.2.2	Bio-Based Hardeners	196
9.2.3	Properties of Cured Bio-Based Epoxy Resins	197
9.3	Bio-Based Epoxy Resins/Clay Nanocomposites	200
9.3.1	Nanocomposites Based on Polyglycerol Polyglycidyl Ether	200
9.3.2	Nanocomposites Based on Epoxidized Vegetable Oil	204
9.4	Conclusion	206
	References	206

10 Electrical Properties and Electromagnetic Interference Shielding Response of Electrically Conducting Thermosetting Nanocomposites 211
Parveen Saini

10.1	Introduction	211
10.2	EMI Shield and Shielding Effectiveness	212
10.2.1	Shielding Effectiveness: Theoretical Expressions	212
10.2.1.1	Shielding Due to Reflection	212
10.2.1.2	Shielding Due to Absorption	213
10.2.1.3	Shielding Due to MIRs	214
10.2.2	Shielding Effectiveness: Experimental Details	214
10.2.3	Materials for EMI Shielding: Polymer-Based Nanocomposites	215
10.2.4	Synthesis of Thermosetting Nanocomposites	216
10.2.5	Electrical Properties of Thermosetting Nanocomposites	217
10.2.5.1	ICP-Based Nanocomposites	218
10.2.5.2	CNT-Based Nanocomposites	223
10.2.5.3	Graphene-Based Nanocomposites	229
10.3	Conclusions	231
	Acknowledgments	232
	References	232

Index 239

Preface

Thermoset polymers are an important class of materials with many superior properties as compared with thermoplastic materials. Generation of thermoset polymer nanocomposites by the incorporation of layered silicates and other nanofillers in the polymer matrices has led to enhancement of the property profiles of the thermoset materials significantly. Nanocomposites with a large variety of thermoset polymers have been explored, and vast knowledge on the synthesis methodologies as well as properties has been generated. The goal of the book is to assimilate these research findings on many thermoset polymer-based nanocomposite systems comprehensively so as to generate better insights into the design, performance, and optimization of thermoset nanocomposites.

Chapter 1 reports the synthesis and properties of epoxy vermiculite nanocomposites. Vermiculite platelets were surface modified to enhance interfacial interactions with the polymer, and the nanocomposites were generated using *in situ* polymerization method. Interesting insights into the permeation properties of the nanocomposites have been reported. Chapter 2 presents photoinduced polymerization as an efficient technique for rapid formation of highly cross-linked networks from liquid epoxy resins. The reaction involves a cationic mechanism and is usually initiated by UV light. Chapter 3 reports the synthesis and properties of polyurethane (PU) nanocomposites, and it was observed that the key to superior properties of nanocomposites is critically dependent on the choice of the organic modifier used to modify the surface of clay as well as the nature of the polymer itself. Chapter 4 reviews recent progresses on thermal properties of formaldehyde-based thermoset/cellulose nanocomposites such as melamine-formaldehyde (MF) resin/clay/cellulose, phenol-formaldehyde (PF) resin/clay/cellulose, or PF resin/multiwalled carbon nanotube (MWCNT)/cellulose nanocomposites, particularly by focusing on thermal curing or degradation kinetics of these nanocomposites. Chapter 5 reports a review of mechanical properties of thermoset polymer nanocomposites. It is reported that the nanoclay particles provide to thermosets interesting mechanical properties when the constituents and the processing parameters are adequately selected. These properties are highly improved in the case of rubbery matrix than in glassy matrix. Chapter 6 demonstrates the unsaturated polyester clay nanocomposite systems. These composites are of high importance as the unsaturated polyester resins are the most widely used thermoset matrix resin in the coatings and composites industry, and constitutes about three-fourth of the total resins used. In Chapter 7, recent advances on applications of hyperbranched polymers as clay surface modifications are presented, with particular reference to the preparation of antimicrobial surface and adsorbents for Cr (VI) water treatment. Chapter 8 focuses on new methods to generate metal- and clay-reinforced nanocomposites. The development of *in situ* methods has clearly facilitated the major advances in the synthesis in a one-pot manner combining polymerization processes leading to the network formation with intercalation/exfoliation or nanoparticle formation, respectively. In Chapter 9, the preparation and properties of the bio-based epoxy resin/hardener/layered silicate nanocomposites are described. The replacement of petroleum-based epoxy resin/hardener with bio-based epoxy resin/hardener is very important from the viewpoint of the conservation of limited petroleum resources and the protection of global environment. In Chapter 10, comprehensive account of electrical and electromagnetic interference (EMI) shielding properties of thermosetting nanocomposites has been provided with special reference to those based on conducting additives like intrinsically conducting polymers and carbonaceous fillers like carbon nanotubes and graphene.

Abu Dhabi *Vikas Mittal*
June 2013

List of Contributors

Concepción Abrusci
Universidad Autónoma de
Madrid-UAM
Departamento de Biología Molecular
Facultad de Ciencias
Cantoblanco
28049 Madrid
Spain

Nourredine Aït Hocine
ENI Val de Loire
Laboratoire de Mécanique et
Rhéologie (LMR)
3, rue de la Chocolaterie
BP 3410, 41034 Blois Cedex
France

Muthukaruppan Alagar
Anna University
Department of Chemical Engineering
AC-Tech Campus
Chennai 600025
India

Muhammad Atif
Politecnico di Torino
Department of Applied Science and
Technology
Corso Duca degli Abruzzi 24
10129 Torino
Italy

Saïd Azem
Université Mouloud Mammeri de
Tizi-Ouzou (UMMTO)
Laboratoire d'Elaboration et de
Caractérisation des Matériaux et
Modélisation (LEC2M)
15000 Tizi-Ouzou
Algeria

Roberta Bongiovanni
Politecnico di Torino
Department of Applied Science and
Technology
Corso Duca degli Abruzzi 24
10129 Torino
Italy

Fernando Catalina
Instituto de Ciencia y Tecnología de
Polímeros
Polymer Photochemistry Group
C.S.I.C. Juan de la Cierva 3
28006 Madrid
Spain

Teresa Corrales
Instituto de Ciencia y Tecnología de
Polímeros
Polymer Photochemistry Group
C.S.I.C. Juan de la Cierva 3
28006 Madrid
Spain

List of Contributors

Kübra Doğan Demir
Istanbul Technical University
Faculty of Science and Letters
Chemistry Department
Maslak
34469 Istanbul
Turkey

Subramani Deveraju
Anna University
Department of Chemical Engineering
AC-Tech Campus
Chennai 600025
India

Kanniyan Dinakaran
Anna University
Department of Chemistry
MIT Campus
Chennai 600044
India

Hanaya Hassan
Centre d'Etude et de Recherche sur les Matériaux Elastomères (CERMEL)
LMR
29, rue des Martyrs
37300 Joué-lès-Tours
France

Manolya Kukut
Istanbul Technical University
Faculty of Science and Letters
Chemistry Department
Maslak
34469 Istanbul
Turkey

Iñigo Larraza
Instituto de Ciencia y Tecnología de Polímeros
Polymer Photochemistry Group
C.S.I.C. Juan de la Cierva 3
28006 Madrid
Spain

Shroff R. Mallikarjuna
National Chemical Laboratory
Division of Polymer Science and Engineering
Dr. Homi Bhabha Road
Pune 411 008
India

Gema Marcelo
Instituto de Ciencia y Tecnología de Polímeros
Polymer Photochemistry Group
C.S.I.C. Juan de la Cierva 3
28006 Madrid
Spain

Vikas Mittal
The Petroleum Institute
Department of Chemical Engineering
Room 2204, Bu Hasa Building
Abu Dhabi 2533
UAE

Byung-Dae Park
Kyungpook National University
Department of Wood Science and Technology
Daegu 702-201
Republic of Korea

Parveen Saini
CSIR-National Physical Laboratory
Conducting Polymers Group
Polymeric and Soft Materials Section
New Rajender Nagar
New Delhi 110 012
India

Marco Sangermano
Politecnico di Torino
Department of Applied Science and Technology
Corso Duca degli Abruzzi 24
10129 Torino
Italy

Said Seghar
Centre d'Etude et de Recherche sur les
Matériaux Elastomères (CERMEL)
LMR
29, rue des Martyrs
37300 Joué-lès-Tours
France

Mitsuhiro Shibata
Chiba Institute of Technology
Department of Life and Environmental
Sciences
Faculty of Engineering
2-17-1, Tsudanuma
Narashino
Chiba 275-0016
Japan

Swaminathan Sivaram
National Chemical Laboratory
Division of Polymer Science and
Engineering
Dr. Homi Bhabha Road
Pune 411 008
India

Mehmet Atilla Tasdelen
Yalova University
Faculty of Engineering
Department of Polymer Engineering
77100 Yalova
Turkey

Yusuf Yagci
Istanbul Technical University
Faculty of Science and Letters
Chemistry Department
Maslak
34469 Istanbul
Turkey

and

King Abdulaziz University
Faculty of Science
Chemistry Department
Jeddah 21589
Saudi Arabia

1
Epoxy–Vermiculite Nanocomposites
Vikas Mittal

1.1
Introduction

Epoxies form a special class of thermosetting polymeric materials having high thermal and environmental stability. They are well known as creep-resistant materials with very high stiffness properties [1–3]. Owing to these properties, a wide spectrum of epoxy applications is available, which includes the use of epoxies as adhesives, coatings, printed circuit boards, electrical insulators, and so on. One of the major areas where epoxy adhesives find tremendous use are packaging laminates where their sole use is to hold together the various polymeric foils used in these commercial packaging laminates. To save the material costs, an overall decrease in the thickness of the packaging laminate can be achieved if the adhesive can also be made to contribute to the properties required for a packaging material apart from its function of being an adhesive. The common properties required being the permeation barrier, mechanical performance, transparency, suitability for food contact applications, ease of printability, and so on. Permeation barrier to oxygen and water vapor form the most important property needed in the packaging materials. This can be achieved by altering the polymer network structure obtained by crosslinking of the epoxide groups with amines or other crosslinking agents [4, 5]. The use of epoxy polymer with stiff rod-like units in the backbone can help to enhance the required properties. The other alternative includes the incorporation of inorganic fillers in the polymer matrix, this approach being easier to monitor and control. As the filler shape, size, and interfacial interactions affect the polymer properties greatly, organically treated plate-like inorganic aluminosilicate particles can be incorporated in the polymer matrix to achieve polymer nanocomposites for improvement in barrier performance. By incorporating impermeable, transparent, plate-like nanoparticles in the polymer matrix, the permeating molecules are forced to wiggle around them in a random walk, hence diffusing through a tortuous pathway [6–8]. Besides, the decrease in transmission rate of the permeant is a function of the aspect ratio of the inclusions, their volume fraction, and orientation.

The synthesis of epoxy–clay nanocomposites has been extensively studied; however, majority of these studies focused on enhancing the mechanical properties

Thermoset Nanocomposites, First Edition. Edited by Vikas Mittal.
© 2013 Wiley-VCH Verlag GmbH & Co. KGaA. Published 2013 by Wiley-VCH Verlag GmbH & Co. KGaA.

with the incorporation of organically modified fillers [9–17], thus largely neglecting the permeation properties. Only a few recent studies have discussed these properties in detail [18–23]. Apart from that, montmorillonite has been the most commonly used aluminosilicate in these studies. Owing to the low charge density (0.25–0.5 equiv mol^{-1}), a larger area per cation is available on the surface, which leads to a lower basal plane spacing in the clay after surface ion exchange with alkyl ammonium ions. On the other hand, minerals with high charge density (1 equiv mol^{-1}), such as mica, and hence subsequent smaller area per cation, do not swell in water and thus do not allow the cation exchange. However, aluminosilicates with medium charge densities of 0.5–0.8 equiv mol^{-1}, such as vermiculite, offer a potential of partial swelling in water and cation exchange, which can lead to a much higher basal plane spacing in the modified mineral if optimum ion exchange is achieved. In the pristine state, vermiculite particles are composed of stacks of negatively charged 2:1 aluminosilicate layers (ca. 0.95 nm thick) with one octahedral sheet sandwiched between two opposing tetrahedral sheets and the resuting regular gap in between (interlayer). The chemical constitution of its unit cell is $(Mg,Al,Fe)_3(Al,Si)_4O_{10}(OH)_2Mg_x(H_2O)_n$ [24, 25]. Due to isomorphic substitutions in the lattice, the layers have permanent negative charges that are compensated mainly by hydrated Mg^{2+} as interlayer cations. Owing to the higher basal plane spacing in the modified mineral, the electrostatic interactions holding the layers together can be expected to be lower than similar montmorillonite counterparts thus increasing the potential of better properties of the hybrid nanocomposites.

The goal of this investigation was to synthesize epoxy–vermiculite nanocomposite coatings and to study their microstructure development as well as their oxygen and water vapor barrier properties in comparison with already reported epoxy–montmorillonite system [18]. Vermiculite platelets modified with two different ammonium ions were prepared for the purpose. The epoxy matrix and the curing agent were chosen to achieve polymer matrix, which meets the requirements of the food and health regulations and has low gas permeability on its own. The nanocomposite coatings were drawn on polyamide and polypropylene substrates and the curing temperatures were kept low in order to avoid the thermal damage to these substrate foils.

1.2
Experimental

1.2.1
Materials

The epoxy resin, bisphenol A diglycidyl ether (4,4'-isopropylidenediphenol diglycidyl ether) with an epoxide equivalent weight 172–176, was supplied by Sigma (Buchs, Switzerland). Tetraethylenepentamine (TEPA) and tetrahydrofuran (THF) were procured from Fluka (Buchs, Switzerland). Benzyldimethylhexadecylammonium chloride (BzC16) was purchased from Acros Organics (Basel, Switzerland). Corona-treated substrate polypropylene (100 μm thick) and polyamide (15 μm thick) foils were supplied by Alcan Packaging (Neuhausen, Switzerland). A surfactant (trade name BYK-307) was used to achieve better wetting and adherence of the neat epoxy coating to the substrate foils and was obtained from Christ Chemie (Reinach, Switzerland). Russian vermiculite with a chemical composition of $(Mg,Al,Fe)_3(Al,Si)_4O_{10}(OH)_2Mg_x(H_2O)_n$ was obtained from Thermax, Greinsfurth, Austria. Benzyl(2-hydroxyethyl)methyloctadecylammonium chloride (BzC18OH) was synthesized by quaternizing the corresponding amine with 1-bromooctadecane as reported earlier [18].

1.2.2
H_2O_2 Treatment

To partially exfoliate the pristine vermiculite, H_2O_2 treatment was performed. Twenty grams of pristine vermiculite mineral was taken in a 2000-ml beaker and 135 ml of H_2O_2 was added to it. The mixture was stirred (glass stirrer) for 5 min at room temperature followed by heating at 50 °C for 10–15 min until the mineral started to foam. The contents were then stirred using a glass rod. The stirring was continued for 15 min with addition of small amounts of water in between to ease the stirring. After 15 min, beaker was removed from the hot plate and was stirred until the mixture stopped foaming. Once the mixture was cooled, it was filtered, washed with water, and finally with methanol and dried under vacuum.

1.2.3
Milling and De-agglomeration

The minerals were wet ground to the desired particle diameter (partially spherical stacks with average particle size of roughly 5 μm) in the wet grinding mill. The grinding operation was performed keeping in mind that only the thickness of the stacks is reduced. For the milling operation, approximately 100 g of the mineral was placed inside the mill. As much as 85 ml of water was subsequently added to the mineral and a homogeneous slurry was generated. It was then milled for 1 h in the milling chamber. The wet slurry was filtered and subsequently washed with methanol. The material was then dried at 70 °C under reduced pressure. The dried powder was sieved in a 160-μm sieve. Particles were also occasionally de-agglomerated by shearing them mechanically in a Teflon vessel containing zirconium oxide beads.

1.2.4
Delamination

In the as-supplied form, vermiculite contained Mg^{2+} ions on the surface. To achieve delamination, 15 g of milled vermiculite was refluxed in 4 M NaCl solution at 170 °C for 12 days followed by filtration and washing with water [26]. The delamination reaction was repeated twice. The resulting Na–vermiculite was dried at 80 °C under reduced pressure.

1.2.5
Cation Exchange Capacity

The cation exchange capacity (CEC) of vermiculite minerals before, during, and after the delamination process was determined as reported earlier [27]. To measure the CEC, 50 g of 5 mM copper sulfate solution was added to 52.5 g of 5 mM triethylenetetramine solution to generate $Cu(trien)^{2+}$ solution. The extinction coefficient (~4.84) of the solution at 255 nm wavelength was measured on a Cary 1E spectrometer (Varian, Palo Alto, CA). Approximately 20–30 mg of the vermiculite mineral was separately suspended (shaking and sonication) in 15 g of water in a polypropylene vial. To this suspension, 12 g of the $Cu(trien)^{2+}$ solution was added and the mixture shaken for 30 min. The suspension was then centrifuged and the supernatant liquid filtered through a cellulose acetate filter with 0.45 µm pore diameter. The concentration of $Cu(trien)^{2+}$ in the filtrate was measured photometrically and the amount of exchanged ions was calculated.

1.2.6
Ion Exchange and Nanocomposite Preparation

The mineral surface was rendered organophilic by exchanging its surface inorganic Na^+ cations with benzyldimethylhexadecylammonium chloride (BzC16) and benzyl(2-hydroxyethyl)methyloctadecylammonium chloride (BzC18OH). To a heated (70 °C) dispersion of unmodified clay in water and ethanol, a solution of ammonium salt corresponding to 110% of the CEC of the mineral in ethanol was added dropwise and was stirred overnight. The modified clay was filtered and washed repeatedly with hot water–ethanol mixture. For nanocomposite synthesis, the required amounts of modified vermiculite and epoxy resin were calculated on the basis of the desired inorganic volume fraction as reported earlier:

$$M_{OV} = M_V + (M_V CEC M_{OC})$$

$$M_{EP} = [(M_V V_{EP} \rho_{EP})/(V_V \rho_V)] - (M_V CEC M_{OC})$$

where M_{OV} is the mass of the modified vermiculite, M_V is the mass of the inorganic aluminosilicate, M_{OC} is the molar mass of the ammonium ion used to modify the vermiculite surface, M_{EP} is the mass of the epoxy resin, V_V is the inorganic volume fraction, ρ_V is the density of sodium vermiculite (2.6 g cm^{-3}), V_{EP} is the epoxy volume fraction, and ρ_{EP} is its density (1.18 g cm^{-3}).

For the composite synthesis, necessary amount of modified montmorillonite was swollen in THF for 2 h followed by sonication (ultrasound horn). The epoxy resin solution was then mixed with the suspension and sonicated. The curing agent, that is, TEPA, was then added and the amine to epoxy mole ratio was maintained at 0.3 : 1. The nanocomposite films were drawn on the corona-treated surface of PP and PA foils with the help of a bar coater (90-µm gap). The coated films were dried at room temperature for 15 min and under reduced pressure for another 15 min, then cured at 70 °C overnight, and postcured at 90 °C for 4 h. It is worth mentioning here that no sedimentation of the inorganic phase was observed for the time scale of the film preparation. Dry-coated films in the thickness range of ca. 10 µm were achieved, and the correct thickness was determined by weighing the samples in air and in ethanol using an analytical balance (Mettler AE 200) and a homemade device similar to the Mettler density kit ME-33360.

1.2.7
Characterization of the Fillers and Composite Films

The mean particle size was measured by light scattering using Malvern mastersizer. As much as 0.1 g of mineral was suspended in 15 ml of deionized water by shaking and sonication. Polydisperse sample analysis mode was employed and particle density of 2.7 g cm^{-3} was used. Laser beam length of 2.4 mm was selected. The sample presentation unit was filled with about 1000 ml of degassed water. Mineral slurry was then added to the sample presentation unit. The slurry was sonicated again for 10 min before performing the light scattering measurements. Apart from mean diameter $D_{0.5}$, aspect ratio of the minerals (largest dimension to the shortest dimension) was also recorded. It should however be noted that these values are only average values as the particle shapes and sizes had a large distribution. The particle size analyzer visualized the mineral particles as spherical in order to calculate the effective mean diameter. Thus, the calculated mean sizes have contributions from length, breadth, and thickness of the particles.

High-resolution (Hi-Res) thermogravimetric analysis (TGA) of the modified clays, in which the heating rate is coupled to the mass loss, that is, the sample temperature is not raised until the mass loss at a particular temperature is completed, was performed on a Q500 thermogravimetric analyzer (TA Instruments, New Castle, DE). All measurements were carried out under an air stream in the temperature range 50–900 °C.

The oxygen (23 °C and 0% RH) and water vapor (23 °C and 100% RH) transmission rate through substrate foils coated with the neat epoxy and the nanocomposites was measured using OX-TRAN 2/20 and PERMATRAN-W 3/31 (Mocon, Minneapolis, MN), respectively. The PP and PA substrates were selected on the basis of their high transmission rate for the specific permeant so that they do not hinder measuring the permeation through the coated film. The transmission rate was normalized with respect to thickness and an average of five values was reported.

Wide-angle X-ray diffraction (WAXRD) patterns were collected on a Scintag XDS 2000 diffractometer (Scintag Inc., Cupertino, CA) using CuK$_\alpha$ radiation (λ = 0.15406 nm) in reflection mode. The instrument was equipped with a graphite monochromator and an intrinsic germanium solid-state detector. The samples were step scanned (step width 0.02° 2θ, scanning rate 0.06° min^{-1}) at room temperature from 1.5 to 15° 2θ. The (001) basal-plane reflection of an internal standard muscovite (very thin platelets, 2θ = 8.84°) was used to calibrate the line position of the reflections.

The microstructure of the filler in the composite films was studied by brightfield TEM using a Zeiss EM 912 Omega (Leo, BRD) microscope. Small pieces of coated foils etched with oxygen plasma were embedded in an epoxy matrix (Epon 812 + Durcopan ACM 3:4, Fluka, Buchs, Switzerland) and 50- to 100-nm-thick sections were carefully microtomed with a diamond knife (Reichert Jung Ultracut E). The sections were supported on 100 mesh grids sputter-coated with a 3-nm-thick carbon layer.

1.3
Results and Discussion

On the one hand, as the lateral dimension of the montmorillonite is limited by the weathering and purification operations, the resulting aspect ratio is also limited to lower values. On the other hand, if vermiculite is optimally modified and hence is made to exfoliate in the polymer matrix, a much higher aspect ratio can be generated. As the permeation properties are significantly affected by exfoliated platelets owing to the enhancement of tortuity in the permeant random path through the membrane, better barrier performance can be expected in the vermiculite system. However, the presence of a greater number of cations in a specific area as compared to montmorillonite also makes it difficult to swell and modify and the platelets are held together by much stronger electrostatic forces. Therefore, only a combined effect of these factors can be expected on the final composite properties.

Table 1.1 shows the mean size ($D_{0.5}$) and cation exchange capacity values of the pristine vermiculite along with mineral after milling, de-agglomeration, H_2O_2

Table 1.1 Mean size, aspect ratio, and cation exchange capacity of vermiculite samples after milling (1 h), de-agglomeration, and delamination.

	$D_{0.5}$ (μm)	Aspect ratio	CEC (μeq g^{-1})
Pristine vermiculite			150
Vermiculite milled, fraction <160 μm	36.87	5.21	240
Vermiculite milled, fraction <160 μm, de-agglomerated	23.59 ($D_{0.9}$ 77.99 μm)		242
Vermiculite milled, fraction 160–250 μm	92.87	6.48	
Vermiculite milled, fraction 160–250 μm, re-milled	29.84 ($D_{0.9}$ 101.74 μm) for fraction <160 μm		
Vermiculite milled, fraction <160 μm, H_2O_2 treatment	35.5		300
Vermiculite milled, fraction >160 μm, H_2O_2 treatment	78.23		200
Delaminated vermiculite	34.11	7.5	1440

treatment as well as delamination. The as-received vermiculite mineral had a low CEC value of 150 µeq g^{-1}. The milled vermiculite was sieved and the fraction with <160 µm size was observed to have a mean diameter of 36.9 µm and a CEC of 240 µeq g^{-1}, which was higher than the pristine mineral. However, after de-agglomeration of the mineral, the size was further reduced to 23.6 µm, but a similar CEC value of 242 µeq g^{-1} was observed, which indicated that the de-agglomeration process was not effective. The $D_{0.9}$ value of 78 µm also indicated the broad distribution in particle size. The fraction in the size range of 160–250 µm had a mean diameter of 92.9 µm. A rough estimate of the aspect ratio was also generated with the particle size analyzer, and the vermiculite fraction with <160 µm fraction had an aspect ratio of 5.21 as compared to 6.48 for the fraction >160 µm. Though the higher sized fraction had slightly higher aspect ratio, it was however a result of higher tactoid or stack thickness as well as higher diameter in the mineral. It should also be noted that due to large distribution in the size of the particles, a distribution in the aspect ratio also existed. The fraction in the range of 160–250 µm was re-milled and an average size of 29.9 µm was measured. A much higher diameter $D_{0.9}$ of 101.8 µm for this fraction indicated that the particle size distribution was still broad. In another trial, the hydrogen peroxide-treated vermiculite was milled and a mean size of 35.5 µm was observed for the fraction <160 µm. A cation exchange capacity value of 300 µeq g^{-1} was obtained for this fraction, indicating that the H_2O_2 treatment of vermiculite prior to milling improved the CEC further as compared to both the pristine vermiculite flakes and the milled vermiculite without H_2O_2 treatment. The fraction above >160 µm had mean size of 78.2 µm and its CEC value of 200 µeq g^{-1} was higher than the pristine mineral. The mean size and aspect ratio values of the particles decreased when the milling time was enhanced to 2 h, which indicated that the lateral dimensions of the minerals were impacted by the longer milling periods.

The findings from Table 1.1 confirm that both milling type and time affected the overall dimensions of the mineral particles. Though the mineral particles have average dimensions in the µm scale and various milling and sieving steps lead to minor changes in the overall size, it should be noted that these particles form precursors to the high aspect ratio primary platelets, thus their handling during the milling and de-agglomeration processes has immense significance. Reduction of lateral dimensions signifies reduced aspect ratio, which does not lead to the optimal enhancements in composite properties when such minerals are incorporated into the polymer matrices. It has been often reported that mechanical and permeation properties of the materials are strongly impacted by filler volume fraction as well as aspect ratio. In fact, only exfoliated platelets were reported to contribute to the barrier performance of the polymer-layered silicate nanocomposites [18]. Pretreatment of the pristine minerals (e.g., H_2O_2 treatment) provided additional advantages in achieving the goal of higher aspect ratio minerals. Enhancement of the CEC of the minerals after treatment suggested that higher extent of mineral surface was exposed due to penetration of the H_2O_2 molecules into the mineral interlayers and possibly reduction in the stack thickness as a consequence. Such treatments are also advantageous to remove any impurities present in the minerals.

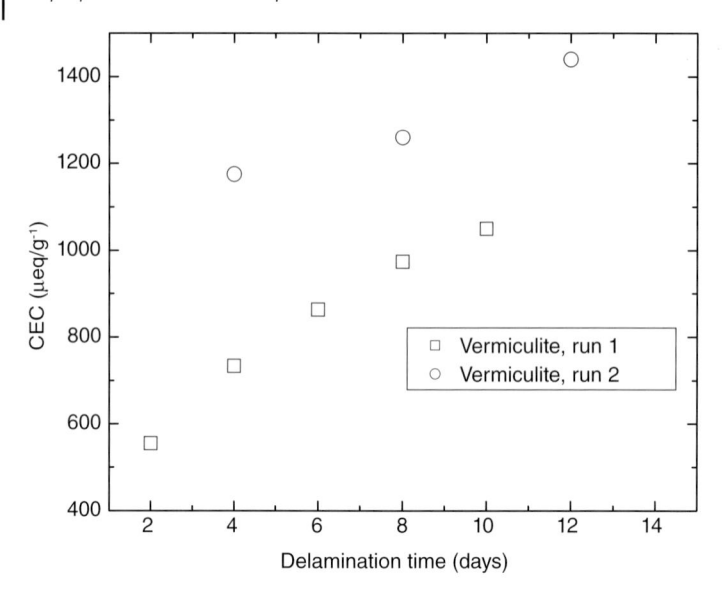

Figure 1.1 CEC values of vermiculite as a function of delamination time with NaCl and LiNO$_3$, respectively.

Figure 1.1 demonstrates the evolution of CEC of vermiculite as a function of number of delamination cycles and delamination time. Vermiculite used for delamination was <160 μm sample, which was treated with H$_2$O$_2$ before milling. The first 10 days of delamination of vermiculite with NaCl allowed partial exchange of the Mg^{2+} ions with Na$^+$ ions generating Na–vermiculite with a CEC of 1050 μeq g^{-1}. Second delamination cycle for further 12 days increased the CEC relatively slowly to 1440 μeq g^{-1}. It signified nearly a tenfold increase in CEC as compared to the pristine mineral. The mean size and aspect ratio of the delaminated minerals in Table 1.1 were similar to the milled materials indicating that although the interlayers were swollen and access to exchangeable cations was enhanced in the delamination process, loose stacks of platelets may still exist. It has also been reported that during the cleaving process, the particle diameter also reduced slightly along with the thickness, thus, justifying the observed values of mean particle size. However, it also confirmed that the mean size of the starting materials was suitable as it allowed the penetration of the NaCl and LiNO$_3$ in the interlayers during the delamination process. Though a partial access was gained to the available surface cations after the cleavage process, the minerals obtained were readily dispersible in water and their enhanced cation exchange capacity was suitable to carry out the surface modification processes to render them compatible with organic polymer matrices.

To achieve compatibility of the mineral and to reduce the surface energy, the sodium ions were exchanged with benzyldimethylhexadecylammonium chloride (BzC16) and benzyl(2-hydroxyethyl)methyloctadecylammonium chloride (BzC18OH).

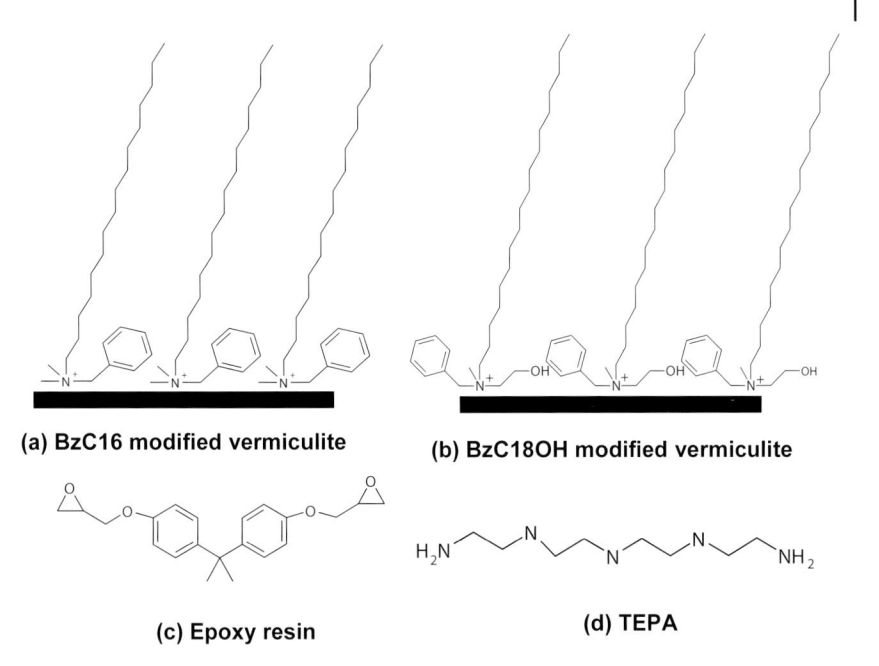

Figure 1.2 (a) and (b) Schematic of modified vermiculites and the chemical architecture of ammonium ions; (c) and (d) chemical structure of the epoxy resin and the crosslinker tetraethylenepentamine.

The two ammonium ions were chosen in order to analyze the effect of different chemical architecture of the ammonium ions on the final composite properties. Depending on the interactions of the swelling solvent and the epoxy prepolymer with the ammonium ion, more chemical interaction of the epoxide groups with the hydroxyl groups present in the ammonium ions can be expected, thus leading to chemical grafting of the epoxy chains on the surface. Similarly, benzyl group present on the surface is also expected to aid in generating stronger van der Waals attraction forces with the epoxy polymer. Figure 1.2 shows the schematic of the modified vermiculites as well as the chemical structures of the system constituents. It is also worth mentioning here that other modifications such as benzyldibutyl(2hydroxyethyl)ammonium chloride, which have been found to significantly exfoliate the montmorillonite clay platelets in the epoxy matrix, were also tried to be exchanged on the vermiculite surface. However, no exchange could be achieved even when the system was constantly stirred for 2 days or higher amounts of ammonium salts corresponding to the mineral CEC were used.

The presence of a local bilayer during the cation exchange was avoided as the presence of excess unattached modifier molecules can negatively interact with the epoxy prepolymer, thus inducing system instability and subsequent loss of composite properties. Figure 1.3 shows the TGA thermograms of the modified vermiculites. Absence of any low-temperature degradation peak in the thermograms

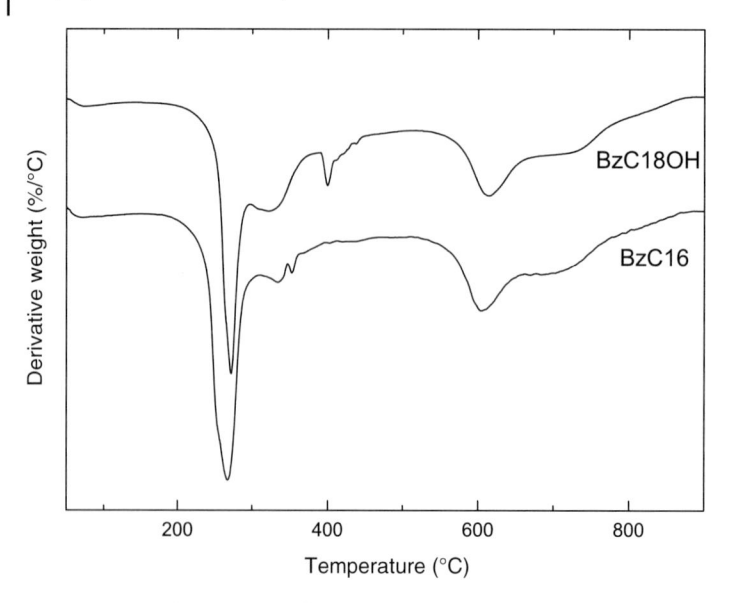

Figure 1.3 TGA thermograms of the surface-modified vermiculites.

Table 1.2 Density of the organo-vermiculites and their 3.5 vol% composites.

Modification	Modified filler Density (g cm^{-3})	Filler weight Fraction (%)	Composite density (g cm^{-3})
BzC18OH	1.43	11.33	1.20
BzC16	1.53	10.80	1.21

indicates that the excess of the modifier molecules could be successfully washed off. However, it was also important to achieve satisfactory extent of surface ion exchange in order to completely organophilize the mineral. Exchange of >95% of the surface ions was achieved in the present system. As the higher CEC of the mineral leads to the presence of a large number of organic chains on the surface leading to a lower density of the modified vermiculite as compared to the unmodified state, therefore these values should be taken into account for calculating the true volume fraction of the filler. Table 1.2 details these values for the modified filler and the composites. The inorganic volume fraction of 3.5% was used for composites as it was found to be an optimum value for achieving the enhancement in the permeation properties.

The basal plane spacing in the nonmodified vermiculite was found to be 1.22 nm, which was enhanced to 3.25 nm for BzC16-modified vermiculite and 3.40 nm for BzC18OH-modified vermiculite. The absence of the peak corresponding to 1.22 nm

1.3 Results and Discussion

Table 1.3 Basal plane spacing of the fillers, their suspensions, and epoxy composites.

Filler	d-spacing filler powder (nm)	d-spacing of filler suspended in DMF (nm)	d-spacing of filler suspended in DMF + epoxy (nm)	d-spacing in epoxy composite (nm)
Na–vermiculite	1.22	1.42	1.42	1.29
BzC18OH	3.40	3.80	3.80	3.96
BzC16	3.25	3.34	3.53	3.68

in the diffractograms of the modified minerals also confirmed that the surface cations were fully exchanged during the exchange process. In contrast, montmorillonite modified with BzC16 and BzC18OH had, respectively, the basal plane spacing of 1.87 and 2.06 nm [18]. It clearly indicates that owing to higher CEC (880 µeq g^{-1} for montmorillonte), the platelets were pushed to much larger distances in vermiculite than in montmorillonite for the same ammonium ions exchanged on the surface. Table 1.3 describes the basal plane spacing values of the dry fillers, the suspension of fillers in DMF, the suspension of fillers in DMF and epoxy prepolymer, and the final composites. DMF has been chosen for these experiments because of its low volatility for the time scale of the X-ray scanning experiments. The presence of an X-ray peak in the solvent suspensions of modified fillers indicates that the fillers were not completely exfoliated in the solvent itself indicating the presence of residual electrostatic forces. The suspensions of fillers in solvent and prepolymer also confirmed that the prepolymer itself was also unable to delaminate the platelets, although there was no decrease in the basal plane spacing as sometimes observed in the montmorillonite system. However, no change in the basal plane spacing on the addition of prepolymer would suggest that no or very weak intercalation of the prepolymer took place. The basal spacing was enhanced to some extent after curing and nanocomposite generation indicating that intercalation occurred during the polymerization leading to further pushing apart of the platelets. However, the presence of diffraction peak in the composite diffractograms also indicated that full exfoliation of the platelets could not be achieved indicating the presence of residual attraction forces. Figures 1.4 and 1.5 show the X-ray diffractograms of the modified fillers and their 3.5 vol% composites. As it is evident from the diffractograms that the intensity of the diffraction peak was reduced after incorporation in the polymer, but owing to the dependency of the intensity on sample preparation and mineral defects, the analysis of the microstructure of the composites was also performed by TEM. Figure 1.6 shows the TEM micrographs of the 3.5 vol% composites. The micrographs showed the presence of mostly mixed morphology with single layers and intercalated tactoids of varying thicknesses. The platelets were also observed to be bent and folded and no specific alignment of the tactoids could be observed at any magnification. One should also note here that the misalignment of platelets leads

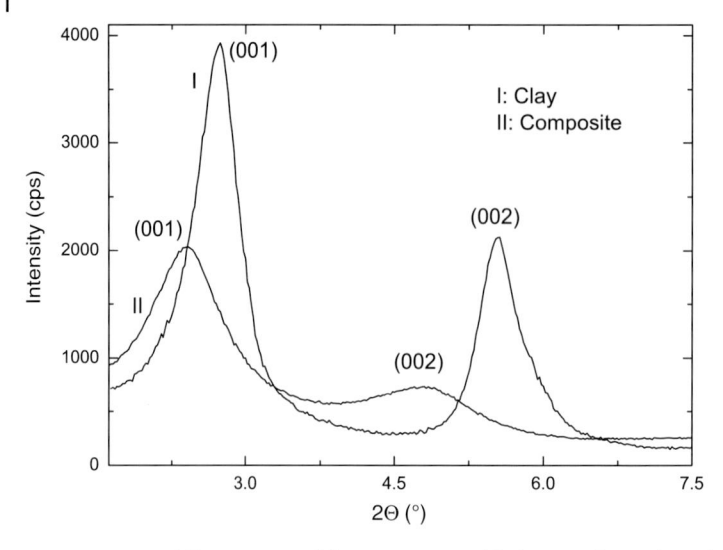

Figure 1.4 X-ray diffractograms of the (I) BzC16-modified vermiculite and (II) the filler in epoxy composite.

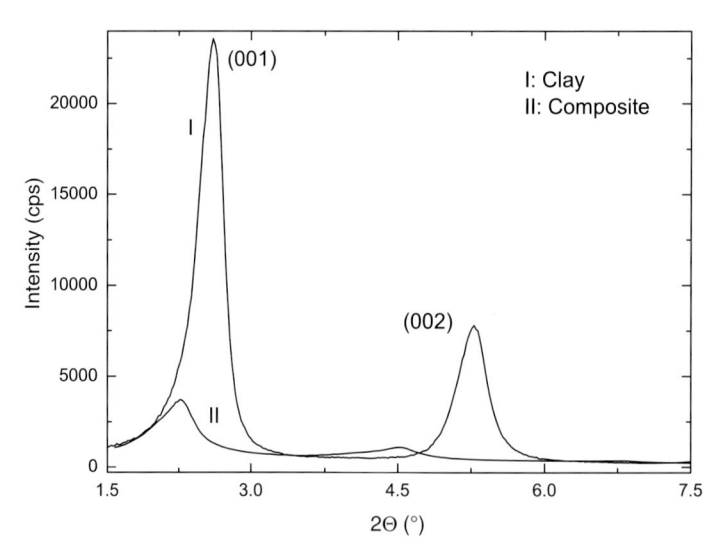

Figure 1.5 X-ray diffractograms of the (I) BzC18OH-modified vermiculite and (II) the filler in epoxy composite.

to reduction in the composite properties, especially gas permeation properties. It has been recently reported by using finite element models of the misaligned platelets that the completely misaligned platelets (as seen in the current system) are roughly one-third effective as barrier materials as compared to the completely aligned platelets thus confirming the need to generate more alignment [28]. The

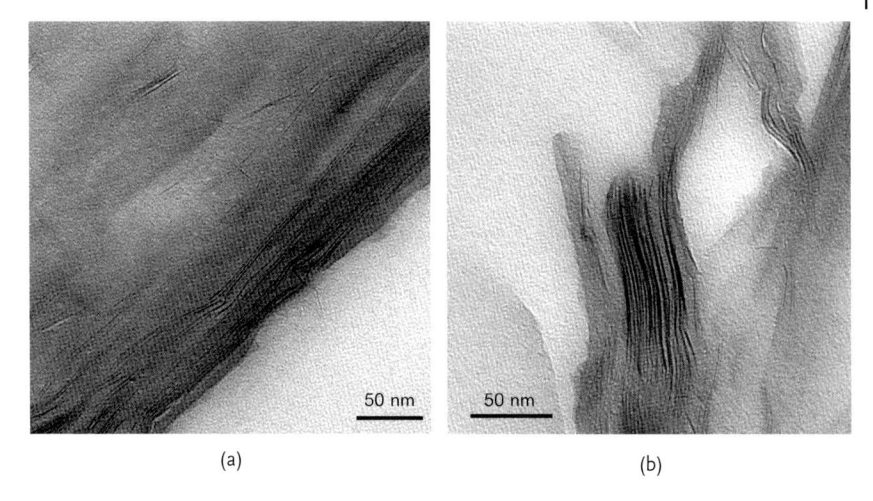

Figure 1.6 TEM micrographs of the 3.5 vol% (a) BzC16–vermiculite epoxy and (b) BzC18OH–vermiculite epoxy composites. The dark lines are cross-sections of aluminosilicate platelets.

Table 1.4 Oxygen and water vapor transmission rates through 3.5 vol% organo-vermiculite nanocomposites at 23 °C.

Composite	Oxygen permeability coefficient[a]	Water vapor transmission rate[a]
	cm^3 μm/(m^2 d mmHg)	g μm/(m^2 d mmHg)
Neat epoxy	2.0	10.0
Na–vermiculite	1.7	37.0
BzC18OH	1.5	7.5
BzC16	1.4	9.7

a) Relative probable error 5%.

TEM micrographs not only confirmed the X-ray findings about the intercalated morphology, but also showed the presence of some exfoliated layers totally missed in the WAXRD.

The gas permeation properties through the composite films containing 3.5% filler volume fraction were measured and are reported in Table 1.4. The oxygen permeation through the pure epoxy films was recorded as 2 cm^3 μm/(m^2 d mmHg), which in itself is much lower than the other adhesives used in the packaging laminates. The oxygen permeation through BzC18OH and BzC16 composites was, respectively, observed as 1.5 and 1.4 cm^3 μm/(m^2 d mmHg), which indicates that in spite of very high oxygen resistance of the matrix, further improvements in the oxygen barrier performance could be achieved by the incorporation of the nanoplatelets. These values for the montmorillonite fillers were reported as 2.2 and 1.6, respectively. It also indicates that the oxygen permeation was unaffected by the chemical nature of the ammonium ion exchanged on the surface and relied more on the polymer filler interactions. The water vapor permeation through the composite films with unmodified vermiculite was observed to be 37 g μm/(m^2 d mmHg), which was much higher than 10 g μm/(m^2 d mmHg) observed for the pure polymer itself. A value of 7.5 and 9.7 g μm/(m^2 d mmHg) was observed for the BzC18OH and BzC16 composites, respectively, indicating that the polarity of the hydrophilic interlayers was significantly reduced after the cation exchange, but interlayers are still partially polar to attract the molecules of water. Thus the composite properties can be represented as a combined effect of positive factors such as potential higher aspect ratio of the platelets, intergallery polymerization, and enhanced basal plane spacing as well as negative factors such as residual electrostatic forces. Thus, attaching much longer alkyl chains as well as increasing the grafting density in the ammonium ions could be expected to further eliminate the electrostatic interactions.

1.4
Conclusions

Vermiculite platelets modified with alkyl ammonium ions contained much higher basal plane spacing than the modified montmorillonites. The clay interlayers were further pushed apart in the composites owing to interlayer polymerization and the polymer chains were not squeezed out during solvent evaporation and curing. However, just having large d-spacing does not constitute the sole requirement for filler exfoliation, because owing to higher number of cations available per unit area in vermiculite, the residual electrostatic forces were still present although the area is covered with a large number of chains. The oxygen permeation through the composite films was further reduced by 30% in 3.5 vol% BzC16 composites, even though the permeation through the pure matrix was also very low. Owing to the polar interlayers, the water vapor permeation through the nonmodified vermiculite composites exponentially increased, which was significantly reduced in modified vermiculite composites. Grafting of large alkyl chains and ammonium ions with higher chain densities can be a possible approach to further reduce the electrostatic forces in these high CEC minerals.

Acknowledgment

Major portion of the presented work has been published in *Journal of Composite Materials*, 2008, 42, 2829, copyright Sage Publishers.

References

1 May, C.A. (1988) *Epoxy Resins Chemistry and Technology*, 2nd edn, Dekker, New York.
2 Lee, H. and Neville, K. (1967) *Handbook of Epoxy Resins*, McGraw-Hill, New York.
3 Ellis, B. (1993) *Chemistry and Technology of Epoxy Resins*, Blackie Academic & Professional, London.
4 Silvis, H.C. (1997) Recent advances in polymers for barrier applications. *Trends Polym. Sci.*, **5** (3), 75–79.
5 Brennan, D.J., Haag, A.P., White, J.E., and Brown, C.N. (1998) High-barrier poly(hydroxy amide ethers): effect of polymer structure on oxygen transmission rates. 2. *Macromolecules*, **31** (8), 2622–2630.
6 Gusev, A.A. and Lusti, H.R. (2001) Rational design of nanocomposites for barrier applications. *Adv. Mater.*, **13** (21), 1641–1643.
7 Eitzman, D.M., Melkote, R.R., and Cussler, E.L. (1996) Barrier membranes with tipped impermeable flakes. *AIChE J.*, **42** (1), 2–9.
8 Fredrickson, G.H. and Bicerano, J. (1999) Barrier properties of oriented disk composites. *J. Chem. Phys.*, **110** (4), 2181–2188.
9 Messersmith, P.B. and Giannelis, E.P. (1994) Synthesis and characterization of layered silicate–epoxy nanocomposites. *Chem. Mater.*, **6** (10), 1719–1725.
10 Lan, T., Kaviratna, P.D., and Pinnavaia, T.J. (1995) Mechanism of clay tactoid exfoliation in epoxy–clay nanocomposites. *Chem. Mater.*, **7** (11), 2144–2150.
11 Zilg, C., Mulhaupt, R., and Finter, J. (1999) Morphology and toughness/stiffness balance of nanocomposites based upon anhydride-cured epoxy resins and layered silicates. *Macromol. Chem. Phys.*, **200** (3), 661–670.
12 Brown, J.M., Curliss, D., and Vaia, R.A. (2000) Thermoset-layered silicate nanocomposites. Quaternary ammonium montmorillonite with primary diamine cured epoxies. *Chem. Mater.*, **12** (11), 3376–3384.
13 Zerda, A.S. and Lesser, A.J. (2001) Intercalated clay nanocomposites: morphology, mechanics, and fracture behavior. *J. Polym. Sci. Part B: Polym. Phys.*, **39** (11), 1137–1146.
14 Kornmann, X., Lindberg, H., and Berglund, L.A. (2001) Synthesis of epoxy–clay nanocomposites: influence of the nature of the clay on structure. *Polymer*, **42** (4), 1303–1310.
15 Kornmann, X., Thomann, R., Mulhaupt, R., Finter, J., and Berglund, L. (2002) Synthesis of amine-cured, epoxy-layered silicate nanocomposites: the influence of the silicate surface modification on the properties. *J. Appl. Polym. Sci.*, **86** (10), 2643–2652.
16 Kong, D. and Park, C.E. (2003) Real time exfoliation behavior of clay layers in epoxy–clay nanocomposites. *Chem. Mater.*, **15** (2), 419–424.
17 Chin, I.J., Thurn-Albrecht, T., Kim, H.C., Russell, T.P., and Wang, J. (2001) On exfoliation of montmorillonite in epoxy. *Polymer*, **42** (13), 5947–5952.
18 Osman, M.A., Mittal, V., Morbidelli, M., and Suter, U.W. (2004) Epoxy-layered silicate nanocomposites and their gas permeation properties. *Macromolecules*, **37** (19), 7250–7257.
19 Triantafyllidis, K.S., LeBaron, P.C., Park, I., and Pinnavaia, T.J. (2006) Epoxy-clay fabric film composites with unprecedented oxygen-barrier properties. *Chem. Mater.*, **18**, 4393–4398.
20 Xu, R., Manias, E., Snyder, A.J., and Runt, J. (2001) New biomedical poly(urethane urea)-layered silicate nanocomposites. *Macromolecules*, **34** (2), 337–339.
21 Tortora, M., Gorrasi, G., Vittoria, V., Galli, G., Ritrovati, S., and Chiellini, E. (2002) Structural characterization and transport properties of organically modified montmorillonite/polyurethane nanocomposites. *Polymer*, **43** (23), 6147–6157.
22 Chang, J.H. and An, Y.U. (2002) Nanocomposites of polyurethane with various organoclays: thermomechanical properties, morphology and gas permeability. *J. Polym. Sci. Part B: Polym. Phys.*, **40** (7), 670–677.

23 Osman, M.A., Mittal, V., Morbidelli, M., and Suter, U.W. (2003) Polyurethane adhesive nanocomposites as gas permeation barrier. *Macromolecules*, **36** (26), 9851–9858.
24 Jasmund, K. and Lagaly, G. (1993) *Tonminerale und Tone*, Steinkopff-Verlag, Darmstadt.
25 Brindley, G.W. and Brown, G. (1980) *Crystal Structure of Clay Minerals and their X-Ray Identification*, Mineralogical Society, London.
26 Osman, M.A. (2006) Organo-vermiculites: synthesis, structure and properties. Platelike nanoparticles with high aspect ratio. *J. Mater. Chem.*, **16** (29), 3007–3013.
27 Osman, M.A. and Suter, U.W. (2000) Determination of the cation-exchange capacity of muscovite mica. *J. Colloid Interface Sci.*, **224** (1), 112–115.
28 Osman, M.A., Mittal, V., and Lusti, H.R. (2004) The aspect ratio and gas permeation in polymer-layered silicate nanocomposites. *Macromol. Rapid Commun.*, **25**, 1145–1149.

2
Polymer Nanocomposites with UV-Cured Epoxies
Roberta Bongiovanni, Muhammad Atif, and Marco Sangermano

2.1
Introduction

Epoxy nanocomposites are heterophasic systems with at least one of the dispersed phase dimensions in the order of a few nanometers, keeping its nature different from the continuous polymeric phase.

Epoxy nanocomposites can be classified according to different criteria:

a) The number of dispersed phase dimensions lying in the nanometric range.[1]
b) The type of interactions between the two phases, that is, the nature of the interface or interphase.
c) The type of process selected to make the nanostructure.

Depending on benchmark (a), three types of nanocomposites can be distinguished:

- Type I: All three dimensions of the dispersed phase are in the order of nanometers.
- Type II: Two dimensions of the dispersed phase are in the nanometer scale.
- Type III: Only one dimension of the dispersed phase is in the nanometer range.

Typical examples of Type I are polymeric matrices with nanoparticles embedded isodimensionally, for example, metallic particles (e.g., Au, Pt, and Ag) or ceramic particles (e.g., CdS, SiO_2, ferrites, and graphite nanoparticles).

For nanocomposites of Type II, typical dispersed phases are made up of particles having an elongated structure, such as nanotubes and nanowhiskers. Examples are sepiolite, palygorskite, and carbon nanotubes (inorganic fillers) and cellulose and chitin nanowhiskers (organic fillers).

Type III nanocomposites contain fillers in the form of sheets. These sheets are one to a few nanometer thick and hundreds to thousands nanometer long. Examples are exfoliated graphite (EG), poly(muconic) acid crystals (organic fillers),

---

1) Colloidal chemistry is the discipline historically dealing with multiphase systems where one phase is in the range of 1μ to 1nm.

Thermoset Nanocomposites, First Edition. Edited by Vikas Mittal.
© 2013 Wiley-VCH Verlag GmbH & Co. KGaA. Published 2013 by Wiley-VCH Verlag GmbH & Co. KGaA.

and the well-known layered silicates, that is, clays and layered double hydroxides. This family of nanocomposites is better known as polymer-layered crystal nanocomposites.

Rendering to class (b)[2], on the basis of the nature of the interface (interphase) between the polymer matrix and the dispersed phase, the nanocomposites are divided into two distinct groups [1]:

1) In Group I, the matrix embeds the dispersed phase. Only weak bonds (hydrogen, van der Waals bonds) are present at the interface and give cohesion to the whole structure

2) In Group II, the two phases are linked together through strong chemical bonds (covalent or ionic-covalent bonds). Obviously, within many Type II materials, organic and inorganic components can also interact via the same kind of weak bonds that define Type I.

Implying to class (c), polymer nanocomposites can be obtained by different approaches. Usually these are a top-down and a bottom-up approach. In a top-down process, the phase to be dispersed in the epoxy matrix is subjected to a mechanical or physicomechanical treatment to reduce its size. In other words, macro- or micro-phases are fragmented into nano. A main drawback of this process is that the addition of fillers at high concentration induces an undesirable viscosity increase, thus influencing the processability. Typically inorganic particles are milled, homogenized, dispersed in the epoxy precursors so that aggregation is lost and primary particles forming the aggregates are adequately separated. The key point is the achievement of a full deagglomeration of the nanoparticles within the epoxy matrix: even though the particles might be well dispersed in the prepolymer solution, aggregation might occur in the matrix, especially during the curing process. To achieve a molecular dispersion and to avoid macroscopic phase separation, the interactions between the organic and the inorganic domains need to be stronger than the agglomeration tendency of the inorganic component [2]. Surface functionalization of the filler is often necessary in order to increase their compatibility, optimize their dispersion, and assure the optimum performance.

Usually a pretreatment of the inorganic fillers is achieved by mixing slurry of the inorganic materials into the solution of a proper surface-modifying agent, obtaining a highly uniform surface treatment. In the case of oxides the coupling agent is an alkoxysilane: it can react with water to form silanol groups, which immediately form covalent bonds by dehydration and condensation on the inorganic particle surface (see Scheme 2.1). The alkoxysilane is selected so that R is compatible with the epoxy matrix and/or contains functional groups, which can participate in the epoxy matrix formation. The surface modification of the ceramic fillers using a coupling agent improves the interfacial adhesion and lowers the viscosity. The hybrids obtained can be Group I or Group II, depending on the nature of R.

2) Commonly these nanocomposites contain an inorganic material as the dispersed phase and are also called organic–inorganic hybrids.

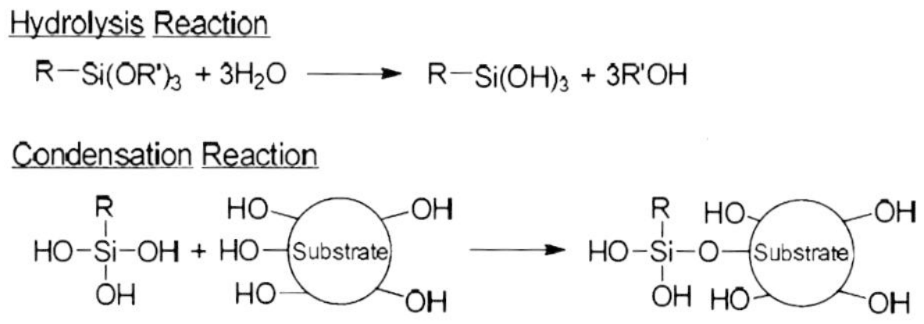

Hydrolysis Reaction

$$R-Si(OR')_3 + 3H_2O \longrightarrow R-Si(OH)_3 + 3R'OH$$

Condensation Reaction

$$\underset{\underset{OH}{|}}{\overset{\overset{R}{|}}{HO-Si-OH}} + \underset{HO}{\overset{HO}{\diagdown}}\!\!\text{(Substrate)}\!\!\underset{OH}{\overset{OH}{\diagup}} \longrightarrow \underset{\underset{OH}{|}}{\overset{\overset{R}{|}}{HO-Si-O}}\!\!-\!\!\underset{HO}{\overset{HO}{\diagdown}}\!\!\text{(Substrate)}\!\!\underset{OH}{\overset{OH}{\diagup}}$$

Scheme 2.1 Surface modification of a ceramic filler.

In the case of clays, ion-exchange reactions with ammonium salts are conventionally employed to make the clays organocompatible: they are referred to as organoclays or organophilic clays.

The bottom-up approach is an elegant way of creating a second phase in the polymeric matrix, through chemical reactions of dispersed precursor molecules. Precursors are usually liquids; therefore, the formulation has a controlled viscosity, while in the top-down synthesis the higher the filler content the higher the viscosity. The reactions of the precursors have to be tuned to obtain nanostructured products, usually inorganic particles. Typical examples of precursors are alkoxides or salts: alkoxides can hydrolyze and condense to form inorganic nano-oxides; salts can be reduced to form metal nanoparticles. During these processes one can control the interface (interphase) of the nanocomposites, eventually making covalent bonds. Therefore, one can choose to obtain hybrids of Group I or Group II.

Concerning the nanocomposites preparation, few more remarks can be useful. If the nanocomposites matrix is a thermoplastic polymer, the dispersed phase can be formed in the neat polymer (either by a top-down strategy or a bottom-up approach); alternatively the polymerization can take place after the second phase formation (either by a top-down strategy or a bottom up-approach). In the field of epoxy nanocomposites, *in situ* polymerization refers either the second phase (top-down approach) is in the monomer (or its solution) or the precursors of the dispersed phase (bottom-up approach) are dissolved in the monomers and react to form the nanostructure during or after polymerization.

In the case of thermoset, as the epoxides usually are, nanocomposites are formed by *in situ* polymerization: the formation of the polymeric network follows or is simultaneous to the process of formation of the inorganic phase. In the case of UV curable epoxies, they are synthesized by an *in situ* polymerization induced by UV light in the presence of a suitable photoinitiator, that is, the curing takes place through irradiation of the reactive epoxides. If the bottom-up approach is preferred, the first step of the synthesis is dispersing the inorganic precursors in the monomers; the following steps can be differently arranged. One method is irradiating (to build the network) and then triggering the processes building the inorganic nanophase; another method is the formation of the nanophase in the monomers and then curing; finally the building of the network by irradiation and the reactions forming the second phase can occur simultaneously.

In the following sections, we will describe the basics on the UV curing of epoxides; then we will review the main works published on UV-cured epoxy nanocomposites, classifying them as bottom-up nanocomposites and top-down nanocomposites.

2.2
Photopolymerization of Epoxides

Photopolymerization is a process induced by light, transforming a monomer into a polymer. Initiation is triggered by radiation: the reaction involves a photosensitive species, namely a photoinitiator that generates either radicals or ions under illumination. A chain reaction follows, leading to the consumption of the monomer, usually a liquid, and the formation of a polymer. Industrial processes usually employ multifunctional monomers, thus obtaining cross-linked systems, and typically the radiation is in the UV region. This is well known as the UV-curing technology. It has got increasing importance in the field of coatings due to its peculiar characteristic: it induces the polymer formation with a fast transformation (no more than minutes) of the liquid monomer into a solid film with tailored physical–chemical and mechanical properties; it can be considered as an environmental friendly technique, due to a solvent-free process, it is usually carried out at room temperature, therefore assuring the energy saving [3].

Most of the work on UV curing has been devoted to free radical polymerization systems, due to the availability of monomers (mainly acrylates and methacrylates) and photoinitiators. The major limits are sensitivity to oxygen that has an inhibition effect, the high shrinkage during the liquid–solid phase change and the irritation effect of some acrylates.

Light-induced cationic polymerization has some advantages like no oxygen inhibition, lower shrinkage; besides that, it is one of the most efficient methods to rapidly cure monomers that are inactive toward radical species, in particular, vinyl ethers and heterocyclic monomers such as epoxides, lactones, and cyclic ethers. The epoxy monomers are the most important and widely used in industrial applications; they can be classified in three families: cycloaliphatic epoxides, aliphatic epoxides, glycidyl ethers (Figure 2.1)

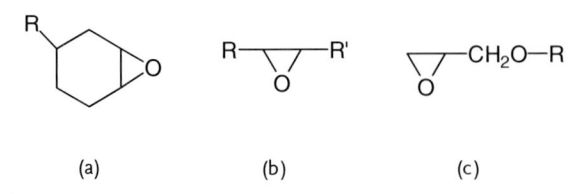

Figure 2.1 Schematic structures of cycloaliphatic epoxy monomer (a), aliphatic epoxy monomer (b), and glycidyl ether (c).

2.2 Photopolymerization of Epoxides

These monomers can be photopolymerized in the presence of cationic photoinitiators. Cationic photoinitiators are mainly onium salts [4]; their light absorption band rests in the UV region. Photopolymerization with sources other than UV lamp has been studied: visible light can promote the reaction in the presence of a suitable photosensitizer, able to absorb it and transfer the energy to the onium salt. Another strategy reported in the literature to conduct the epoxide polymerization under visible light is the radical induced cationic polymerization [5].

The mechanism for cationic ring-opening polymerization of epoxides is based on a multistep process (Scheme 2.2) [6]. The initiation mechanism involves the photoexcitation of diaryliodonium or triarylsulfonium salts and then the decay of the resulting excited singlet state with both heterolytic and homolytic cleavages. Cations and aryl cations are generated and produce strong Bronsted acids, which are the actual initiators of the cationic polymerization [7].

As per Scheme 2.2 , protonation of the monomer involves the reversible electrophilic attack of the photo-generated proton onto the nucleophilic oxygen of the epoxy ring and a secondary onium ion (oxonium ion) is formed. In the subsequent steps, the protonated monomers activate the cationic chain growth propagation. The ability of a given acid to protonate a specific monomer is directly related to its inherent acidity as well as the basicity of the monomer. Since epoxy monomers are only weak bases, strong acids are required to shift the equilibrium significantly toward the chain growth propagation.

Initiation

$$Ph_3S^+X^- \xrightarrow{h\nu} [Ph_3S^+X^-]^1 \longrightarrow \begin{bmatrix} Ph_2S^{+\bullet}X^- + Ph^\bullet \\ Ph_2S + Ph^+X^- \end{bmatrix} \rightleftarrows H^+X^-$$

$$\triangleleft O + H^+X^- \longrightarrow H-\overset{+}{O}\triangleleft \quad X^-$$

Propagation

$$H-\overset{+}{O}\triangleleft + \triangleleft O \longrightarrow HOCH_2CH_2-\overset{+}{O}\triangleleft$$

$$HOCH_2CH_2-\overset{+}{O}\triangleleft + n\, \triangleleft O \longrightarrow H(OCH_2CH_2)_{n+1}-\overset{+}{O}\triangleleft$$

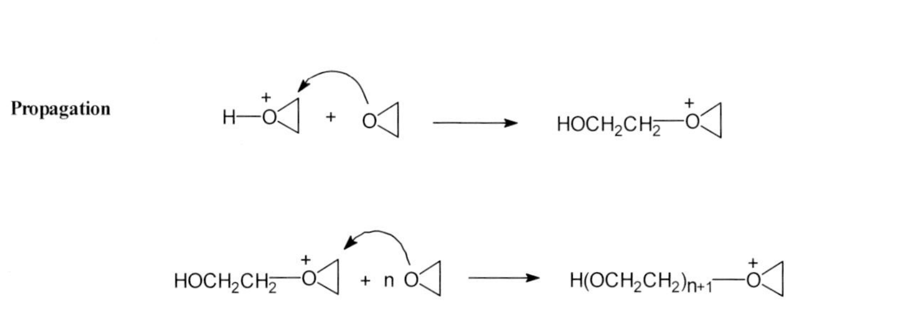

Scheme 2.2 UV-induced cationic ring-opening polymerization.

In addition to the mechanism shown in Scheme 2.2, Penczek and Kubisa described an alternative mechanism for the polymerization of epoxides, called activated monomer (AM) mechanism. It takes place when the cationic polymerization of epoxides is carried out in the the presence of alcohols [8].

As depicted in Scheme 2.3, during the polymerization, the growing ionic chain end undergoes a nucleophilic attack by the alcohol to give protonated ether. Deprotonation of this latter species by the epoxy monomers results in the termination of the growing chain. The proton transfers to the monomer and can start a new chain. The polymer has now an alcohol fragment as an end group. The effect of the chemical structure of the alcohol added on the occurrence of this chain transfer reaction has been deeply reported and discussed in the literature [9].

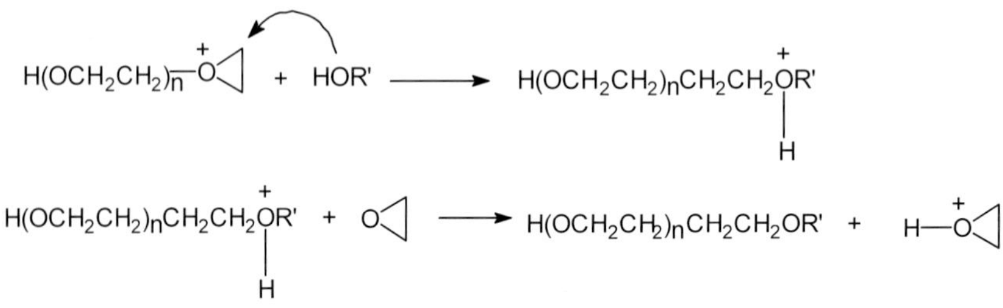

Scheme 2.3 Activated monomer mechanism.

In contrast to radical-initiated polymerization, in a cationic mechanism, the propagating chains are not interacting, so that the polymerization has a "living" character and will continue to develop after the UV exposure [10]. The oxonium species are essentially nonterminating, leading to very long life active centers (hours or even days). Thus there is a beneficial postcure effect that can be enhanced further by a thermal treatment. The postcure phase usually permits to achieve higher conversion and better material properties: the duration of this period depends on the thickness of the specimen and on the chemistry of the epoxide selected. The long-living species are also able to penetrate into nonilluminated parts: the shadow curing can be relatively long as said before; however, it can permit the curing of objects of complex geometries.

Termination still occurs, either by chain transfers producing inactive species, or by reaction of the propagating chains with nucleophilic impurities, in particular, traces of water.

If multifunctional epoxides or epoxy-substituted polymers are used as the starting materials, cross-linking readily occurs to generate a tridimensional polymer network. In the UV-curing field, only cycloaliphatic epoxies have reached commercial significance due to their good reactivity in cationic photopolymerization and also due to excellent adhesion, chemical resistance, and mechanical properties of resulting polymers. In particular, the monomers containing epoxycyclohexane rings are widely employed: they display higher rates of polymerization than other epoxides as glycidyl ethers. In fact, there is a high level of ring strain and its opening in the presence of an acid is favored.

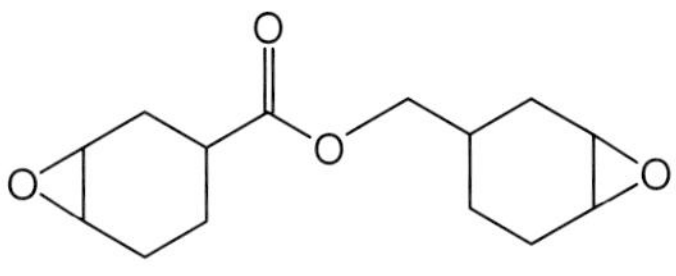

Figure 2.2 Structure of 3,4-epoxycyclohexylmethyl-3′,4′-epoxycyclohexane carboxylate (CE).

Structural parameters are, however, very important. In fact, 3,4-epoxycyclohexylmethyl-3′,4′-epoxycyclohexane carboxylate (CE, Figure 2.2) is less reactive than could be predicted from the reactivity of model compounds such as cyclohexane oxide.

The reason for its lower reactivity can be attributed to the interaction of the epoxy group with the ester carbonyl group, when dialkoxycarbenium ions are formed in epoxide ring-opening polymerization (Scheme 2.4) [11]. In general, the presence of an ester and an ether groups in epoxy monomers can limit their reactivity.

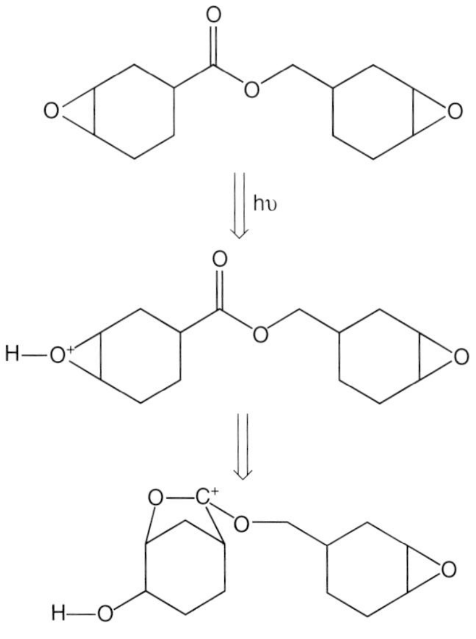

Scheme 2.4 The interaction of the epoxy group with the ester carbonyl group when dialkoxycarbenium ions are formed in epoxy ring-opening polymerization.

Unfortunately, the rate of photopolymerization of difunctional epoxides is by one order of magnitude lower than that of diacrylates, which are in fact the most employed monomers in photocuring. Attraction of the epoxides is due to their limited shrinkage (always lower compared to acrylics), which assures reduction of internal stresses and good adhesion.

2.3
Limits in Curing Epoxy Composites by UV Irradiation

As discussed earlier, UV-curing processes rely on the absorption of radiation passing through the matter; therefore, the thickness of the material is usually limited and shadowed areas cannot be processed. Fillers and reinforcements introduced in the epoxides enhance the problem as they can interfere with the light: as a consequence, the thickness of the material is even more limited. For this reason, nowadays the main applications of UV curing are commonly in the field of varnishes, paints, printing inks, adhesives, printing plates, microcircuits, and other devices in the form of thin polymer films. In the case of epoxies, the existence of long-living species and frontal mechanisms partially solve the problems. The long-living characteristic of epoxies during photopolymerization has been described earlier. On the contrary, for systems requiring radical polymerization, curing starts at the exposed surface and unreacted monomers remain below in nonexposed areas.

In a frontal polymerization the epoxide is converted into the cross-linked polymer by means of propagation through the reaction vessel. The photo-frontal polymerization requires the localized reaction (front) driven by an external UV source. The front is then supported by the combination of thermal diffusion and temperature-dependent reaction rates [12].

Alternatively, layer-by-layer curing allows UV polymerization: this technique is currently used in dental prosthetics and rapid prototyping. In the processing of traditional composites where the typical reinforcements are glass fibers, chopped glass fiber mats, woven glass fiber fabric, the UV curing is limited to open mold processes and thin laminates, with simple shapes. Crivello succeeded in the polymerization of glass fiber-reinforced epoxy resin to fabricate a real kayak and showed that conversion was good at a greater thickness than the neat resin: light scattering and wave-guiding effects of fibers enhanced the light penetration. Even sunlight gave excellent curing, thanks to the use of photosensitizers able to absorb at longer wavelength than the onium salt and to transfer the excitation to the photoinitiator [13]. The kayak was constructed by ordinary wet layup techniques by impregnating glass cloth with epoxy containing triarylsulfonium salt in a mold and then cured by exposure to direct solar irradiation. The irradiation time was 30 min. The upper and lower portions of the kayak were made and then joined at the midseam using a resin-impregnated tape and once again subjected to irradiation with sunlight.

If carbon materials or aramid fibers, which are opaque to UV, are incorporated in the matrix, after photocuring composites are usually subjected to thermal postcuring cycle (photocuring is for B-stage composites).

Although there is severe limitation of thickness and shadowed areas, composite preparation techniques have been adapted to UV curing as the curing time is in the order of magnitudes of minutes while thermal curing takes hours. Wet layup techniques, vacuum infusion type processes, filament winding, prepreg processes, and so on, have been adapted to UV curing [14]. Moreover, new procedures have been developed: an interesting example is the "wet edge lamination," the components are cured through a mask, a second layer is overlapped onto the uncured part, and by a further irradiation step a joint is formed. Application of this technique is the bonding of stiffeners onto shell structures. Repairs of localized damage are also suggested.

In an industrial production environment, a limitation of UV curing is represented by the potential hazard of radiation for operational personnel: the main organs affected by UV exposure are the skin and the eyes. Safety regulations impose shielding of the sources and suitable protective cloths and face shields. With respect to this point, we are witnessing an increasing interest in the use of LED lamps, with many scientific efforts in the development of photoinitiating systems active at longer wavelength.

2.4
Top-Down UV-Cured Epoxy Nanocomposites

Fillers mostly used in top-down photocured epoxy composites are those successfully employed in epoxy matrices and other polymers since the beginning of the nanocomposites era. We can mention clays, oxides, metal particles, and carbon in its different allotropic forms (CNT, graphite particles, graphenes, fullerenes, carbon black) and with different oxidation degree. The combination at the nanosize level of the inorganic fillers mentioned before and an epoxy matrix has made possible to improve properties such as mechanical performance, scratch resistance, abrasion resistance, heat stability as well as other characteristics [15]. Eventually, multifunctional and smart materials have been obtained as mentioned in the following.

It is well known that these materials can be obtained by preparing epoxy nanocomposites by conventional curing. Photopolymerization is particularly attractive because the high rate of the process may allow a polymer network formation much more rapidly than phase separation or macroscopic aggregation [16]. The rapid initiation and kinetics allow the medium to quickly solidify around the dispersed particles: in other words, by UV curing it is possible to freeze the degree of dispersion achieved in the starting prepolymer.

The use of lamellar silicates to prepare nanocomposites became popular in the late 1990s [17]. The most frequently used material is montmorillonite, often partially modified by introducing organic ammonium cations, bearing long alkyl chains, which increase the basal spacing, improve the compatibility between the filler and the epoxy matrix, and enhance the intercalation of the epoxides in the nanoclay galleries.

Epoxy nanocomposites containing exfoliated clays could show improved mechanical and barrier properties. Therefore, it is of high interest to investigate clay dispersion into UV-curable epoxy resins. Decker et al. described the UV-cured epoxy nanocomposites [18] containing montmorillonite clay platelets either packed together in a disordered arrangement (intercalated morphology) or dispersed as isolated nanoparticles (exfoliated morphology). The highly cross-linked epoxy nanocomposites were quite resistant to organic solvents, moisture, and weathering, as well as to mechanical aggression.

Bongiovanni *et al.* compared the kinetics of photopolymerization of cycloaliphatic diepoxides in the presence of different organophilic clays [19]. Interestingly, a difference in the photopolymerization rate was observed depending on the amount of water present in the clay: only after a thermal treatment to make the filler anhydrous, the induction time disappeared. When the modifier was methyl-tallow-bis-2-hydroxyethyl ammonium chloride (Cloisite 30B) the reaction rate was higher than the neat resin and the overall conversion was greater. The positive effect on curing was attributed to the presence of the hydroxyl groups of the clay modifier: in agreement with the previous discussion on the AM mechanism, they can act as chain transfer agents. Similar findings were reported in the literature [20]. XRD diffraction studies and TEM investigations evidenced, respectively, an increase of the interlayer distances and the presence of tactoids. As an example, the d_{001} spacing found for Cloisite 30B embedded in a CE matrix was 34.6A, that is, nearly twice the basal distance for the solid filler, indicating a relevant intercalation. XRD were recorded both before and after irradiating the sample: interestingly, no improvement of the basal distance was due to the photopolymerization reaction, confirming that once a dispersion is formed and the swelling of the clay by the epoxide is achieved, the curing just freezes the morphology of the systems. Further investigations on this system (CE and Cloisite 30B) regarded the effect of the addition of a diol, ethylene glycol, to the polymeric matrix. The diol changed the kinetics of photopolymerization as expected and the properties of the cured films. Moreover, it increased the state of dispersion of the organoclays inside the epoxy matrix as confirmed by rheological methods. The results of the linear viscoelasticity measurements indicated that while the pure resin CE displays negligible elastic properties and behaves as a Newtonian fluid, in the presence of the clay at low frequencies G' and G'' are at a plateau as for a solid-like material. This behavior is attributed to the existence of strong interactions between the organoclays and the resin monomer, which leads to formation of a physical network. Moreover, on addition of organoclays, a shear-thinning behavior is observed. The increase of the linear viscoelastic dynamic parameter is due to two contributions: the enhanced hydrodynamic effects caused by the mere presence of solid particles and the pronounced colloidal interactions, which slow the relaxation modes of the aggregates [21]. As the polymerization process does not affect further the morphology of the systems, the rheological characterization performed before curing is a valid tool to foresee the ability of the photoreactive monomer to produce nanostructured materials. Using Cloisite 30B, a quasi-exfoliated morphology in a photocured epoxy matrix (obtained from the CE monomer) was achieved by further modification of the organophilic clay with maleinized polybutadienes [22], according to the reaction in Figure 2.3. The morphology of the nanocomposite is shown in Figure 2.4a, while the XRD spectra (Figure 2.4b) shows the improvement obtained by the innovative modi-

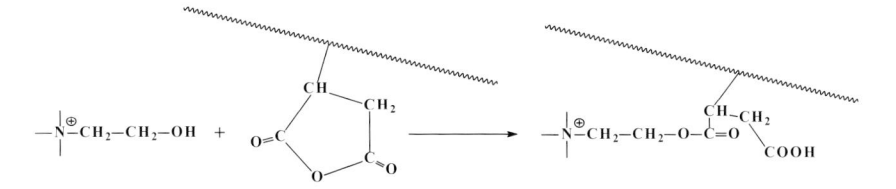

Figure 2.3 Modification of the reaction between the ammonium salt of the organophilic clay and maleinized polybutadiene.

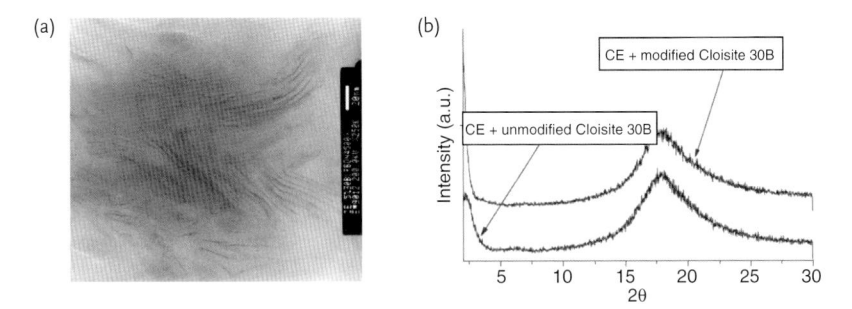

Figure 2.4 (a) TEM image of an epoxy nanocomposite containing modified Cloisite 30B; (b) comparison of XRD spectra before and after maleinization.

fication of the clay: whereas with Cloisite 30B the basal distance is 3.5 nm, after maleinization the reflection is no more present in the accessible 2θ range. An interesting feature of the quasi-exfoliated nanocomposites was the better thermal resistance (by TGA measurements) than the pure polymer.

An alternative modification relies on the use of alkoxysilanes. The silanization reaction is reported in the literature both for natural clays [23] and for organophilic clays [24]. Keeping in mind the preparation of epoxy-based photocured nanocomposites, silanization was conducted with glycidoxypropyl-trimethoxysilane (GPTS) [25], therefore introducing functional groups onto the clay, which can co-react with the epoxy monomer. The nanocomposite coating obtained by photopolymerization of CE in the presence of 5% w/w or 10% w/w of modified clay showed a mixed intercalated/exfoliated structure. The coatings maintained their optical transparency (Figure 2.5a) and were characterized by better thermal and mechanical properties with respect to the pure CE coatings. In particular, the scratch resistance of the films was enhanced as well as the resistance to the formation of cracks in the material (Figure 2.5b). Data of the critical load (e.g., the force needed to start the scratching, CL1) was encouraging: for the nanocomposites coating when the filler amount was 10 wt%, CL1 doubled.

The use of pristine Na–montmorillonite as nanofiller for epoxies is also possible. However, the intercalation of the reactive monomer inside the galleries of

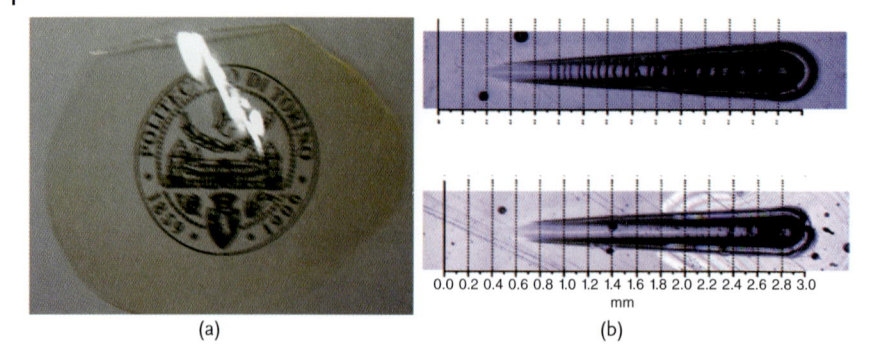

Figure 2.5 (a) Aspect of the nanocomposite under visible light; (b) scratch resistance test on the neat epoxide (top) and on the corresponding nanocomposite.

the neatly layered montmorillonite (without surfactants) can take place only under specific conditions. The hydration degree of the Na–montmorillonite was found to be a critical prerequisite in order to have intercalative polymerization and to obtain a nanoscale material. In the absence of water, that is, when the clay structure is collapsed, no intercalation occurs [26].

Yagci and co-workers have developed an effective route to allow polymer molecules to grow inside the clay galleries on irradiation and consequently to form covalent bonds between organic and inorganic phases [27]. Direct photoinitiation method was set to promote the propagation and the exfoliation processes, assuring the formation of homogeneous clay–polymer nanocomposites.

Charged ammonium ions are known to intercalate between the clay layers through ion exchange process. Thus, any photoinitiator with a quaternary ammonium group can easily be incorporated into the clay and used as a photoinitiator. On irradiation, the salts decompose and photopolymerization through the interlayer galleries of clay takes place, forming exfoliation/intercalation structures (Scheme 2.5).

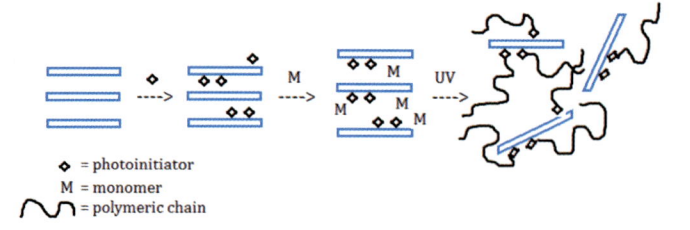

◆ = photoinitiator
M = monomer
∩∩ = polymeric chain

Scheme 2.5 Direct photoinitiation method for clay nanocomposites.

Besides montmorillonites, other clays were tested as nanofillers for epoxies. Corcione et al. [28] investigated the rheological behavior of several formulations based on a cycloaliphatic epoxy resin, containing an organo-modified boehmite. A general increase in viscosity was observed with increase in the volume fraction of the filler. A quasi-Newtonian behavior was found at relatively high shear rates when organophilic boehmite was added.

Similarly, Sangermano and co-workers [29] dispersed boehmite nanoparticles, by ultrasonification, in an epoxy resin in the range 1–5 wt% and the formulations were UV cured. An increase of T_g values was observed: this could be due to a good dispersion of the filler within the polymeric network that can hinder the segmental motion of the polymeric chains and decrease the free volume. TEM analysis showed boehmite aggregates ranging from 100–200 nm to 30–40 µm. Larger particle aggregation was avoided, allowing a certain degree of nanostructuration of the samples.

The addition of boehmite was also able to improve significantly the electrical properties of epoxy resin, particularly the space charge accumulation. This phenomenon occurs in almost all the polymers due to the huge amount of available localized state, particularly located in the amorphous region, in which electronic (or ionic) charge can be trapped even for a long time, depending on the trap depth.

Among the oxides, silica and titania nanoparticles have been widely studied in top-down nanocomposites. To obtain silica nanocomposite epoxides, Sangermano and co-workers added it to the reference diepoxide CE in the range between 5 and 15 wt% [30]. Similar investigation is reported elsewhere [31]. In the case of titania the highest concentration was 5% w/w [32]. The influence of the presence of the oxides on the rate of polymerization was investigated by real-time FT-IR. In the case of TiO_2 it was evidenced that by increasing its amount in the photocurable formulations, a decrease in the photopolymerization rate and epoxy group conversion was induced. These results were explained taking into account the important UV absorption phenomena undertaken by TiO_2 in the form of Anatase: this effect competed with the UV absorption of the photoinitiator, preventing its decomposition and therefore reducing the epoxy group conversion. In agreement with the FT-IR data, a decrease in T_g by increasing the TiO_2 content was found, indicating a decrease in the cross-linking density of the network.

In the case of silica, the kinetics improved and the result was explained on the basis of a chain transfer mechanism involving the silanol groups. The SiO_2 nanofiller, dispersed in the prepolymer by sonication, induced both a bulk and a surface modification of UV-cured coatings with an increase on T_g values, elastic modulus, and surface hardness as a function of the amount of silica. In general, the increase on T_g values and elastic modulus is attributed to a good adhesion between the polymer and the filler. The nanometer-sized particles can restrict the segmental motion of the polymeric chains with a consequent increase on glass transition temperature. Strong interaction between the nanoparticles and the polymeric chains were probably due to the chemical linkages achieved by the chain transfer reactions involving the silanol SiOH groups, in accordance with the kinetics data. TEM investigations confirmed that the silica filler keeps a size distribution ranging between 5 and 20 nm, without macroscopic agglomeration.

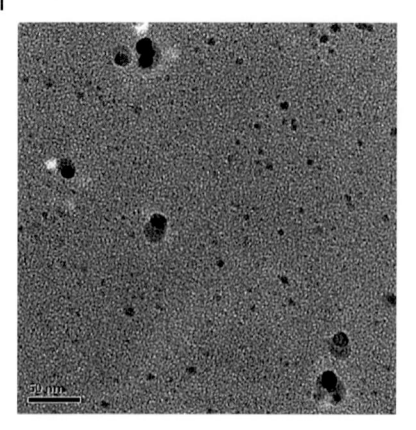

Figure 2.6 Bright field TEM micrograph for dicycloaliphatic epoxide photocured film containing 10 wt% of iron oxide nanoparticles.

In the case of titania, besides a mechanical and thermal reinforcement, the optical and electronic properties of the epoxides can be modified: TiO_2 imparts to the polymer matrix a high refractive index. When it is in the specific crystalline form anatase, strong photocatalytic properties can be exploited. The photocatalytic efficiency of TiO_2–epoxy nanocomposites was evaluated by studying the degradation of some organic target compounds under UV light [33]. Methylene blue was employed as model molecule to investigate the photoactivity toward organic molecules directly adsorbed on the epoxy surface. Under irradiation, complete dye degradation was achieved within 90 min; phenol and 3,5-dichlorophenol also underwent complete degradation.

Another system embedded in an epoxy network was iron oxide. Nanoparticles were functionalized by means of sol–gel chemistry (Scheme 2.1): using the GPTS, reactive epoxy group functionalities were covalently linked on the oxides surface. The functionalized nanoparticles were used in content up to 10 wt% [34]. All the samples were transparent to visible light; TEM showed that the inorganic particles were not macroscopically aggregated and their size distribution ranged from 5 to 10 nm, as shown in Figure 2.6. The control of the phase separation in the nanoscale range offers the possibility to obtain transparent composites coatings with improved mechanical properties. Moreover, due to the magnetism of the filler, the nanocomposites resulted to have magnetic susceptibility. The cured films also showed an important increase in T_g values. In this case, the inorganic component with small size easily infiltrates into the free volume of the organic phase with a consequent decrease on free-volume available, which will lead to a further increase on T_g values.

More recently the effects of several carbon fillers were compared: the curing process and the physical properties of the CE epoxy resin were studied, in the presence of exfoliated graphite (EG), functionalized graphene sheets (FGS), mul-

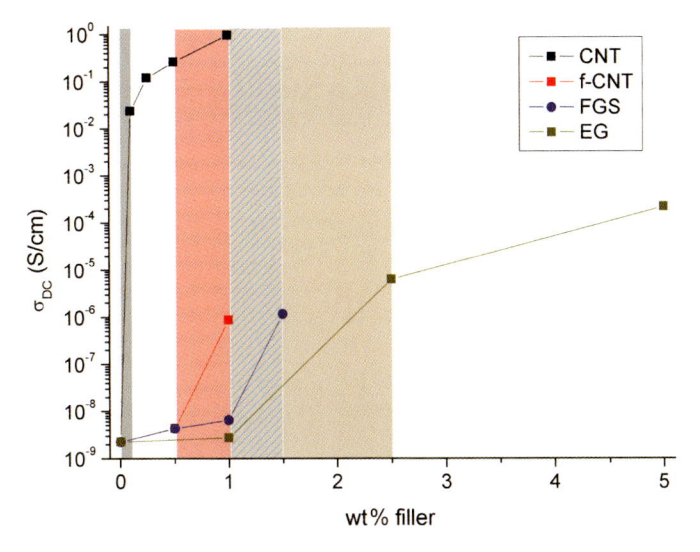

Figure 2.7 Values of the DC electrical conductivity of the epoxy nanocomposites as a function of the nanofiller content.

tiwalled carbon nanotubes (CNT), and oxidized and functionalized carbon nanotubes (f-CNT) [35]. It was found that all the nanofillers delayed the curing reaction due to a shielding effect as well as to an increase of the resin viscosity. Nevertheless, fully cured films were made by using a higher light intensity. Figure 2.7 shows DC conductivity for the nanocomposites studied as a function of nanofiller type and content. All the systems showed an electrical percolation threshold, but with CNTs, it was attained at a lower concentration (<0.1 wt%). In addition, FGS showed the best response in terms of the dynamic mechanical and microindentation performances. An increase of more than 20 °C in the T_g was observed with the addition of 1 wt% of FGS.

Also, Datta *et al.* prepared electrically conductive cationic UV-cured composites using exfoliated graphite platelets (EG). Electrical resistivity measurements of the composites as a function of EG concentration showed that at low filler concentration the chemistry of the resin can influence the electrical percolation behavior [36]. The formulations were based on cycloaliphatic epoxy resin, polyalcohol, and a cationic photoinitiator. It was observed that by incorporating different types of polyols in the system it is possible to change the properties of the binder system. While the polyols affect the electrical resistivity of the system at low filler content, at higher filler content all the formulations showed comparable DC electrical resistivity values.

Antistatic coatings were obtained in the presence of only 0.025 wt% of CNTs into the epoxy monomer [37]. An extended percolative structure forming a conductive CNT network was clearly evidenced by TEM analysis.

2.5
Bottom-Up UV-Cured Epoxy Nanocomposites

An interesting method to obtain nanoparticles is the sol–gel process, which allows synthesis of inorganic domains within the polymeric network. The process involves a series of hydrolysis and condensation reactions starting from a hydrolysable multifunctional metal-alkoxide as a precursor of the inorganic domain formation as reported in Scheme 2.6 [38].

$$M(OR)_4 + 4 H_2O \rightleftharpoons M(OH)_4 + 4 ROH \quad (1)$$

$$\equiv M\text{-}OH + HO\text{-}M\equiv \rightleftharpoons \equiv M\text{-}O\text{-}M\equiv + H_2O \quad (2a)$$

$$\equiv M\text{-}OH + RO\text{-}M\equiv \rightleftharpoons \equiv M\text{-}O\text{-}M\equiv + ROH \quad (2b)$$

Scheme 2.6 Sol–gel process to form nanoparticles through hydrolysis (1) and condensation reactions (2a and 2b).

The sol–gel preparation of nanoparticles can be combined with a photopolymerization process; this allows joining the advantages of both curing methods. The use of a suitable coupling agent permits to obtain a strictly interconnected network preventing macroscopic phase separation. The coupling agent provides bonding between the organic and the inorganic phases, therefore well-dispersed nanostructured phases may result [39].

The first authors to propose the use of sol–gel process in cationic UV curing were Wu et al. [40], who used reactive diluents modified by tetraethyl orthosilicate (TEOS). The resulting additives were used to formulate cationic UV coatings with CE. The coatings showed greater tensile modulus, lower elongation, and higher glass transition temperature. In addition, the siloxane functionalized polyols also effectively reduced the viscosity of the coating formulations.

The sol–gel reactions and the photopolymerization are combined by researchers in different ways as sketched below. In Scheme 2.7a the sol–gel reactions take place before photocuring of the reactive monomers, therefore the inorganic domains are formed first and then embedded in the polymers built up during

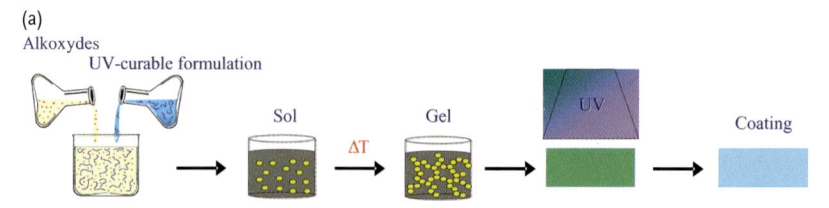

Scheme 2.7a Bottom-up preparation of UV-cured nanocomposites: sol–gel process comes before UV curing.

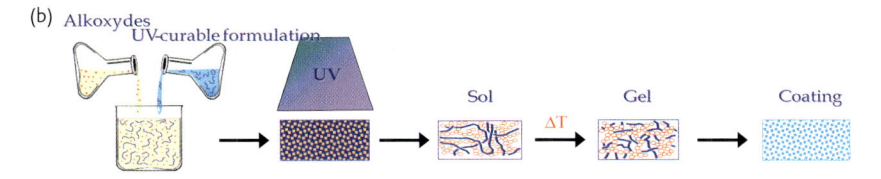

Scheme 2.7b Bottom-up preparation of UV-cured nanocomposites: sol–gel process follows UV curing.

irradiation. In Scheme 2.7b an alternative strategy is followed: the polymeric matrix is formed by UV curing, then the sol–gel process is triggered and the inorganic phase is formed inside the cross-linked matrix. Usually the latter method is preferred when the T_g of the photopolymer is low and the network is in the rubbery state, therefore the mobility is high enough so that the diffusion of the inorganic precursors is not hindered and the sol–gel process kinetics is good.

The research group of Mulhouse demonstrated that the organic–inorganic networks could be formed simultaneously, because the photoinitiator forms upon irradiation a strong acid, which is able to catalyze the sol–gel process in the presence of the moisture of air. The authors proved that the diffusion of water originated from the atmosphere is sufficient to induce the hydrolysis of the alkoxysilane functions in the presence of the photo-generated acid. Therefore, the photoacid catalysis of the alkoxysilane groups occurs simultaneously to the cationic ring-opening polymerization of epoxy monomer, leading to organic–inorganic epoxy networks via a single UV-curing process [41]. These hybrid coatings showed enhanced friction and wear properties, as measured on a ball-on disk tribometer [42].

Organic–inorganic hybrid coatings are described in [43]: the inorganic domains are formed by using tetraethoxy-orthosilicate (TEOS), as inorganic precursor for the silica network, and GPTS, as coupling agent. The morphology of the systems is in Figure 2.8.

DMTA showed an increase in T_g and in the storage modulus above T_g as a function of TEOS content (Figure 2.9). Both, the increase of T_g values and of modulus, can be attributed to strong and extensive interfacial interactions between the organic and inorganic phases. In fact, the inorganic silica particles, formed during the sol–gel process, can restrict the segmental motion of the polymeric chains and increase the cross-linking density of the polymer network.

Titania [44] and zirconia [45] inorganic domains were formed within epoxy matrices using the same technique with different inorganic precursors: optical properties were deeply modified and the transparent coatings obtained had high refractive index. These new materials could find advanced applications such as antireflective coatings.

Also metallic nanoparticles can be synthesized *in situ* by a bottom-up approach following a redox process conducted simultaneously to UV curing. Silver epoxy nanocomposites based on photoinduced electron transfer and cationic polymerization processes were proposed [46]. In this approach, electron donor alkoxybenzyl

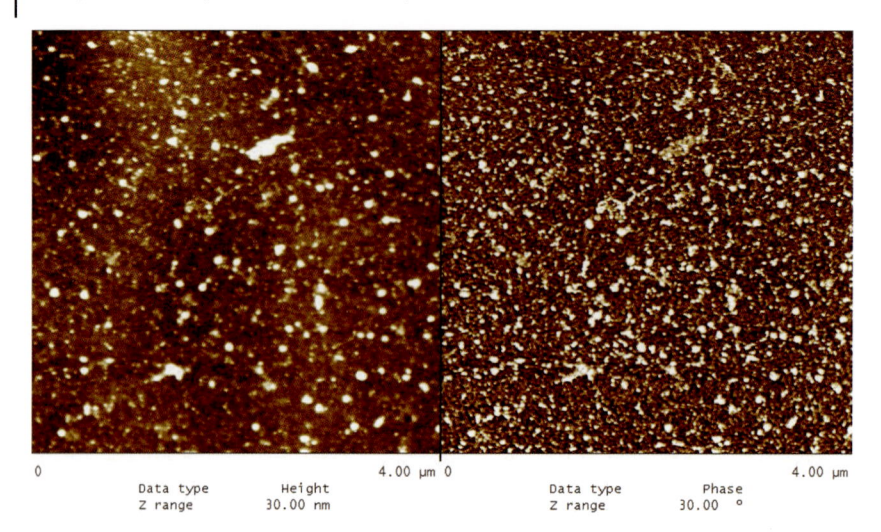

Figure 2.8 AFM image of the nanocomposite formed by an epoxy UV-cured matrix and silica domains obtained *in situ*.

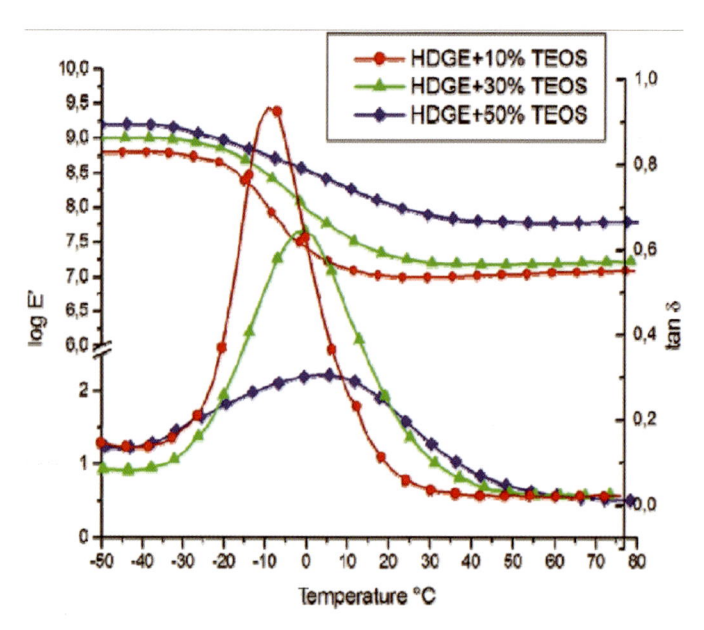

Figure 2.9 LogE' and tanδ curves for hybrid organic–inorganic UV-cured epoxy coatings based on hexanedioldiglycidyl ether (HDGE) containing increasing content of TEOS as inorganic precursor.

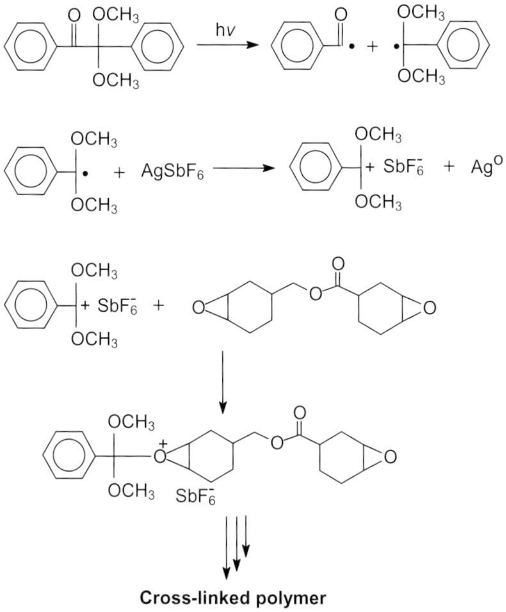

Scheme 2.8 Mechanism for bottom-up preparation of UV-cured metal nanocomposites.

radicals, formed from the photoinduced cleavage of 2,2-dimethoxy-2-phenyl acetophenone, are oxidized to the corresponding carbocations. At the same time, silver ions are reduced to the metallic state. The carbocations induce the cationic chain growth propagation of CE, forming the epoxy matrix. The reaction mechanism is in Scheme 2.8.

It was demonstrated that the nanoparticles are homogeneously distributed in the network without macroscopic agglomeration. This method is applicable not only to silver nanoparticles, but also to other metals, such as gold, whose salts undergo similar redox reactions with photochemically generated electron donor radicals [47]. The cured products showed the typical plasmon effect imparting a purple color.

2.6
Conclusions

Photoinduced polymerization is a proficient technique for rapid formation of highly cross-linked networks from liquid epoxy resins. The reaction involves a cationic mechanism and is usually initiated by UV light. Nanocomposites based on UV-cured epoxy matrix are reported in the literature: the polymer is combined with many nanofillers, formed either by a top-down method or a bottom-up strategy. The epoxy nanocomposites are often in the form of transparent coatings, characterized by better mechanical performance, improved thermal stability, eventually peculiar electrical and optical properties.

References

1. Judeinstein, P. and Sanchez, C. (1996) *J. Mater. Chem.*, **6**, 511.
2. Sanchez, C. (2001) *Chem. Mater.*, **13**, 3061.
3. Davidson, R.S. (1998) *Exploring the Science Technology and Applications of UV and EB Curing*, STA Tech. Ltd, London; Fouassier, J.P. and Rabek, J.C. (1993) *Radiation Curing in Polymer Science and Technology*, Elsevier, London.
4. Crivello, J.V. (1984) *Adv. Polym. Sci.*, **62**, 3.
5. Yagci, Y., Jockusck, S., and Turro, N.J. (2010) *Macromolecules*, **43**, 6245.
6. Crivello, J.V., Colon, D.A., Olson, D.R., and Webb, K.K. (1986) *J. Radiat. Curing*, **13**, 3.
7. Crivello, J.V. (1999) *J. Polym. Sci. Part A, Polym. Chem.*, **37**, 4241; Fouassier, J.P., Burr, D., and Crivello, J.V. (1994) *J. Macromol. Sci.*, **A31**, 677.
8. Penczek, S. and Kubisa, P. (1993) *Ring Opening Polymerization* (ed. P.J. Brunelle), Hanser, Munich, p. 17.
9. Crivello, J.V. and Liu, S. (2000) *J. Polym. Sci. Part A, Polym. Chem.*, **38**, 389; Crivello, J.V. and Acosta Ortiz, R. (2002) *J. Polym. Sci. Part A, Polym. Chem.*, **40**, 2298; Sangermano, M., Malucelli, G., Morel, F., Decker, C., and Priola, A. (1999) *Eur. Polym. J.*, **35**, 636; Bongiovanni, R., Malucelli, G., Sangermano, M., and Priola, A. (2002) *Macromol. Symp.*, **187**, 481.
10. Ficek, B.A., Tiesen, A.M., and Scranton, A.B. (2008) *Eur. Polym. J.*, **44**, 98.
11. Crivello, J.V. and Varlemann, U. (1995) *J. Polym. Sci. Part A, Polym. Chem.*, **33**, 2473; Crivello, J.V. and Varlemann, U. (1995) *J. Polym. Sci. Part A, Polym. Chem.*, **33**, 2463.
12. Mariani, A., Bidali, S., Fiori, S., Sangermano, M., Malucelli, G., Bongiovanni, R., and Priola, A. (2004) *J. Polym. Sci. Part A, Polym. Chem.*, **42**, 2066; Crivello, J.V., Falk, B., and Zonca, M.R. (2004) *J. Polym. Sci. Part A, Polym. Chem.*, **42**, 1630.
13. Crivello, J.V. and Lam, J.H.W. (1979) *J. Polym. Sci. Polym. Chem. Ed.*, **17**, 977.
14. Endruweit, A., Johnson, M.S., and Long, A.C. (2006) *Polym. Compos.*, **27**, 119.
15. Njuguna, J., Pielichowski, K., and Desai, S. (2008) *Polym. Adv. Technol.*, **19**, 947; Weifeng, Z., Haiquan, W., Haitao, T., and Guohua, C. (2006) *Polymer*, **47**, 8401; Decker, C., Keller, L., Zahouily, K., and Benfarhi, S. (2005) *Polymer*, **46**, 6640.
16. Clapper, J.D., Sievens-Figueroa, L., and Guymon, C.A. (2008) *Chem. Mater.*, **20**, 768.
17. Alexandre, M. and Dubois, P. (2000) *Mater. Sci. Eng.*, **28**, 1; Le Baron, P.C., Wang, Z., and Pinnavaia, T.J. (1999) *Appl. Clay. Sci.*, **15**, 11.
18. Decker, C., Keller, L., Zahouily, K., and Benfarhi, S. (2005) *Polymer*, **46**, 6640; Decker, C., Zahouily, K., Keller, L., Benfarhi, S., Bendaikha, T., and Baron, J. (2002) *J. Mater. Sci.*, **37**, 4831; Benfarhi, S., Decker, C., Keller, L., and Zahouily, K. (2004) *Eur. Polym. J.*, **40**, 493.
19. Bongiovanni, R., Turcato, E.A., Di Gianni, A., and Ronchetti, S. (2008) *Prog. Org. Coat.*, **62**, 336.
20. Oral, A., Tasdelen, M.A., Demirel, A.L., and Yagci, Y. (2009) *J. Polym. Sci. Part A, Polym. Chem.*, **47**, 5328.
21. Ceccia, S., Turcato, E.A., Maffettone, P.L., and Bongiovanni, R. (2008) *Prog. Org. Coat.*, **63**, 110.
22. Malucelli, G., Bongiovanni, R., Sangermano, M., Ronchetti, S., and Priola, A. (2007) *Polymer*, **48**, 7000.
23. Herrera, N.N., Letoffe, J.M., Putaux, J.M., David, L., and Bourgeat-Lami, E. (2004) *Langmuir*, **20**, 1564.
24. Chen, G.X. and Yoon, J.S. (2005) *Macromol. Rapid Commun.*, **26**, 899.
25. Di Gianni, A., Amerio, E., Monticelli, O., and Bongiovanni, R. (2008) *Appl. Clay Sci.*, **42**, 116.

26 Bongiovanni, R., Mazza, D., Ronchetti, S., and Turcato, E.A. (2006) *J. Colloid Interface Sci.*, **296**, 515.
27 Nese, A., Sen, S., Tasdelen, M.A., Nugay, N., and Yagci, Y. (2006) *Macromol. Chem. Phys.*, **207**, 820.
28 Corcione, C.E., Frigione, M., and Acierno, D. (2009) *J. Appl. Polym. Sci.*, **112**, 1302.
29 Sangermano, M., Deorsola, F.A., Fabiani, D., Montanari, G.C., and Rizza, G. (2009) *J. Appl. Polym. Sci.*, **114**, 2541.
30 Sangermano, M., Malucelli, G., Amerio, E., Priola, A., Billi, E., and Rizza, G. (2005) *Prog. Org. Coat.*, **54**, 134.
31 Isin, D., Kayaman-Apohan, N., and Gungor, A. (2009) *Prog. Org. Coat.*, **65**, 477.
32 Sangermano, M., Malucelli, G., Amerio, E., Bongiovanni, R., Priola, A., Di Gianni, A., Voit, B., and Rizza, G. (2006) *Macromol. Mater. Eng.*, **291**, 517.
33 Calza, P., Rigo, L., and Sangermano, M. (2011) *Appl. Catal. B*, **106**, 657.
34 Sangermano, M., Priola, A., Kortaberria, G., Jimeno, A., Garcia, I., Mondragon, I., and Rizza, G. (2007) *Macromol. Mater. Eng.*, **292**, 956.
35 Martin-Gallego, M., Hernandez, M., Lorenzo, V., Verdejo, R., Lopez-Manchado, M.A., and Sangermano, M. (2012) *Polymer*, **53**, 1831.
36 Datta, S., Htet, M., and Webster, D.C. (2011) *Macromol. Mater. Eng.*, **296**, 70.
37 Sangermano, M., Pegel, S., Pötschke, P., and Voit, B. (2008) *Macromol. Rapid Comm.*, **5**, 396.
38 Bandyopadhyay, A., Bhowmick, A.R., and De Sarkar, M. (2004) *J. Appl. Polym. Sci.*, **93**, 2579.
39 Ajayan, P.M., Schadler, L.S., and Braun, P.V. (2003) *Nanocomposite Science and Technology*, John Wiley & Sons, Inc., New York, p. 112.
40 Wu, S., Sears, M.T., Soucek, M.D., and Simonsick, W.J. (1999) *Polymer*, **40**, 5675; Wu, S., Sears, M.T., and Sucek, M.D. (1999) *Prog. Org. Coat.*, **36**, 89.
41 Chemtob, A., Versace, D.L., Belon, C., Croutxé-Barghorn, C., and Rigolet, S. (2008) *Macromolecules*, **41**, 7390; Belon, C., Chemtob, A., Croutxé-Barghorn, C., Rigolet, S., Schmitt, M., BIstac, S., Le Houréou, V., and Gauthier, C. (2010) *Polym. Int.*, **59**, 1175.
42 Belon, C., Schmitt, M., BIstac, S., Croutxé-Barghorn, C., and Chemtob, A. (2011) *Surf. Sci.*, **257**, 6618.
43 Amerio, E., Sangermano, M., Malucelli, G., Priola, A., and Voit, B. (2005) *Polymer*, **46**, 11241; Amerio, E., Sangermano, M., Malucelli, G., Priola, A., and Rizza, G. (2006) *Macromol. Mater. Eng.*, **291**, 1287.
44 Sangermano, M., Malucelli, G., Amerio, E., Bongiovanni, R., Priola, A., Di Gianni, A., Voit, B., and Rizza, G. (2006) *Macromol. Mater. Eng.*, **291**, 517.
45 Sangermano, M., Voit, B., Sordo, F., Eichhorn, K.J., and Rizza, G. (2008) *Polymer*, **49**, 2018.
46 Sangermano, M., Yagci, Y., and Rizza, G. (2007) *Macromolecules*, **40**, 8827.
47 Yagci, Y., Sangermano, M., and Rizza, G. (2008) *Macromolecules*, **41**, 7268–7270.

3
Influence of Organic Modification and Polyurethane Structure on Clay Dispersion in Polyurethane–Clay Nanocomposites

Shroff R. Mallikarjuna and Swaminathan Sivaram

3.1
Polymer Nanocomposites: An Introduction

Materials are key components of our civilization. We ascribe the major historical landmarks of our society to materials with epithets such as the stone age, bronze age, iron age, steel age (the industrial revolution), polymer age, silicon age, and so on. This reflects the importance of materials in the evolution of our society. It is becoming even more apparent that future technologies will unfold through a better understanding of structure–property relationship in materials and their functions. The emerging importance of nanoscale materials affords significant opportunities to create novel materials with unique and useful properties. Such materials augur new applications by exploiting the unique synergism that arises between constituents only when the length scales of the morphology and the critical lengths associated with the fundamental physics of a given property converge.

Polymer nanocomposites (PNCs) are a novel class of composites that are particle-filled polymers in which at least one dimension of the dispersed particle is in the nanometer range. PNCs have been an area of intense industrial and academic research for the past twenty years. PNCs represent an alternative to conventional filled polymers or polymer blends, which are ubiquitous in the polymer industry. In contrast to conventional composites, where dimensions of the reinforcing agents are of the order of microns, PNCs are characterized by discrete particles whose size is only about of a few nanometers. Thus nanocomposites possess structural inhomogeneities in the scale range of nanometers.

One can distinguish three types of nanocomposites depending on the number of dimensions of the dispersed particles in the nanometer range. When all the three dimensions are of the order of nanometers, we are dealing with isodimensional nanoparticles, such as spherical silica nanoparticles obtained by *in situ* sol–gel methods [1, 2] or by polymerization promoted directly from their surface [3], nanoclusters based on semiconductors [4]. When two dimensions are in the nanometer scale and the third is larger, forming an elongated structure, we speak about nanotubes or whiskers as, for example, carbon nanotubes

Thermoset Nanocomposites, First Edition. Edited by Vikas Mittal.
© 2013 Wiley-VCH Verlag GmbH & Co. KGaA. Published 2013 by Wiley-VCH Verlag GmbH & Co. KGaA.

[5] or cellulose whiskers [6, 7], which are extensively studied as reinforcing nanofillers, leading to composite materials with exceptional properties. The third type of nanocomposites is characterized by materials whose only dimension is in the nanometers range. In this case, the filler is present in the form of layered sheets or platelets of one to a few nanometers thick to hundreds to thousands nanometers long. This family of composites is termed as polymer-layered nanocomposites.

Among all the potential nanocomposite precursors, those based on clay and layered silicates have been more widely investigated, probably because the starting clay materials are easily available and their intercalation chemistry has been well explored for a long time [8, 9]. Silicates have a characteristic distance between galleries of 1 nm; the basal spacing of a gallery is also approximately 1 nm. The negatively charged galleries are held together by electrostatic forces, typically alkali metal cations. The important characteristics of clay minerals in polymer nanocomposites are their rich intercalation chemistry, high strength and stiffness and high aspect ratio of individual platelets, natural abundance, and relatively low cost. Their unique layered structure and high intercalation capabilities allow them to be chemically modified to be compatible with polymers, which make them particularly attractive in the development of clay-based polymer nanocomposites. In addition, their relatively low layer charge implies a weak force between adjacent layers, making the interlayer cations exchangeable and rendering facile exfoliation or delamination of layers. Therefore, intercalation of inorganic and organic cations and other molecules into the interlayer space is feasible, which is an important consideration in the preparation of polymer nanocomposites. Exchanging hard alkali metal cations with softer alkylammonium ions renders the clay galleries more hydrophobic, which facilitates intercalation of polymer molecules into clay. Among the smectite clays, montmorillonite (MMT) and hectorite are the most commonly used. Smectite clays are not natural nanoparticles; however, they can be exfoliated or delaminated into nanometer platelets of thickness of nearly 1 nm, an aspect ratio of 100–1500, and surface area of 700–800 m^2/g. Each platelet has very high strength and stiffness and can be regarded as a rigid inorganic polymer whose molecular weight (ca. 1.3×10^8) is much larger than that of a typical polymer. Therefore, very low loading of clays is required to achieve equivalent properties compared with the conventional composites. The best properties of nanocomposites are achieved when the surface area of the nanofilter is the largest; consequently, preventing aggregation of nanoparticles and promoting exfoliation of clays enhances the properties of a nanocomposites. Usually, exfoliation of clay layers in a polymer matrix requires matching of polarity of the clay surface and the polymer. Owing to the nanometer-sized particles obtained by dispersion, these nanocomposites exhibit markedly improved mechanical, thermal, optical, and physicochemical properties when compared with the pristine polymer or conventional (microscale) composites. There are several reviews and books that highlight the major developments in the area of polymer/clay nanocomposites and discuss their techniques of preparation, characterization, properties, and potential and applications [10–25]. The reader is referred to them for a detailed treatment of the subject.

3.2 Polyurethane–Clay Nanocomposites

3.2.1 Polyurethane Products and Chemistry

Polyurethanes (PUs) are a class of polymers containing carbamate (or urethane) link in the backbone of polymer chains. The carbamate link is formed by a reaction of isocyanate group with hydroxyl group. Polyurethanes, in general, contain three components: a polyol, diisocyante, and a chain extender. Polyols used for this purposes are generally macromonomers made of oligomers or polymers with more than two reactive hydroxl groups. The diisocyantes used can be aromatic diisocyantes such as 4,4′-diphenylmethane diisocyanate (MDI) or toluene diisocyanate (TDI) or aliphatic diisocyanates such as hexamethylene diisocyante or isophorone diisocynate (IPDI). Sometimes compounds such as polyisocyanates that contain more than two isocyanates are also used (e.g., polydiphenylmethane diisocyanate, which has two, three, and four or more isocyanate groups with an average functionality of 2.7). The chain extenders used can be short chain diols such as ethylene glycol, 1,4-butanediol, or can be a triol such as trimethylol propane (TMP) or pentaerythritol. Addition of small quantites of water during polymerization reaction between isocyanates and alcohols can result in the formation of urea linkages and the release of CO_2. This process is used in the production of PU foams. External foaming agents, such as pentane, are also widely used. Uncontrolled reaction of isocyanates can lead to formation of isocyanurates and allophanates, which further results in cross-linking and network formation during polyurethane synthesis.

Given the range of monomers that can be used to produce polyurethanes, it is not surprising that the basic chemistry yields polyurethane products ranging from soft elastomers to rigid thermoplastics and thermosets that are widely used in diverse applications.

3.2.2 Polyurethane Clay Composites

Polyurethanes (PUs) are versatile polymeric materials exhibiting a wide range of physical and chemical properties and find wide applications in coatings, adhesives, foams, rubbers, thermoplastic elastomers, and composites [26–28]. Nevertheless, PU suffers from a few weaknesses, such as low thermal stability and poor mechanical strength. Chemical modifications of PU or use of inorganic filler may in specific instances mitigate these problems. Many properties of PU are reported to be improved by incorporation of fillers. Calcium carbonate, aluminum hydroxide, kaolin, titanium dioxide, zinc oxide were used to improve the mechanical properties [29–34]. However, use of inorganic particulate fillers also has a deleterious effect on the fatigue property of PU and reduces its elongation at break [35]. Nanostructured PU/clay composites have been extensively studied in recent years [36]. Various methods of preparation, such as melt blending, solution blending, high shear mixing, sonication, and *in situ* polymerization in the presence or absence of solvents, and melt intercalation methods, are employed in the preparation of PU/clay nanocomposite. In the case of *in situ* polymerization studies, clay particles are intercalated either by soft polyols prior to reaction with diisocyanates, or by prepolymer terminated with diisocyanates are mixed with organoclay followed by chain extension with short chain diols [37–40]. In general, tensile strength and thermal stability of PU-nanocomposites showed improvements without loss of elongation [41–59] except for one study in which a decrease of tensile modulus was observed [41].

3.2.2.1 Thermoplastic Polyurethane Elastomers and their Composites

Thermoplastic polyurethane elastomers (TPUs) are produced from three component systems consisting of macrodiols (soft block) and diisocyantes and short chain diol chain extenders (hard block). The properties of thermoplastic elastomers are controlled by suitably adjusting the ratio of hard block to soft block and the choice of monomers. Thermoplastic elastomer/clay nanocomposites are prepared by solvent casting, melt mixing, or *in situ* polymerization [60–89]. The dispersion of clay in the polyurethane matrix depended on the type of clay, clay modification, method of preparation of nanocomposite, and the type of polyurethane. In general, TPU/clay nanocomposites show improvement in mechanical properties such as tensile strength, tensile modulus, impact strength, enhancement in thermal properties such as T_g, fire retardancy and char content, higher decomposition temperatures, and decreased gas permeabilities. TPU with semicrystalline soft block showed enhanced crystallization rates than the pristine polymer. The level of enhancement in properties depend on the extent of dispersion of clay in the polymer matrix. Polymer chain confinement in the intralayer gallery is believed to be the cause of such property improvements.

3.2.2.2 Thermoset Polyurethanes and Their Composites

Thermoset polyurethanes are produced when the monomeric components contain cross-linkable units. The cross-linkable units are introduced either by having macromonomers with hydroxyl functionality more than two or by having chain extenders with more than two hydroxyl groups or use of polyamines as curing agents. Sometimes, UV curable formulations containing acrylate units such as hexane diol diacrylates in presence of photoinitiators are also used to produce cross-linked PU networks. Organoclay nanocomposites of thermoset polyurethanes were produced by introducing suitable organoclay before the curing process [90–

103]. The nanocomposites obtained had high abrasion and solvent resistance, hardness and heat resistance, stiffness, and barrier resistance to gases and vapors.

3.2.2.3 Polyurethane Foams and Their Composites

Thermoset polyurethanes have wide applications in soft and rigid foams. They are produced by addition of calculated amounts of water to react with isocyantes to form ureas and release of CO_2. The gas released is trapped in the polymer matrix to form polyurethane foams. The clay is found to induce gas bubble nucleation resulting in reduced cell size and narrow cell size distribution, improved mechanical properties, decreased coefficient of thermal conductivity, enhanced glass transition temperature, decomposition temperature, wear resistance, barrier properties, fire retardancy, and increased char contents [104–115]. The clay is believed to act as a heterogeneous catalyst for the foaming and polymerization reactions. Polyurethane nanocomposite foams exhibit partially exfoliated clay structurees in the polymer matrix.

3.2.2.4 Polyurethanes Dispersions and Their Composites

Polyurethane dispersions are produced for a wide range of applications in adhesives, coatings, and inks. Nanocomposite films based on polyurethane dispersions and clay showed enhanced abrasion resistance, weather resistance, improved thermal and mechanical properties and improved barrier properties [116–139]. Clay/nanocomposite formulations based on superhydrophobic coatings showed enhanced adhesion strength along with good antiwetting, antifouling, antiaging, and self-cleaning properties compared with the pristine formulation.

3.3
Influence of Organic Modification of Clay and Structure of PU on PU/Clay Nanocomposites Structure

3.3.1
Reactive and Nonreactive Modifiers

Naturally available clays are generally hydrophilic. To render these hydrophilic clays compatible with organic polymers, it is necessary to modify the surface of the layered silicate and make them organophilic. Usually this is accomplished by replacing the exchangeable Na^+ (hard) cations present in the interlayer gallery with organoammonium (soft) cations (organomodifiers). Surface of clays can also be made organophilic by reacting the edge hydroxyls of the clay with organosilanes. The nature of the clay modifier plays an important role in improving the compatibility of clay layers with the polymer matrix, leading to significant enhancement in properties. Organic modifiers can also possess reactive functionalities, which can covalently link to polymer chains during nanocomposite formation (reactive modifiers). In case of polyurethane nanocomposites, reactive modifiers are usually those containing hydroxyl, amine, isocyanate, or UV curable groups.

Table 3.1 Interlayer distances measured from wide angle X-ray diffraction (WAXD) for organoclays after swelling in various diols and polyols.

Organoclay[a]	Air dried	Glycol (62)[b]	PEG300 (300)[b]	PPG2000 (2000)[b]	V230-238[c] (700)[b]	V230-112[c] (1500)[b]	V230-056[c] (3000)[b]
C12A-CWC	22.2	33.9	32.0	33.1	32.3	32.9	32.9
C18A-CWC	23.0	36.7	36.6	37.8	37.1	38.0	38.7

a) C12A and C18A denote $C12H25NH3(+)$ and $C18H37NH3(+)$ exchange cations, respectively.
b) Molecular weight of polyol.
c) Glycerol propoxylates, which are polyols with hydroxyl functionality of 3.

3.3.2
Effect of Nature of Modifier on Clay Dispersion in PU

The effect of chain length of the organic modifier on the extent of intercalation was studied by Wang and Pinnavaia [140]. The nanocomposites were prepared by *in situ* polymerization in the absence of solvents. It was shown that irrespective of the molecular weight of the short chain diol, or the macrodiol (e.g., polyethylene glycol [PEG], polypropylene glycol [PPG]) or macro triols used for swelling the clays, the extent of intercalation depends only on the chain length of the organic modifier and was consistently higher when the organoclay used was modified by C18 alkyl ammonium cation than when the organoclay used was modified by C12 alkyl ammonium cation (Table 3.1). It was also shown that during polymerization the interlayer distance increased by reacting with isocyanates after swelling the clays with macrodiols or triols as showed in Figure 3.1.

The role of carboxyl group or amine group in the modifier on the extent of intercalation/exfoliation was studied by Chen et al. [44]. The nanocomposites were prepared by solvent casting method. The extent of intercalation/exfoliation was better when the organoclay used contained a modifier bearing as carboxyl group compared with an organoclay that was modified using benzidine.

Pattanayak and Jana [83–85, 141] showed that in addition to tethering PU chains on to the clay surface by reacting the isocyanate with hydroxyl functionalities of the modifier, shear stress of mixing also helped in better exfoliation of clay in the resultant nanocomposite. The nanocomposites were prepared by *in situ* bulk polymerization in the absence of solvents. Three types of clays were used with different exchangeable cations (modifiers) as shown in Table 3.2.

Two methods of mixing were employed as shown schematically in Figure 3.2. The schematic shows the order of mixing of various components in method 1 and method 2. Method 2 induce higher shear stress than method 1. Clay 2 is capable of tethering the polymer chains by forming a urethane link. Lowered shear stress of mixing or absence of tethering reactions did not result in exfoliation as

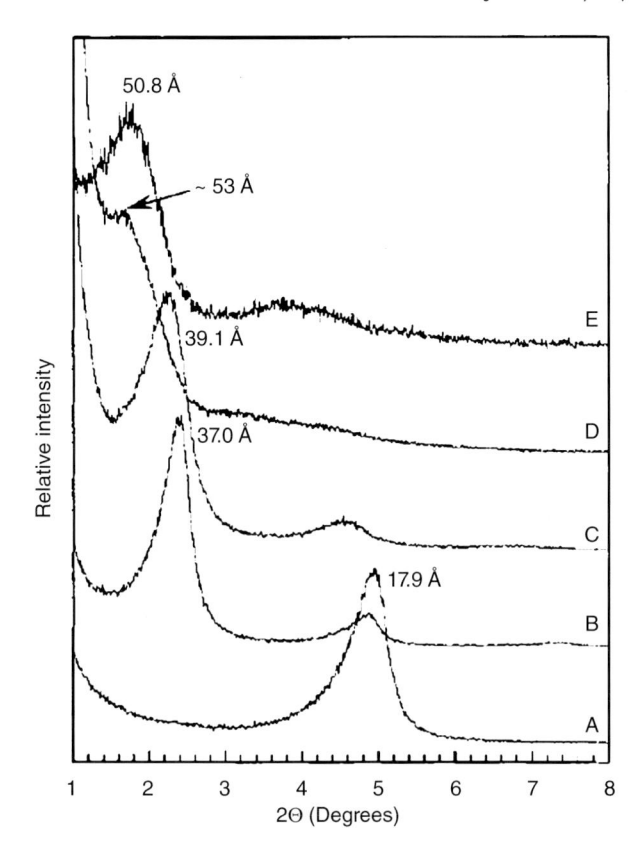

Figure 3.1 X-ray diffraction patterns of C18-Swy montmorillonite: (a) unsolvated clay and (b) solvated by polyol V230-238. The remaining curves are for mixtures of organoclay (10 wt%), polyol, and Rubinate diisocyanate R9272 after curing at (c) 95 °C for 10 min, (d) 95 °C for 25 min, and (e) 95 °C for 10 h. (Reproduced from Ref. [140] with permission from Americal Chemical Society.)

Table 3.2 Clays and their exchangeable cations (modifiers).

Clay	Exchangeable cation
Clay 1	Na^+
Clay 2	N,N-bis(hydroxyethyl)-N-(hydrogenated-tallow)-N-methyl ammonium cation
Clay 3	Dodecylammonium cation

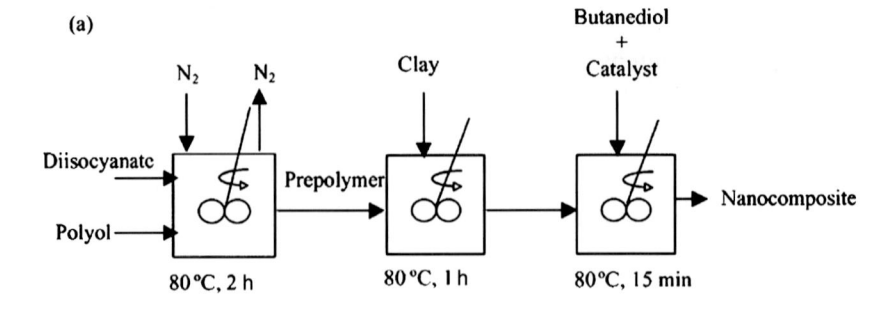

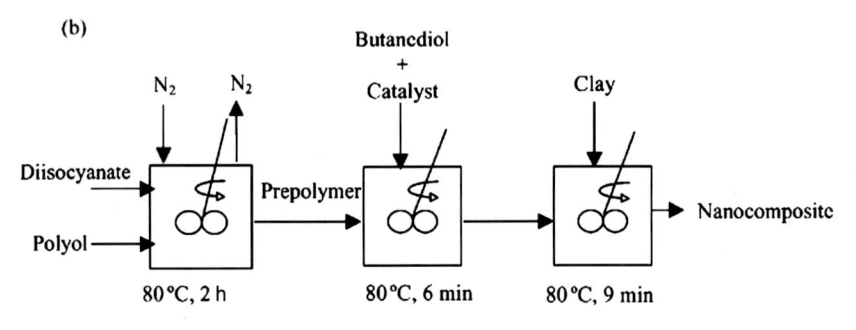

Figure 3.2 Schematic of composite preparation (a) method I and (b) method II. (Reproduced from Ref. [85] with permission from Elsevier.)

evidenced by transmission electron microscope (TEM) (Figure 3.3). TEM (Figure 3.3) image showed that among the clays, clay 2, which can tether polymer on to the clay surface, dispersed well in the polymer matrix, and between the two methods employed, method 2, which induces higher shear, resulted in better dispersion of clay in the PU matrix. Although the extent of tethering to the clay surface was similar for both polyether- and polyester-based thermoplastic urethane (TPU), the latter exhibited better exfoliation due to higher shear stress during mixing and polymerization.

Dan et al. [80] showed that exfoliated nanocomposites can be obtained by melt compounding of polyester or polyether-based TPU when the organoclay has been modified with hydroxyl group containing modifiers (Cloisite 30B, all Cloisites are organic modified quarternary ammonium ion exchanged MMT clays, produced by Southern Clay Products, USA, www.nanoclay.com), while only intercalated nanocomposites were obtained when no hydroxyl group was present in the modifier (Cloisite 25A or Cloisite 15 A, see Table 3.4). The improved dispersion with Cloisite 30B was attributed to stronger interaction of PU chains with the clay surface by formation of hydrogen bonding between carbonyl of urethane with hydroxyl of Cloisite 30B.

Effect of organoclay structure on the dispersion of TPU nanocomposites prepared by melt mixing was studied by Chavarria and Paul [60]. Organoclays and their respective modifiers used for this study are shown in Table 3.3. The structure of modifiers are shown in Figure 3.4.

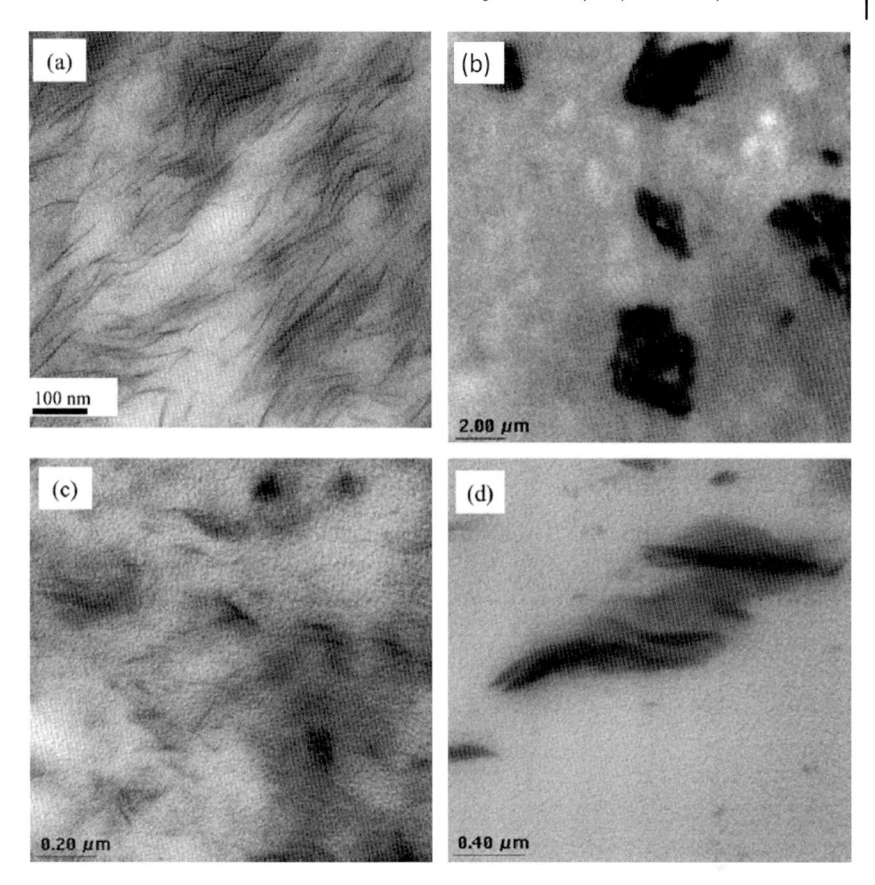

Figure 3.3 TEM images of nanocomposites of 5 wt% clay (a) method II, clay 2; (b) method I, clay 2; (c) method II, clay 3 and (d) method clay 1. (Reproduced from Ref. [85] with permission from Elsevier.)

The TEM images of the nanocomposites obtained are shown as in Figure 3.5. It was concluded that ammonium ion having one alkyl tail rather than two, presence of hydroxyethyl groups rather than methyl groups on the nitrogen, and a longer alkyl tail along with hydroxyethyl group rather than shorter alkyl chain along with hydroxyethyl group lead to better clay dispersion. The study also showed that TPU with harder segments, that is, with more polar functionalities, favors better polymer–clay interaction.

Xia et al. [142] examined the extent of intercalation and exfoliation in a PU matrix by employing a mixture of two organoclays, namely Cloisite 30B (which contains reactive hydroxyethyl group in the modifier) and Cloisite 25A (which does not contain reactive group). Higher ratios of Cloisite C30B in the mixture resulted in a higher degree of exfoliation of the resulting nanocomposites.

Exfoliated PU/clay nanocomposites based on a polyol and polymeric methane diisocyanate were prepared by *in situ* polymerization method in presence of

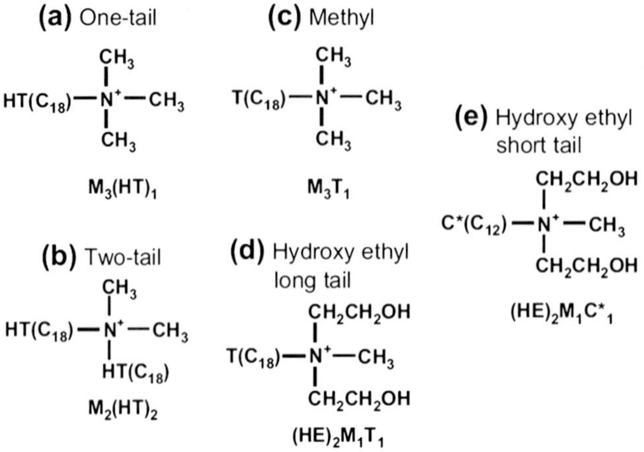

(a) One-tail

$$HT(C_{18})-\overset{\overset{\displaystyle CH_3}{|}}{\underset{\underset{\displaystyle CH_3}{|}}{N^+}}-CH_3$$

$M_3(HT)_1$

(b) Two-tail

$$HT(C_{18})-\overset{\overset{\displaystyle CH_3}{|}}{\underset{\underset{\displaystyle HT(C_{18})}{|}}{N^+}}-CH_3$$

$M_2(HT)_2$

(c) Methyl

$$T(C_{18})-\overset{\overset{\displaystyle CH_3}{|}}{\underset{\underset{\displaystyle CH_3}{|}}{N^+}}-CH_3$$

M_3T_1

(d) Hydroxy ethyl long tail

$$T(C_{18})-\overset{\overset{\displaystyle CH_2CH_2OH}{|}}{\underset{\underset{\displaystyle CH_2CH_2OH}{|}}{N^+}}-CH_3$$

$(HE)_2M_1T_1$

(e) Hydroxy ethyl short tail

$$C^*(C_{12})-\overset{\overset{\displaystyle CH_2CH_2OH}{|}}{\underset{\underset{\displaystyle CH_2CH_2OH}{|}}{N^+}}-CH_3$$

$(HE)_2M_1C^*_1$

Figure 3.4 Structure of the modifiers and their notations. (Reproduced from Ref. [60] with permission from Elsevier.)

Table 3.3 Organoclay notations and the modifiers.

Organoclay	Modifier
$M_3(HT)_1$	Trimethyl hydrogenated-tallow ammonium chloride
$M_2(HT)_2$	Dimethylbis(hydrogenated-tallow) ammonium chloride
M_3T_1	Trimethyltallow quaternary ammonium chloride
$(HE)_2M_1T_1$	Bis(2-hydroxyethyl) methyl tallow ammonium chloride
$(HE)_2M_1C_1$	Bis(2-hydroxyethyl) methyl coco ammonium chloride

polymeric 4,4′-diphenyl methane diisocyanate (PMDI)-modified clay [143]. It was shown that PMDI is anchored covalently to the clay surface and contains additional reactive isocyanate groups, which can undergo further polymerization.

PU nanocomposites were prepared using organoclays that are modified with a macrodiol containing protonated amine moiety, capable of exchange with Na$^+$ ions of the clay, polyethylene glycol units, and urethane groups in the backbone [144]. The structure of the organic modifier used is shown in Figure 3.6. This modifier was quaternized by reaction with HCl and dispersed in clay in N,N'-dimethyl formamide. Sonication was used to ensure effective dispersion. PU nanocomposites obtained using this organoclay was found to have intercalated structures with some disorder and the extent of intercalation improved with increasing dispersability of the organoclay as a consequence of sonication. The barrier property, tensile strength, and Young's modulus increased with increasing clay dispersion.

A comparative study of UV-active silane grafted and ion-exchanged organoclay for applications in photo-curable urethane acrylate nano and micro composites has been reported by Dean et al. [99]. They employed organically modified clays containing either methacrylate or acrylate functionality, which are capable of react-

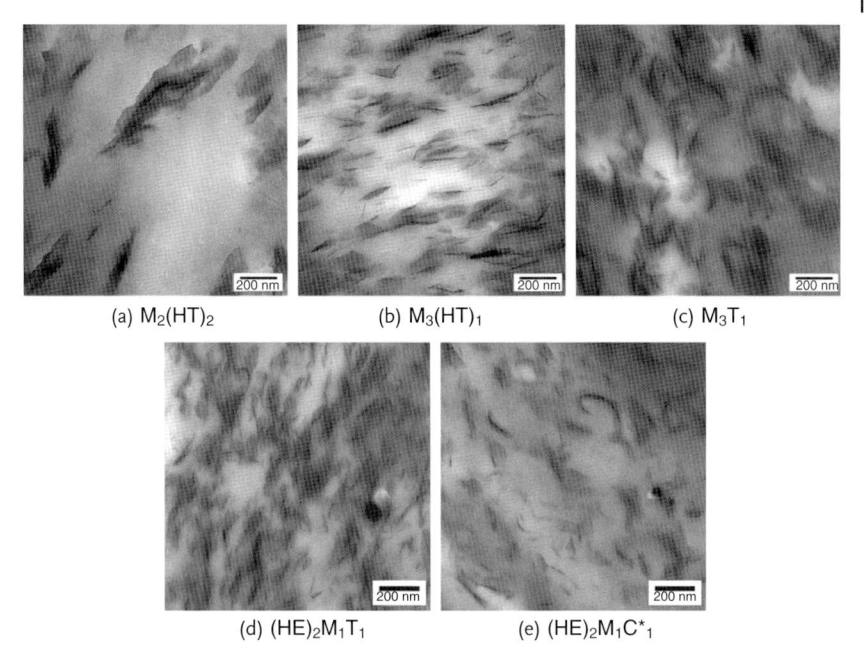

Figure 3.5 TEM images for nanocomposites based on M-H TPU and organoclays of concentration in the range 4.6–6%: (a) $M_2(HT)_2$, (b) $M_3(HT)_1$, (c) $M3T_1$, (d) $(HE)_2M_1T_1$, (e) $(HE)_2M_1C_1$. (Reproduced from Ref. [60] with permission from Elsevier.)

Figure 3.6 Structure of the organic modifier. (Reproduced from Ref. [144] with permission from Elsevier.)

ing with acrylic groups in the urethane acrylate matrix. [2-(Acryloyloxy)ethyl]trimethyl ammonium ion or [2-(methacryloyloxy)ethyl]trimethyl ammonium ion were exchanged with sodium ion in MMT. Silane grafting was undertaken using [3-(acryloxy)propyl]dimethylmethoxysilane or [3-(methacryloxy)propyl]dimethylmethoxysilane. Better dispersion was obtained when silane-grafted MMT containing acrylate or methacrylate were used. The structure of these modifiers are shown in Figure 3.7.

Tan et al. [145] prepared photopolymerized PU/organoclay nanocomposites based on polyurethane acrylates using a series of organoclays containing reactive methacrylate groups attached to the ammonium center by varying alkyl chain lengths. The structure of modifiers and WAXD patterns of the nanocomposites prepared are as shown in Figure 3.8. The resultant nanocomposites showed intercalated structures, and the extent of intercalation decreased with increase of alkyl chain length in the methacrylate.

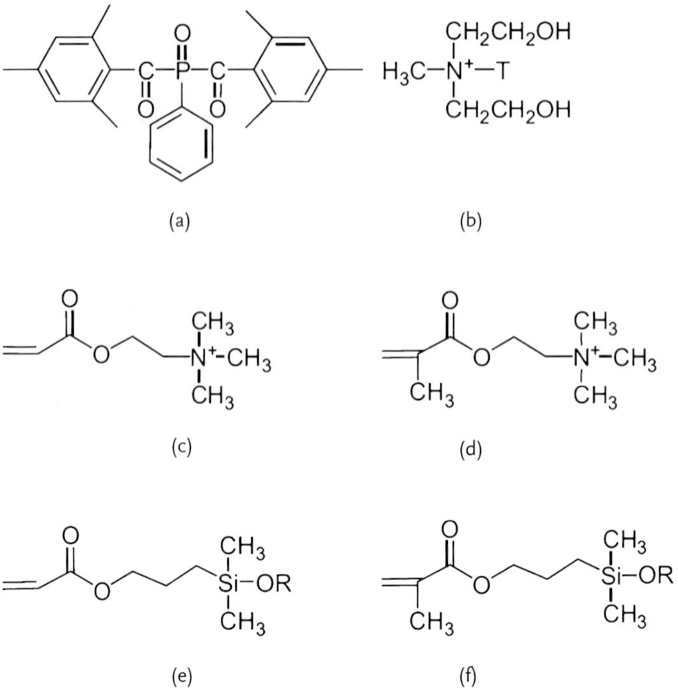

Figure 3.7 Structure of: (a) UV initiator acyphosphine oxide-Irgacure 819; and clay modifiers such as (b) bishydroxyethyl methyl tallow ammonium ion; (c) [2-(acryloyloxy) ethyl]trimethyl ammonium ion (AOETMA); (d) [2-(methacryloyloxy)ethyl]trimethyl ammonium ion (MAOETMA); (e) [3-(acryloxy)propyl]dimethylmethoxysilane (APDMMS); (f) [3-(methacryloxy)propyl] dimethylmethoxysilane (MAPDMMS). (Reproduced from Ref. [99] with permission from Elsevier.)

Exfoliated PU/clay nanocomposites based on polytetramethylene oxide (Mn:1800), MDI and 1,4-butanediol were effectively prepared by solvent casting in N,N-dimethyl formamide by introducing a cationic groups in PU [146] using dimethyl bis(2-hydroxyethyl) ammonium iodide as chain extender along with butane diol. The resulting nanocomposite contain polyurethane with a cationic group. The WAXD patterns (Figure 3.9) shows that when there is a cationic group in the polyurethane backbone the clay layers are completely exfoliated while only intercalated structures are obtained in the absence of cationic groups.

Effect of different dispersing medium (such as N,N-dimethyl acetamide and tetrahydrofuran) on clay dispersion and properties of TPU/clay nanocomposites prepared via solution mixing was studied by Dan et al. [77]. The polyether and polyester-based TPUs were used along with organically modified clays such as Cloisite 30B, Cloisite 25A, and Cloisite 15A and pristine montmorillonite. Affinity between the solvent and clay becomes more important when there is no strong interaction between the clay and the polymer. For example, Cloisite 25A dispersed in N,N-dimethyl acetamide is better than Cloisite 15A, while tetrahydrofuran is a

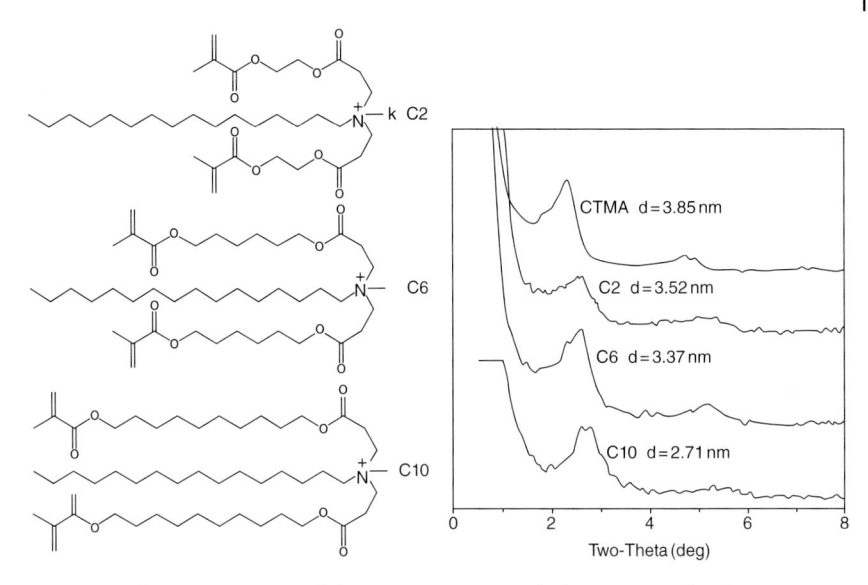

Figure 3.8 The WAXD patterns of the nanocomposites with the structure of surfactants. (Reproduced from Ref. [145] with permission from John Wiley and Sons.)

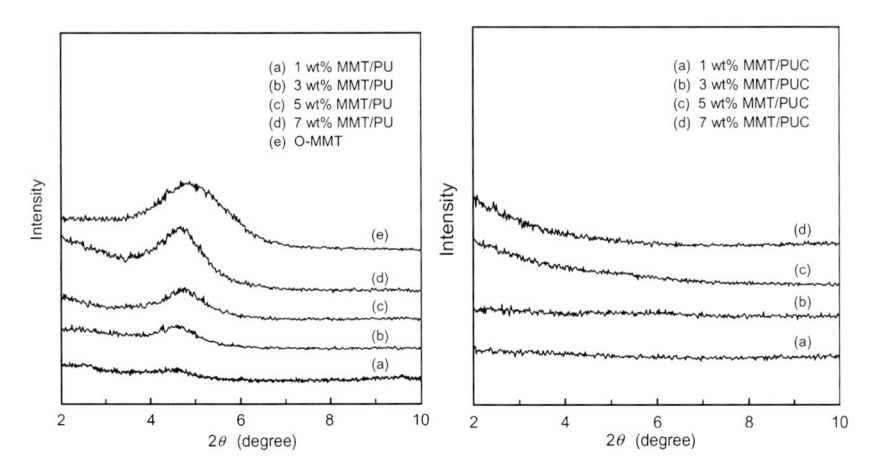

Figure 3.9 WAXD patterns of MMT/PU nanocomposites and MMT/PUC nanocomposites. (Reproduced from Ref. [146] with permission from Elsevier.)

good dispersing medium for the latter. Compatibility between clays and polymers becomes dominating if there exists a specific interaction such as hydrogen bonding between Cloisite 30B and TPU. Irrespective of the nature of the solvent used, Cloisite 30B dispersed well in the TPU matrix.

The effect of the length of hydroxy alkyl groups in the organo modifier on the dispersion of clay in TPU by solvent mixing in N,N-dimethyl acetamide as solvent was studied by Kim *et al.* [76]. They employed TPUs based on polyesters and polyethers for this purpose. The structure of modifier with longer alkyl chain is shown in Figure 3.10. The dispersion was shown to be superior when the organoclays possessed longer hydroxyalkyl group in the modifier than the organoclays containing shorter hydroxyalkyl chains (such as Cloisite 30B). TEM images of the nanocomposites (Figure 3.11) showed that the number of clay layers per tactoid is lower when the organoclay contains a longer hydroxyalkyl group.

Figure 3.10 Structure of modifiers with long hydroxyl alkyl group. (Reproduced from Ref. [76] with permission from John Wiley and Sons.)

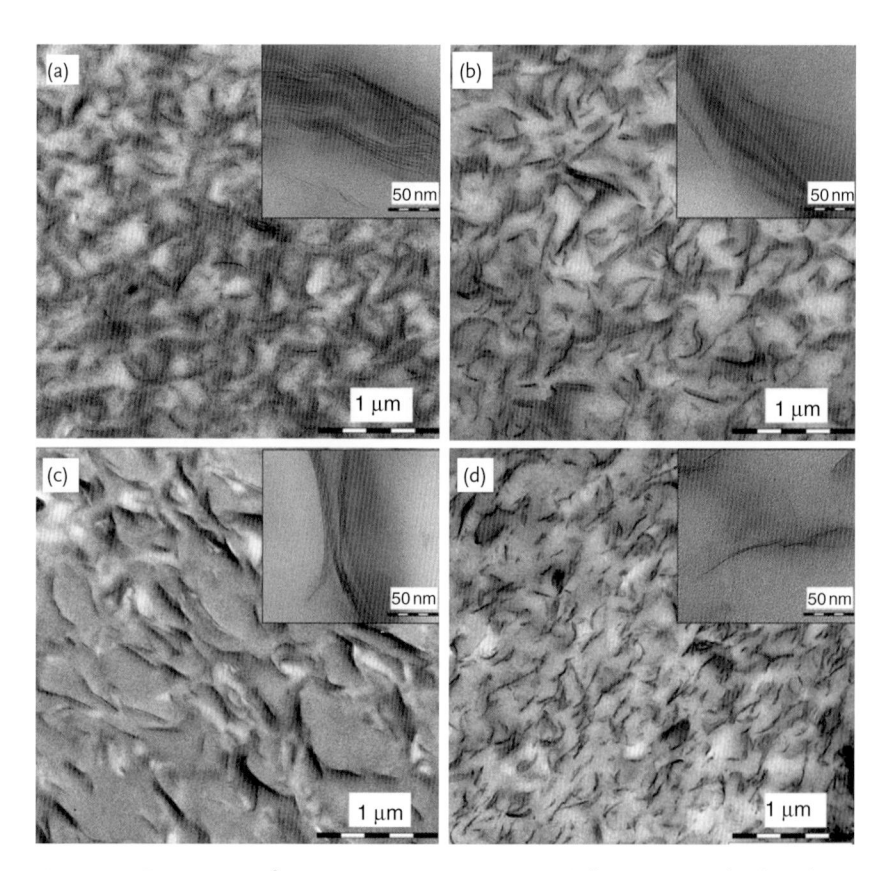

Figure 3.11 TEM images of nanocomposites containing 5wt% clay: (a) C30B/ether based TPU; (b) MMT-OH/ether based TPU; (c) C30B/ester based TPU; (d) MMT-OH/ester based TPU. (Reproduced from Ref. [76] with permission from John Wiley and Sons.)

The effect of different types of organomodifiers, nature of branching in the polyol used, and conditions used for preparation, such as temperature and shear stress, during mixing were investigated by Xia and Song [103]. They showed that presence of branched polyol, higher temperature of mixing and higher shear stress during mixing, and presence of reactive hydroxyl group in clay helped in better dispersion of clay in the PU matrix. Exfoliated PU/clay nanocomposites, based on polyether polyol and hydrogenated MDI, were obtained whenever the functionality of the polyol used was greater than two [103].

Plummer et al. showed that exfoliated polyurethane thermoset/clay nanocomposites could be prepared by incorporating hydroxyl-terminated hyperbranched polyesters in the polyol precursors along with MMT before curing [147]. The effect of hard segment content on the extent of intercalation/exfoliation of clay in PU matrix was studied by Hu et al. The extent of intercalation improved with increase in the content of hard segments [49].

Tien and Wei [47] showed that PU based on poly(tetramethylene glycol) and 4,4′-methylene diphenyl diisocyanate can be ionically tethered to the surface of the clay by choosing modifiers having reactive hydroxyl groups. They used short chain hydroxyl containing modifiers. The structure of the modifiers are shown in Figure 3.12. The hydroxyl groups in the modifier of the clay were reacted with isocyanate terminated PU-prepolymer. The presence of three hydroxyl groups per modifier favored the formation of exfoliated PU nanocomposites as evidenced by TEM (Figure 3.13).

Moon et al. [148] used the same organoclay to prepare PU nanocomposites based on poly(butylene succinate) (PBS). Poly(ethylene glycol) was employed as the soft segment and 1,4-butanediol and hydrogenated MDI as hard segments. The nanocomposites obtained exhibited an intercalated structure. Inability to form exfoliated structures was attributed to the poorer swelling of organoclay by PBS.

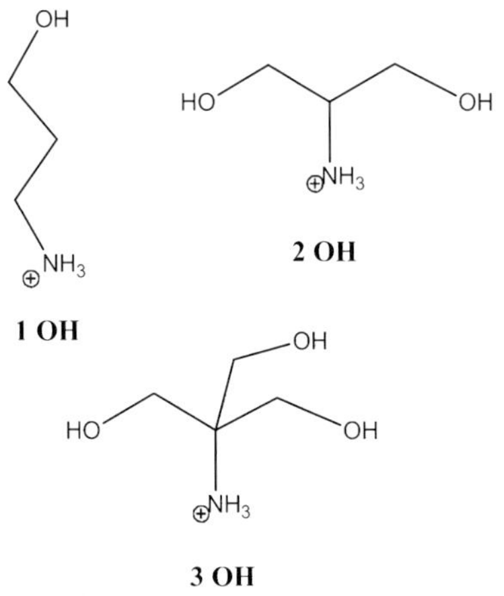

Figure 3.12 Structure of hydroxyl containing organic modifiers. (Reproduced from Ref. [47] with permission from American Chemical Society.)

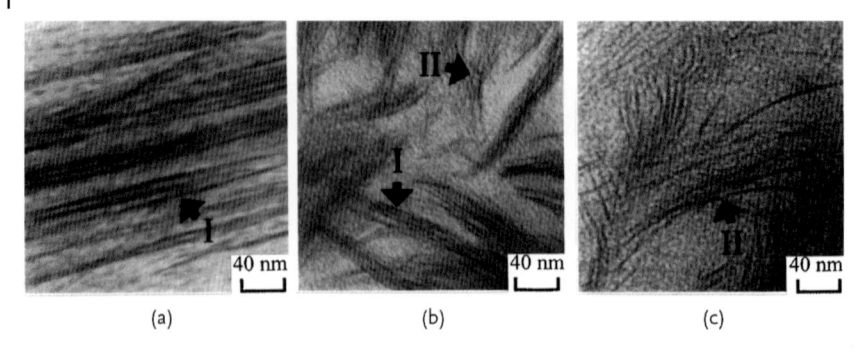

Figure 3.13 TEM images of the cross-sectional views of PU containing (a) 1 wt% 1OH-Mont (b) 1 wt% 2OH-Mont, and (c) 1 wt% 3OH-Mont: (I) indicated intercalated structure; (II) indicated exfoliated structure. (Reproduced from Ref. [47] with permission from American Chemical Society.)

In spite of several studies, the dependence of the structure of nanocomposites obtained (intercalated or exfoliated) on the structure and composition of the poly(urethane)s are still poorly understood. An exfoliated nanocomposite results when a branched polyol is used [103]. This is supported by the theoretical modeling study by Singh and Balaz [149]. Similarly, when a modifier containing three methylol groups was ionically anchored to the clay and reacted with an isocyanate-terminated prepolymer (based on poly(tetramethylene glycol) and MDI), an exfoliated structure was obtained [47] while only intercalated structures were obtained when the PU matrix was based on PBS [148]. More recently, Mallikarjuna and Sivaram [150] studied the effect of composition of PU and the nature of modifier functionality on the structure of PU nanocomposites when prepared by *in situ* polymerization in the presence of toluene as a solvent. The importance of tethering groups and the effect of branched and cross-linked PU structures on the dispersion of polyurethane/clay nanocomposites were examined. Trimethylol propane (TMP) was used as a branching agent in the preparation of cross-linked PUs. Organoclays used in this study are shown in the Table 3.4. When Cloisite 30B containing reactive hydroxyl in the modifiers was used, an increase in the content of the branching agent caused an improvement in dispersion of clay in the polymer matrix, and at higher contents of TMP, an exfoliated structure was obtained. When Cloisite 25A was used as organoclay, increasing the content of branching agent (TMP) did not have an effect on dispersion of organoclay in the PU matrix as evidenced by WAXD patterns (Figure 3.14) and TEM (Figure 3.15). When the PU nanocomposites were prepared using organoclays containing modifiers with increasing number of hydroxyl functionality (1-OH(MMT), 2-OH(MMT), and 3-OH(MMT), Table 3.4) no exfoliation was observed in the absence of branching agent. Thus, it is apparent that presence of either tethering groups in the organoclay or branch points in PU alone does not result in exfoliated structures (Figure 3.16).

Very recently hydroxyl functional quarternary ammonium ions have been used as catalytic modifiers for the preparation of exfoliated PU elastomer nanocomposites. In this study, vermiculite was chosen as the clay. The PU elastomer had 30%

3.3 Structure of PU on PU/Clay Nanocomposites Structure

Table 3.4 List of organoclays used and the structure of their modifiers.

Organoclay	Structure of the modifier	Name of the modifier
Cloisite 25A	$H_3C-N^+(CH_3)_2-TH$ Where TH = hydrogenated tallow	N-(hydrogenated tallow)-N,N,N-trimethyl ammonium cation
Cloisite 30B	$HO-CH_2CH_2-N^+(CH_3)(TH)-CH_2CH_2-OH$ Where TH = hydrogenated tallow	N,N-bis(hydroxyethyl)-N-(hydrogenated tallow)-N-methyl ammonium cation
1-OH(MMT)	$H_3C-N^+(CH_3)(C_{16}H_{33})-CH_2CH_2-OH$	N-(2-hydroxyethyl)-N,N-dimethyl-N-hexadecylammonium cation
2-OH(MMT)	$HO-CH_2CH_2-N^+(CH_3)(C_{16}H_{33})-CH_2CH_2-OH$	N,N-bis(2-hydroxyethyl)-N-methyl-N-hexadecylammonium cation
3-OH(MMT)	$HO-CH_2CH_2-N^+(CH_3)(CH_2CH(OH)C_{16}H_{33})-CH_2CH_2-OH$	N,N-bis(2-hydroxyethyl)-N-methyl-N-(2-(hydroxymethyl)octadecyl)ammonium cation

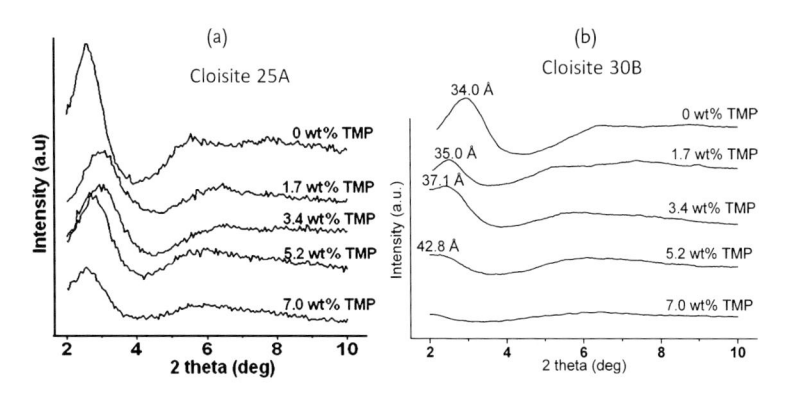

Figure 3.14 WAXD pattern of PU/clay nanocomposites with varying amount of trimethylol propane (TMP) and with organoclay (a) Cloisite 25A and (b) Cloisite 30B. (Reproduced from Ref. [150] with permission from John Wiley and Sons.)

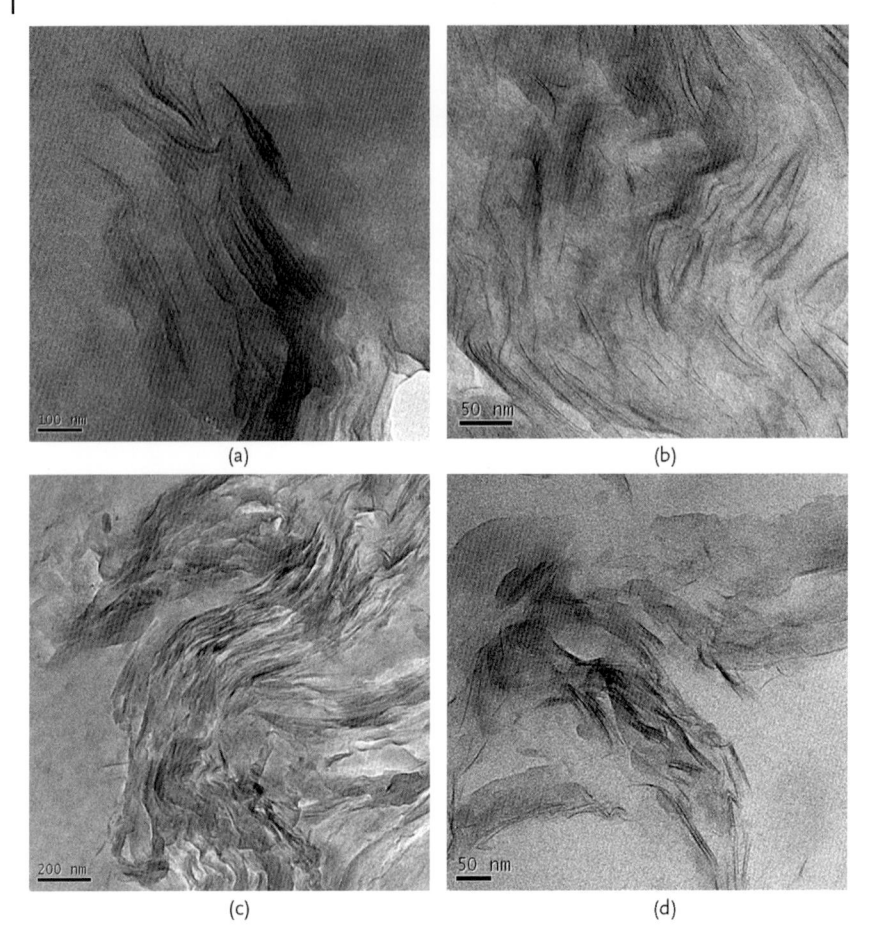

Figure 3.15 TEM images of PU/Cloisite 30B nanocomposite with (a) 1.7 wt% TMP (b) 7.0 wt% TMP and TEM images of PU/Cloisite 25A nanocomposite with (c) 1.7 wt% TMP and (d) 7.0 wt% TMP. (Reproduced from Ref. [150] with permission from John Wiley and Sons.)

hard block and consisted of a trifunctional polyol, 1,4-butanediol and MDI. Commercial PU catalysts were modified by quarternization with 1-bromohexadecane (Figure 3.17). The catalyst bearing the hydroxyl group tethered the polymer to the polymer chains, thus restricting its mobility and promoting reactions predominantly in the intragallery space of the clay. This in turn improved the tensile modulus and decreased carbon dioxide permeability [151].

3.3.3
Effect of Nature of Modifier on Properties of PU Nanocomposites

PU/clay nanocomposites exhibit various property enhancements. In general, when the organic modifier has a reactive group that can tether the polymer chains

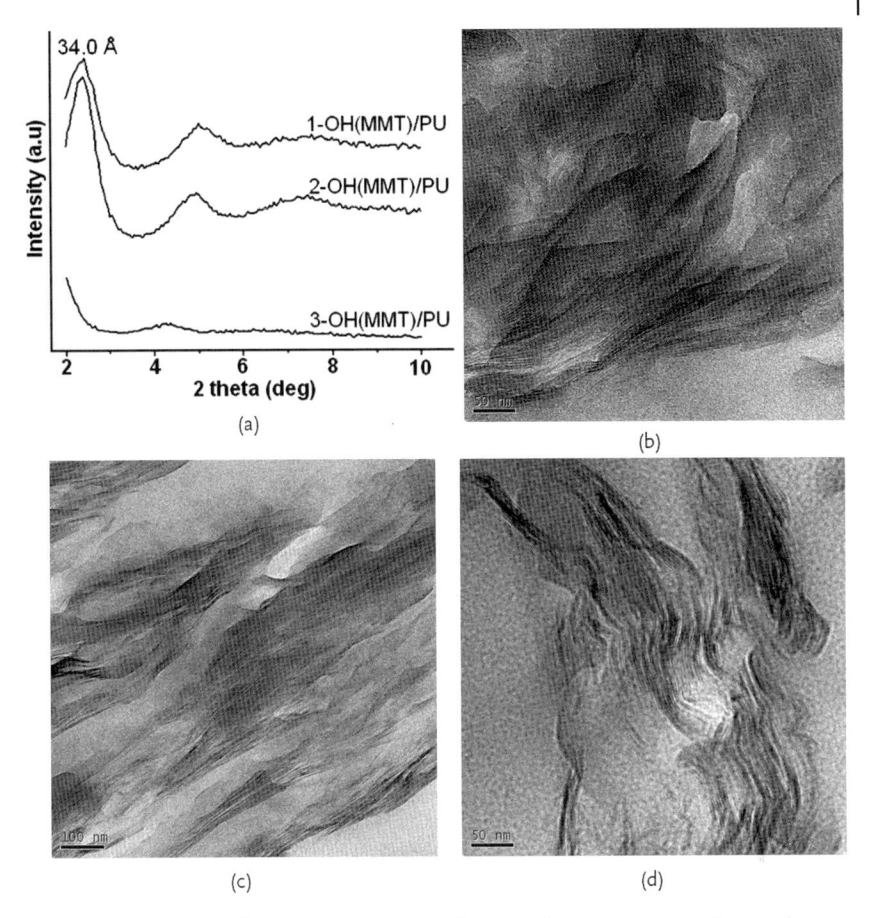

Figure 3.16 (a) WAXD of PU nanocomposites with organoclays containing varying number of hydroxyl groups, and TEM images of (b) 1-OH(MMT)/PU; (c) 2-OH(MMT)/PU; (d) 1-OH(MMT)/PU. (Reproduced from Ref. [150] with permission from John Wiley and Sons.)

on the intragallery clay surface, improved clay dispersion is observed. This, in turn, results in enhanced mechanical properties such as improved tensile strength, tensile modulus [39, 60, 76, 77, 85], and storage modulus [7]. The percentage elongation at break improved with the increasing content of clay; however, beyond a critical content (3%), the property deteriorated [85, 142, 143]. Similarly, the adhesive strength of PU nanocomposites prepared with organoclay having an organic modifier with a reactive group increased with the content of the clay, while the composites prepared with unmodified clay did not show any improvement in adhesive strength compared with the pristine polymer [60]. The water absorption capacity of the nanocomposites were reduced when clays used contained organic modifiers having reactive groups [39, 44]. Similarly nanocomposites showed enhanced abrasion resistance when the clay used contained reactive organic modifiers [142]. When the organoclay used contain reactive modifiers, the transparency

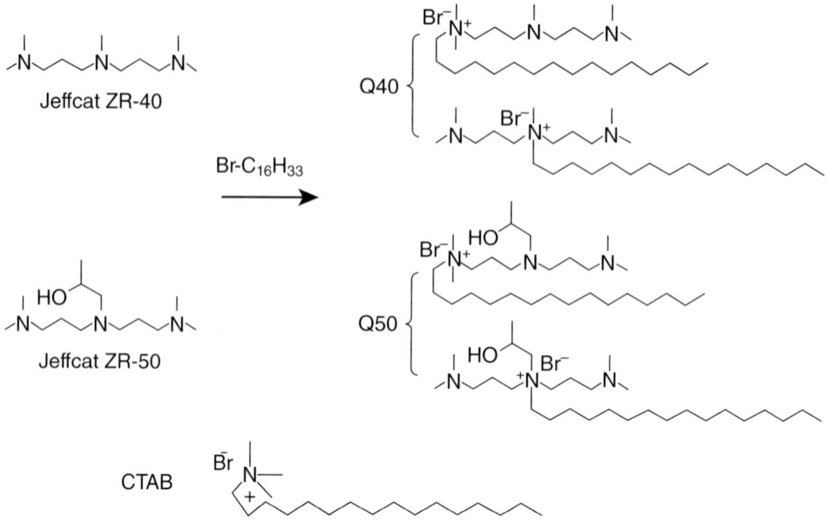

Figure 3.17 Structures of functional amine catalysts bearing quarternary ammonium cations. (Reproduced from Ref. [151] with permission from Elsevier.)

of the nanocomposites were better due to effective nanolayer dispersion of clay in the polymer matrix [85]. Among the organoclay modifiers used, those which induced better dispersion of clay in the polymer matrix and stronger interaction between the clay layers and polymer chains resulted in better property enhancements [76, 150].

3.4
Conclusions

The key to superior properties of PU nanocomposites is critically dependent on the choice of the organic modifier used to modify the surface of clay as well as the nature of the polymer itself. A judicious choice of organic modifiers with reactive functionalities and nature of monomers used is necessary to promote compatibility between clay particles and the polymer matrix.

A number of structural features of the organic modifiers are recognized as being critical. These are the length of the alkyl chain attached to the ammonium ion, the nature of functionality attached to the alkyl ammonium ion and the degree of functionality. Equally important is the nature of monomers used, degree of branching of the polyol, and the polarity of the polyol. The method of nanocomposite preparation also plays an important role. *In situ* polymerization, where the low-molecular-weight monomers react inside the clay gallery with the organic modifier, resulting in polymers tethered to the surface of the clay that lead to most effective clay dispersions and more pronounced property improvements. There are several reports showing that the presence of reactive groups in the organic modifier can form urethane links with polymer chains as well as improved hydrogen bonding between clay layers and the polymer [85, 99, 142]. Therefore, the combined effect of increased dispersion of clay layers and improved interaction between the polymer chains and the clay layers show positive effects. Our understanding of the interface between clay surface and the polymers is still rudimentary. Therefore, all approaches have been somewhat empirical. The goal of rational tailoring of the interface in an optimal way still remains elusive.

Glossary

MDI	4,4′-diphenyl methane diisocyanate
MMT	montmorillonite
PBS	poly(butylene succinate)
PEG	polyethylene glycol
PMDI	polymeric 4,4′-diphenyl methane diisocyanate
PPG	polypropylene glycol
PU	polyurethane
TEM	transmission electron microscope
TMP	trimethylol propane
TPU	thermoplastic polyurethane elastomers
WAXD	wide-angle X-ray diffraction

References

1 Mark, J.E. (1996) Ceramic-reinforced polymers and polymer-modified ceramics. *Polym. Eng. Sci.*, **36**, 2905–2920.

2 Reynaud, E., Gauthier, C., and Perez, J. (1999) Nanophases in polymers. *Rev. Metall./Cah. Inf. Tech.*, **96** (2), 169–176.

3 von Werne, T. and Patten, T.E. (1999) Preparation of structurally well-defined polymer nanoparticle hybrids with controlled/living polymerizations. *J. Am. Chem. Soc.*, **121**, 7409–7410.

4 Herron, N. and Thorn, D.L. (1998) Nanoparticles: uses and relationships to molecular cluster compounds. *Adv. Mater.*, **10**, 1173–1184.

5 Calvert, P. (1997) Potential applications of nanotubes, in *Carbon Nanotubes* (ed. T.W. Ebbesen), CRC Press, Boca Raton, FL, p. 277.

6 Favier, V., Canova, G.R., Shrivastava, S.C., and Cavaille, J.Y. (1997) Mechanical percolation in cellulose whiskers nanocomposites. *Polym. Eng. Sci.*, **37**, 1732–1739.

7 Chazeau, L., Cavaille, J.Y., Canova, G., Dendievel, R., and Boutherin, B. (1999) Viscoelastic properties of plasticized PVC reinforced with cellulose whiskers. *J. Appl. Polym. Sci.*, **71**, 1797–1808.

8 Theng, B.K.G. (1974) *The Chemistry of Clay-Organic Reactions*, John Wiley & Sons, Inc., New York.

9 Ogawa, M. and Kuroda, K. (1997) Preparation of inorganic-organic nanocomposites through intercalation of organoammonium ions into layered silicates. *Bull. Chem. Soc. Jpn.* **70**, 2593–2618.

10 Pinnavaia, T.J. and Beall, G.W. (2000) *Polymer Clay Nanocomposites*, John Wiley and Sons Ltd, Chichester, UK.

11 Koo, J.H. (2006) *Polymer Nanocomposites: Processing, Characterization and Applications*, McGraw-Hill Publications, New York.

12. Mai, Y.-W. and Yu, Z.-Z. (2006) *Polymer Nanocomposites, Woodhead Publishing in Materials*, CRC Press, Boca Raton, FL.
13. LeBaron, P.C., Wang, Z., and Pinnavaia, T.J. (1999) Polymer-layered silicate nanocomposites: an overview. *Appl. Clay Sci.*, **15**, 11–29.
14. Alexandre, M. and Dubois, P. (2000) Polymer-layered silicate nanocomposites: preparation, properties and uses of a new class of materials. *Mater. Sci. Eng. R: Rep.*, **28**, 1–63.
15. Ray, S.S. and Okamoto, M. (2003) Polymer/layered silicate nanocomposites: a review from preparation to processing. *Prog. Polym. Sci.*, **28**, 1539–1641.
16. Viswanathan, V., Laha, T., Balani, K., Agarwal, A., and Seal, S. (2006) Challenges and advances in nanocomposite processing techniques. *Mater. Sci. Eng. R: Rep.*, **54**, 121–285.
17. Nguyen, Q.T. and Baird, D.G. (2006) Preparation of polymer–clay nanocomposites and their properties. *Adv. Polym. Technol.*, **25**, 270–285.
18. Leszynska, A., Njuguna, J., Pielichowski, K., and Banerjee, J.R. (2007) Polymer/montmorillonite nanocomposites with improved thermal properties: part I. Factors influencing thermal stability and mechanisms of thermal stability improvement. *Thermochim. Acta*, **453**, 75–96.
19. Leszynska, A., Njuguna, J., Pielichowski, K., and Banerjee, J.R. (2007) Polymer/montmorillonite nanocomposites with improved thermal properties: part II. Thermal stability of montmorillonite nanocomposites based on different polymeric matrixes. *Thermochim. Acta*, **454**, 1.
20. Vaia, R.A. and Maguire, J.F. (2007) Polymer nanocomposites with prescribed morphology: going beyond nanoparticle-filled polymers. *Chem. Mater.*, **19**, 2736–2751.
21. Chen, B., Evans, J.R.G., Greenwell, H.C., Boulet, P., Coveney P.V., Bowden, A.A., and Whiting, A. (2008) A critical appraisal of polymer–clay nanocomposites. *Chem. Soc. Rev.*, **37**, 568–594.
22. Esfandiari, A., Nazokdast, H., Rashidi, A.S., and Yazdanshenas, M.-E. (2008) Review of polymer–organoclay nanocomposites. *J. Appl. Sci.*, **8**, 545–561.
23. Yeh, J.-M. and Chang, K.-C. (2008) Polymer/layered silicate nanocomposite anticorrosive coatings. *J. Ind. Eng. Chem.*, **14**, 275–291.
24. Ciardelli, F., Coiai, S., Passaglia, E., Pucci, A., and Ruggeri, G. (2008) Reactive blending of polyolefins to nanostructured functional thermoplastic materials. *Polym. Int.*, **57**, 805–836.
25. Paul, D.R. and Robson, L.M. (2008) Polymer nanotechnology: nanocomposites. *Polymer*, **49**, 3187–3204.
26. Meckel, W., Goyert, W., and Wieder, W. (1987) *Thermoplastic Elastomers*, Hanser, Munich.
27. Frisch, K.C. (1980) Recent developments in urethane elastomers and reaction injection molded (RIM) elastomers. *Rubber Chem. Technol.*, **53**, 126–140.
28. Bayer, O., Muller, E., Petersen, S., Piepenbrink, H.F., and Windemuth, E. (1950) Polyurethanes: VI. New highly elastic synthesis. "Vulcollans". *Angew. Chem.*, **62**, 57–66.
29. Goda, H. and Frank, C.W. (2001) Fluorescence studies of the hybrid composite of segmented-polyurethane and silica. *Chem. Mater.*, **13**, 2783–2787.
30. Furukawa, M. and Yokoyama, T. (1994) Mechanical properties of organic–inorganic polyurethane elastomers. I. Al(OH)$_3$–polyurethane composites based on PPG. *J. Appl. Polym. Sci.*, **53**, 1723–1729.
31. Dolui, S.K. (1994) Unusual effect of filler (CaCO$_3$) on thermal degradation of polyurethane. *J. Appl. Polym. Sci.*, **53**, 463–465.
32. Feldman, D. and Lacasse, M.A. (1994) Polymer–filler interaction in polyurethane kraft lignin polyblends. *J. Appl. Polym. Sci.*, **51**, 701–709.
33. Otterstedt, E.A., Ekdahl, J., and Backman, J. (1987) Fine colloidal silica as reinforcing filler in polyurethane polymers. *J. Appl. Polym. Sci.*, **34**, 2575.
34. Nunes, R.C.R., Fonesca, J.L.C., and Pereira, M.R. (2000) Polymer–filler

interactions and mechanical properties of a polyurethane elastomer. *Polym. Test.*, **19**, 93–103.
35 Hepburn, C. (1982) *Polyurethane elastomer*, Applied Science Publishers, London.
36 Khudyakov, I.V., Zopf, D.R., and Turro, N.J. (2009) Polyurethane nanocomposites. *Des. Monomers Polym.*, **12**, 279–290.
37 Kojima, Y., Usuki, A., Kawasumi, M., Okada, A., Fukushima, Y., and Kurauchi, T. (1993) Mechanical properties of nylon 6–clay hybrid. *J. Mater. Res.*, **8**, 1185–1189.
38 Fukushima, Y., Okada, A., Kawasumi, M., Kurauchi, T., and Kamigaito, O. (1998) Swelling behavior of montmorillonite by poly-6-amide. *Clay Miner.*, **23**, 27–34.
39 Ni, P., Li, J., Suo, J.S., and Li, S.B. (2004) Novel polyether polyurethane/clay nanocomposites synthesized with organic-modified montmorillonite as chain extenders. *J. Appl. Polym. Sci.*, **94**, 534–541.
40 Ni, P., Wang, Q.L., Li, J., Suo, J.S., and Li, S.B. (2006) Novel polyether polyurethane/clay nanocomposites synthesized with organicly modified montmorillonite as chain extenders. *J. Appl. Polym. Sci.*, **99**, 6–13.
41 Zilg, C., Thomann, R., Muelhaupt, R., and Finter, J. (1999) Polyurethane nanocomposites containing laminated anisotropic nanoparticles derived from organophilic layered silicates. *Adv. Mater.*, **11**, 49–52.
42 Petrovic, Z.S., Javni, I., Waddon, A., and Banhegyi, G. (2000) Structure and properties of polyurethane–silica nanocomposites. *J. Appl. Polym. Sci.*, **76**, 133–151.
43 Chen, T.K., Tien, Y.I., and Wei, K.H. (1999) Synthesis and characterization of novel segmented polyurethane/clay nanocomposite via poly(ε-caprolactone)/clay. *J. Polym. Sci. Part A: Polym. Chem.*, **37**, 2225–2233.
44 Chen, T.K., Tien, Y.I., and Wei, K.H. (2000) Synthesis and characterization of novel segmented polyurethane/clay nanocomposites. *Polymer*, **41**, 1345–1353.
45 Xu, R., Manias, E., Snyder, A.J., and Runt, J. (2001) New biomedical poly(urethane urea)–layered silicate nanocomposites. *Macromolecules*, **34**, 337–339.
46 Ma, J., Zhang, S., and Qi, Z. (2001) Synthesis and characterization of elastomeric polyurethane/clay nanocomposites. *J. Appl. Polym. Sci.*, **82**, 1444–1448.
47 Tien, Y.I. and Wei, K.H. (2001) High tensile-property layered silicate/polyurethane nanocomposites by using reactive silicates as pseudo chain extenders. *Macromolecules*, **34**, 9045–9052.
48 Tien, Y.I. and Wei, K.H. (2001) Hydrogen bonding and mechanical properties in segmented montmorillonite/polyurethane nanocomposites of different hard segment ratios. *Polymer*, **42**, 3213–3221.
49 Hu, Y., Song, L., Xu, J., Yang, L., Chen, Z., and Fan, W. (2001) Synthesis of polyurethane/clay intercalated nanocomposites. *Colloid Polym. Sci.*, **279**, 819–822.
50 Yao, K.J., Song, M., Hourston, D.J., and Luo, D.Z. (2002) Polymer/layered clay nanocomposites: 2 polyurethane nanocomposites. *Polymer*, **43**, 1017–1020.
51 Tien, Y.I. and Wei, K.H. (2002) Effect of nanosized silicate layers from montmorillonite on glass transition, dynamic mechanical, and thermal degradation properties of segmented polyurethane. *J. Appl. Polym. Sci.*, **86**, 1741–1748.
52 Chang, J.H. and An, Y.U. (2002) Nanocomposites of polyurethane with various organoclays: thermomechanical properties, morphology, and gas permeability. *J. Polym. Sci. Part B: Polym. Phys.*, **40**, 670–677.
53 Tortora, M., Gorrasi, G., Vittoria, V., Galli, G., Ritrovati, S., and Chiellini, E. (2002) Structural characterization and transport properties of organically modified montmorillonite/polyurethane nanocomposites. *Polymer*, **43**, 6147–6157.
54 Zhang, X., Xu, R., Wu, Z., and Zhou, C. (2003) The synthesis and characterization

of polyurethane/clay nanocomposites. *Polym. Int.*, **52**, 790–794.

55 Song, M., Hourston, D.J., Yao, K.J., Tay, J.K.H., and Ansarifar, M.A. (2003) High performance nanocomposites of polyurethane elastomer and organically modified layered silicate. *J. Appl. Polym. Sci.*, **90**, 3239–3243.

56 Mishra, J.K., Kim, I., and Ha, C.S. (2003) New millable polyurethane/organoclay nanocomposite: preparation, characterization and properties. *Macromol. Rapid Commun.*, **24**, 671–675.

57 Chen, X., Wu, L., Zhou, S., and You, B. (2003) In situ polymerization and characterization of polyester-based polyurethane nano-silca composites. *Polym. Int.*, **52**, 993–998.

58 Rhoney, I., Brown, S., Hudson, N.E., and Pethrik, R.A. (2003) Influence of processing method on the exfoliation process for organically modified clay systems. I. Polyurethanes. *J. Appl. Polym. Sci.*, **91**, 1335–1343.

59 Osman, M.A., Mittal, V., Morbidelli, M., and Suter, U.W. (2003) Polyurethane adhesive nanocomposites as gas permeation barrier. *Macromolecules*, **36**, 9851–9858.

60 Chavarria, F. and Paul, D.R. (2006) Morphology and properties of thermoplastic polyurethane nanocomposites: effect of organoclay structure. *Polymer*, **47**, 7760–7773.

61 Naguib, H.F., Abdel, A., Mohamed, S., Sherif, S.M., and Saad, G.R. (2012) Thermal properties of biodegradable poly(PHB/PCL-PEG-PCL) urethanes nanocomposites using clay/poly(caprolactone) nanohybrid based masterbatch. *Appl. Clay Sci.*, **57**, 55–63.

62 Koo, J.H., Ezekoye, O.A., Lee, J.C., Ho, W.K., and Bruns, M.C. (2011) Rubber–clay nanocomposites based on thermoplastic elastomers, in *Rubber-Clay Nanocomposites* (ed. M. Galimberti), John Wiley & Sons, Inc., New York, pp. 489–521.

63 Kaushik, A., Ahuja, D., and Salwani, V. (2011) Synthesis and characterization of organically modified clay/castor oil based chain extended polyurethane nanocomposites. *Compos., Part A: Appl. Sci. Manuf.*, **42A** (10), 1534–1541.

64 Aslzadeh, M.M., Sadeghi, G.M.M., and Abdouss, M. (2012) Synthesis and characterization of chlorine-containing flame-retardant polyurethane nanocomposites via *in situ* polymerization. *J. Appl. Polym. Sci.*, **123** (1), 437–447.

65 Sheng, D., Tan, J., Liu, X., Wang, P., and Yang, Y. (2011) Effect of organoclay with various organic modifiers on the morphological, mechanical, and gas barrier properties of thermoplastic polyurethane/organoclay nanocomposites. *J. Mater. Sci.*, **46** (20), 6508–6517.

66 Barick, A.K. and Tripathy, D.K. (2011) Effect of organically modified layered silicate nanoclay on the dynamic viscoelastic properties of thermoplastic polyurethane nanocomposites. *Appl. Clay Sci.*, **52** (3), 312–321.

67 Mishra, A.K., Rajamohanan, P.R., Nando, G.B., and Chattopadhyay, S. (2011) Structure-property of thermoplastic polyurethane–clay nanocomposite based on covalent and dual-modified laponite. *Adv. Sci. Lett.*, **4** (1), 65–73.

68 Tabuani, D., Bellucci, F., Terenzi, A., and Camino, G. (2012) Flame retarded thermoplastic polyurethane (TPU) for cable jacketing application. *Polym. Degrad. Stab.*, **97** (12), 2594–2601.

69 Spirkova, M., Pavlicevic, J., Strachota, A., Poreba, R., Bera, O., Kapralkova, L., Baldrian, J., Slouf, M., Lazic, N., and Budinski-Simendic, J. (2011) Novel polycarbonate-based polyurethane elastomers: composition-property relationship. *Eur. Polym. J.*, **47** (5), 959–972.

70 Cho, T.-W. and Kim, S.-W. (2011) Morphologies and properties of nanocomposite films based on a biodegradable poly(ester)urethane elastomer. *J. Appl. Polym. Sci.*, **121** (3), 1622–1630.

71 Mishra, A.K., Chattopadhyay, S., Rajamohanan, P.R., and Nando, G.B. (2011) Effect of tethering on the structure–property relationship of TPU-dual modified Laponite clay nanocomposites prepared by *ex-situ* and

in-situ techniques. *Polymer*, **52** (4), 1071–1083.

72 Mondal, M., Chattopadhyay, P.K., Chattopadhyay, S., and Setua, D.K. (2010) Thermal and morphological analysis of thermoplastic polyurethane–clay nanocomposites: comparison of efficacy of dual modified laponite vs. commercial montmorillonites. *Thermochim. Acta*, **510** (1–2), 185–194.

73 Xu, B., Fu, Y.Q., Huang, W.M., Pei, Y.T., Chen, Z.G., De Hosson, J.T.M., Kraft, A., and Reuben, R.L. (2010) Thermal–mechanical properties of polyurethane-clay shape memory polymer nanocomposites. *Polymers*, **2** (2), 31–39.

74 Koo, J.H., Nguyen, K.C., Lee, J.C., Ho, X., Wai, K., Bruns, M.C., and Ezekoye, O.A. (2010) Flammability studies of a novel class of thermoplastic elastomer nanocomposites. *J. Fire Sci.*, **28** (1), 49–85.

75 Meng, X., Wang, Z., Yu, H., Du, X., Li, S., Wang, Y., Jiang, Z., Wang, Q., and Tang, T. (2009) A strategy of fabricating exfoliated thermoplastic polyurethane/clay nanocomposites via introducing maleated polypropylene. *Polymer*, **50** (16), 3997–4006.

76 Kim, W., Chung, D., and Kim, J.H. (2008) Effect of length of hydroxyalkyl groups in the clay modifier on the properties of thermoplastic polyurethane/clay nanocomposites. *J. Appl. Polym. Sci.*, **110** (5), 3209–3216.

77 Dan, C.H., Kim, Y.D., Lee, M., Min, B.H., and Kim, J.H. (2008) Effect of solvent on the properties of thermoplastic polyurethane/clay nanocomposites prepared by solution mixing. *J. Appl. Polym. Sci.*, **108** (4), 2128–2138.

78 Dan, C.H., Kim, W.T., and Kim, J.H. (2007) Effect of solvent on the properties of thermoplastic polyurethane/clay nanocomposites. *PMSE Preprints*, **96**, 641–642.

79 Choi, M.Y., Anandhan, S., Youk, J.H., Baik, D.H., Seo, S.W., and Lee, H.S. (2006) Synthesis and characterization of *in situ* polymerized segmented thermoplastic elastomeric polyurethane/layered silicate clay nanocomposites. *J. Appl. Polym. Sci.*, **102** (3), 3048–3055.

80 Dan, C.H., Lee, M.H., Kim, Y.D., Min, B.H., and Kim, J.H. (2006) Effect of clay modifiers on the morphology and physical properties of thermoplastic polyurethane/clay nanocomposites. *Polymer*, **47** (19), 6718–6730.

81 Dan, C.H., Lee, M.H., Min, B.H., and Kim, J.H. (2006) The morphology and physical properties of thermoplastic polyurethane/clay nanocomposites. *PMSE Prepr.*, **94**, 532–533.

82 Pattanayak, A. and Jana, S.C. (2005) Thermoplastic polyurethane nanocomposites of reactive silicate clays. 63rd Annual Technical Conference–Society of Plastics Engineers, pp. 1933–1937.

83 Pattanayak, A. and Jana, S.C. (2005) Thermoplastic polyurethane nanocomposites of reactive silicate clays: effects of soft segments on properties. *Polymer*, **46** (14), 5183–5193.

84 Pattanayak, A. and Jana, S.C. (2005) Properties of bulk-polymerized thermoplastic polyurethane nanocomposites. *Polymer*, **46** (10), 3394–3406.

85 Pattanayak, A. and Jana, S.C. (2005) Synthesis of thermoplastic polyurethane nanocomposites of reactive nanoclay by bulk polymerization methods. *Polymer*, **46** (10), 3275–3288.

86 Marchant, D., Koo, J.H., Blanski, R.L., Weber, E.H., Ruth, P.N., Lee, A., and Schlaefer, C.E. (2004) Flammability and thermophysical characterization of thermoplastic elastomer nanocomposites. *PMSE Prepr.*, **91**, 32–33.

87 Pattanayak, A. and Jana, S.C. (2004) Polyurethane-clay nanocomposites via bulk polymerization methods. 62nd (Vol. 2) Annual Technical Conference–Society of Plastics Engineers, pp. 1962–1966.

88 Finnigan, B., Martin, D., Halley, P., Truss, R., and Campbell, K. (2004) Morphology and properties of thermoplastic polyurethane nanocomposites incorporating hydrophilic layered silicates. *Polymer*, **45** (7), 2249–2260.

89 Pattanayak, A. and Jana, S.C. (2003) A study on intercalation and exfoliation of layered silicate nanoparticles in thermoplastic polyurethanes. 61st (Vol. 2) Annual Technical Conference–Society of Plastics Engineers, pp. 1424–1428.

90 Joulazadeh, M. and Navarchian, A.H. (2011) Study on elastic modulus of crosslinked polyurethane/organoclay nanocomposites. *Polym. Adv. Technol.*, **22** (12), 2022–2031.

91 Zia, K.M., Zuber, M., Barikani, M., Hussain, R., Jamil, T., and Anjum, S. (2011) Cytotoxicity and mechanical behavior of chitin-bentonite clay based polyurethane bio-nanocomposites. *Int. J. Biol. Macromol.*, **49** (5), 1131–1136.

92 Deka, H. and Karak, N. (2011) Bio-based hyperbranched polyurethane/clay nanocomposites: adhesive, mechanical, and thermal properties. *Polym. Adv. Technol.*, **22** (6), 973–980.

93 Corcione, C.E., Maffezzoli, A., and Cannoletta, D. (2009) Effect of a nanodispersed clay fillers on glass transition of thermosetting polyurethane. *Macromol. Symp.*, **286**, 180–186.

94 Esposito, C.C., Mensitieri, G., and Maffezzoli, A. (2009) Analysis of the structure and mass transport properties of nanocomposite polyurethane. *Polym. Eng. Sci.*, **49** (9), 1708–1718.

95 Deka, H. and Karak, N. (2009) Vegetable oil-based hyperbranched thermosetting polyurethane/clay nanocomposites. *Nanoscale Res. Lett.*, **4** (7), 758–765.

96 Maji, P.K., Guchhait, P.K., and Bhowmick, A.K. (2009) Effect of the microstructure of a hyperbranched polymer and nanoclay loading on the morphology and properties of novel polyurethane nanocomposites. *ACS Appl. Mater. Interfaces*, **1** (2), 289–300.

97 Jia, Q., Shan, S., Wang, Y., Gu, L., and Li, J. (2008) Tribological performance and thermal behavior of epoxy resin nanocomposites containing polyurethane and organoclay. *Polym. Adv. Technol.*, **19** (7), 859–864.

98 Tan, H. and Nie, J. (2007) Photopolymerization and characteristics of polyurethane/organoclay nanocomposites. *Macromol. React. Eng.*, **1** (3), 384–390.

99 Dean, K.M., Bateman, S.A., and Simons, R. (2007) A comparative study of UV active silane-grafted and ion-exchanged organo-clay for application in photocurable urethane acrylate nano- and micro-composites. *Polymer*, **48** (8), 2231–2240.

100 Sreedhar, B., Chattopadhyay, D.K., and Swapna, V. (2006) Thermal and surface characterization of polyurethane–urea clay nanocomposite coatings. *J. Appl. Polym. Sci.*, **100** (3), 2393–2401.

101 Jia, Q., Zheng, M., Shen, R., and Chen, H. (2006) Synthesis, characterization and properties of organo clay-modified polyurethane/epoxy interpenetrating polymer network nanocomposites. *Polym. Int.*, **55** (3), 257–264.

102 Li, J. (2006) High performance epoxy resin nanocomposites containing both organic montmorillonite and castor oil-polyurethane. *Polym. Bull.*, **56** (4–5), 377–384.

103 Xia, H. and Song, M. (2006) Intercalation and exfoliation behavior of clay layers in branched polyol and polyurethane/clay nanocomposites. *Polym. Int.*, **55** (2), 229–235.

104 Harikrishnan, G., Patro, T.U., Unni, A.R., and Khakhar, D.V. (2011) Clay nanoplatelet induced morphological evolutions during polymeric foaming. *Soft Matter*, **7** (15), 6801–6804.

105 Valizadeh, M., Rezaei, M., and Eyvazzadeh, A. (2011) Effect of nanoclay on the mechanical and thermal properties of rigid polyurethane/organoclay nanocomposite foams blown with cyclo and normal pentane mixture. *Key Eng. Mater.*, **471–472** (Pt. 1, Composite Science and Technology), 584–589.

106 Zainuddin, F., Ahmad, S., Rasid, R., and Adam, S.N.F.S. (2010) Preparation and characterizations of palm oil based rigid nanocomposite polyurethane foam. *Solid State Sci. Technol.*, **18** (1), 121–128.

107 Kim, S.H., Lee, M.C., Kim, H.D., Park, H.C., Jeong, H.M., Yoon, K.S., and Kim, B.K. (2010) Nanoclay reinforced rigid polyurethane foams. *J. Appl. Polym. Sci.*, **117** (4), 1992–1997.

108 Harikrishnan, G., Lindsay, C.I., Arunagirinathan, M.A., and Macosko, C.W. (2009) Probing nanodispersions of clays for reactive foaming. *ACS Appl. Mater. Interfaces*, **1** (9), 1913–1918.

109 Patro, T.U., Harikrishnan, G., Misra, A., and Khakhar, D.V. (2008) Formation and characterization of polyurethane–vermiculite clay nanocomposite foams. *Polym. Eng. Sci.*, **48** (9), 1778–1784.

110 Han, M.S., Kim, Y.H., Han, S.J., Choi, S.J., Kim, S.B., and Kim, W.N. (2008) Effects of a silane coupling agent on the exfoliation of organoclay layers in polyurethane/organoclay nanocomposite foams? *J. Appl. Polym. Sci.*, **110** (1), 376–386.

111 Xu, Z., Tang, X., Gu, A., Fang, Z., and Tong, L. (2007) Surface-modifiers of clay on mechanical properties of rigid polyurethane foams/organoclay nanocomposites. *J. Appl. Polym. Sci.*, **105** (5), 2988–2995.

112 Mondal, P., and Khakhar, D.V. (2007) Rigid polyurethane–clay nanocomposite foams: preparation and properties. *J. Appl. Polym. Sci.*, **103** (5), 2802–2809.

113 Harikrishnan, G., Patro, T.U., and Khakhar, D.V. (2006) Polyurethane foam–clay nanocomposites: nanoclays as cell openers. *Ind. Eng. Chem. Res.*, **45** (21), 7126–7134.

114 Cao, X., Lee, L.J., Widya, T., and Macosko, C. (2005) Polyurethane/clay nanocomposites foams: processing, structure and properties. *Polymer*, **46** (3), 775–783.

115 Cao, X., Lee, L., Widya, T., and Macosko, C. (2004) Structure and properties of polyurethane/clay nanocomposites and foams. 62nd (Vol. 2) Annual Technical Conference – Society of Plastics Engineers, pp. 1896–1900.

116 Noh, H.H., Lee, J.K., Liu, X., and Choi, Y.M. (2012) Synthesis and characterization of siloxane-modified polyurethane dispersion/clay nanocomposites. *Adv. Mater. Res.*, **415–417** (Pt. 2, Advanced Materials), 1196–1199.

117 Delpech, M.C., Miranda, G.S., Espirito, X., and Santo, W.L. (2011) Aqueous dispersions based on nanocomposites of polyurethanes and hydrophilic Brasilian clays: synthesis and characterization. *Polimeros: Ciencia e Tecnologia*, **21** (4), 315–320.

118 Horta Pinto, F.C., Silva-Cunha, A., Pianetti, G.A., Ayres, E., Orefice, R.L., and Da Silva, G.R. (2011) Montmorillonite clay-based polyurethane nanocomposite as local triamcinolone acetonide delivery system. *J. Nanomater.*, **2011**, article ID: 528628.

119 Steele, A., Bayer, I., Yeong, Y.H., De Combarieu, G., Lakeman, C., Deglaire, P., and Loth, E. (2011) Adhesion strength and superhydrophobicity in polyurethane/organoclay nanocomposites (eds M. Laudon and B. Romanowicz). Nanotech Conference & Expo 2011: An Interdisciplinary Integrative Forum on Nanotechnology, Biotechnology and Microtechnology, Boston, MA, United States, June 13–16, 2011, 1 399–402.

120 Ashhari, S., Sarabi, A.A., Kasiriha, S.M., and Zaarei, D. (2011) Aliphatic polyurethane–montmorillonite nanocomposite coatings: preparation, characterization, and anticorrosive properties. *J. Appl. Polym. Sci.*, **119** (1), 523–529.

121 Zuber, M., Zia, K.M., Mahboob, S., Hassan, M., and Bhatti, I.A. (2010) Synthesis of chitin-bentonite clay based polyurethane bio-nanocomposites. *Int. J. Biol. Macromol.*, **47** (2), 196–200.

122 Jin, H., Wie, J.J., and Kim, S.C. (2010) Effect of organoclays on the properties of polyurethane/clay nanocomposite coatings. *J. Appl. Polym. Sci.*, **117** (4), 2090–2100.

123 Gao, C., Ni, J., Zheng, Q., Chen, S., Lin, W., and Tu, J. (2010) Preparation and properties of waterborne polysiloxaneurethane/clay nanocomposite. *Youjigui Cailiao*, **24** (1), 12–18.

124 Huh, J.H., Rahman, M.M., and Kim, H.-D. (2009) Properties of waterborne polyurethane/clay nanocomposite adhesives. *J. Adhes. Sci. Technol.*, **23** (5), 739–751.

125 Rahman, M.M., Kim, H.-D., and Lee, W.-K. (2008) Preparation and characterization of waterborne polyurethane/clay nanocomposite: effect

on water vapor permeability. *J. Appl. Polym. Sci.*, **110** (6), 3697–3705.
126 Yeh, J.-M., Yao, C.-T., Hsieh, C.-F., Lin, L.-H., Chen, P.-L., Wu, J.-C., Yang, H.-C., and Wu, C.-P. (2008) Preparation, characterization and electrochemical corrosion studies on environmentally friendly waterborne polyurethane/Na$^+$–MMT clay nanocomposite coatings. *Eur. Polym. J.*, **44** (10), 3046–3056.
127 Madbouly, S.A., Otaigbe, J.U., Nanda, A.K., and Wicks, D.A. (2008) Synthesis of polyurethane–urea/clay nanocomposite from aqueous prepolymer dispersions. Abstracts of Papers, 235th ACS National Meeting, New Orleans, LA, United States, April 6–10, 2008. PMSE-491.
128 Ayres, E. and Orefice, R.L. (2007) Nanocomposites derived from polyurethane aqueous dispersion and clay: role of clay on morphology and mechanical properties. *Polimeros: Ciencia e Tecnologia*, **17** (4), 339–345.
129 Subramani, S., Lee, J.-Y., Choi, S.-W., and Kim, J.H. (2007) Waterborne trifunctional silane-terminated polyurethane nanocomposite with silane-modified clay. *J. Polym. Sci. Part B: Polym. Phys.*, **45** (19), 2747–2761.
130 Subramani, S., Choi, S.-W., Lee, J.-Y., and Kim, J.H. (2007) Aqueous dispersion of novel silylated (polyurethane-acrylic hybrid/clay) nanocomposite. *Polymer*, **48** (16), 4691–4703.
131 Rahman, M.M., Yoo, H.-J., Mi, C.J., and Kim, H.-D. (2007) Synthesis and characterization of waterborne polyurethane/clay nanocomposite–effect on adhesive strength. *Macromol. Symp.*, **249/250** (Advanced Polymers for Emerging Technologies), 251–258.
132 Ayres, E., Orefice, R.L., and Sousa, D. (2006) Influence of bentonite type in waterborne polyurethane nanocomposite mechanical properties. *Macromol. Symp.*, **245/246** (World Polymer Congress–MACRO 2006), 330–336.
133 Subramani, S., Lee, J.-Y., Kim, J.H., and Cheong, I.W. (2007) Crosslinked aqueous dispersion of silylated poly(urethane-urea)/clay nanocomposites. *Compos. Sci. Technol.*, **67** (7–8), 1561–1573.
134 Lee, H.-T., Hwang, J.-J., and Liu, H.-J. (2006) Effects of ionic interactions between clay and waterborne polyurethanes on the structure and physical properties of their nanocomposite dispersions. *J. Polym. Sci. Part A Polym. Chem.*, **44** (19), 5801–5807.
135 Lee, H.-T. and Lin, L.-H. (2006) Waterborne polyurethane/Clay nanocomposites: novel effects of the clay and its interlayer ions on the morphology and physical and electrical properties. *Macromolecules*, **39** (18), 6133–6141.
136 Ma, C.-C.M., Kuan, H.-C., Chuang, W.-P., and Su, H.-Y. (2004) Hydrogen bonding, mechanical and physical property, and surface morphology of waterborne polyurethane/clay nanocomposite. Composites Technologies for 2020, Proceedings of the Asian-Australasian Conference on Composite Materials (ACCM-4), 4th, Sydney, Australia, July 6–9, 2004, pp. 731–735.
137 Kuan, H.-C., Chuang, W.-P., Ma, C.-C.M., Chiang, C.-L., and Wu, H.-L. (2005) Synthesis and characterization of a clay/waterborne polyurethane nanocomposite. *J. Mater. Sci.*, **40** (1), 179–185.
138 Jeong, H.M., Jang, K.H., and Cho, K. (2003) Properties of waterborne polyurethanes based on polycarbonate diol reinforced with organophilic clay. *J. Macromol. Sci. Part B: Phys.*, **B42** (6), 1249–1263.
139 Kim, B.K., Seo, J.W., and Jeong, H.M. (2002) Morphology and properties of waterborne polyurethane/clay nanocomposites. *Eur. Polym. J.*, **39** (1), 85–91.
140 Wang, Z. and Pinnavaia, T.J. (1998) Nanolayer reinforcement of elastomeric polyurethane. *Chem. Mater.*, **10**, 3769–3771.
141 Pattanayak, A. and Jana, S.C. (2005) High-strength and low-stiffness composites of nanoclay-filled thermoplastic polyurethanes. *Polym. Eng. Sci.*, **45**, 1532–1539.

142 Xia, H., Shaw, S.J., and Song, M. (2005) Relationship between mechanical properties and exfoliation degree of clay in polyurethane nanocomposites. *Polym. Int.*, **54** (10), 1392–1400.

143 Seo, W.J., Sung, Y.T., Han, S.J., Kim, Y.H., Ryu, O.H., Lee, H.S., and Kim, W. (2006) Synthesis and properties of polyurethane/clay nanocomposite by clay modified with polymeric methane diisocyanate. *J. Appl. Polym. Sci.*, **101** (5), 2879–2883.

144 Choi, W.J., Kim, S.H., Kim, Y.J., and Kim, S.C. (2004) Synthesis of chain-extended organifier and properties of polyurethane/clay nanocomposites. *Polymer*, **45** (17), 6045–6057.

145 Tan, H., Ma, G., Xiao, M., and Nie, J. (2009) Photopolymerization and characteristics of reactive organoclay–polyurethane nanocomposites. *Polym. Compos.*, **30** (5), 612–618.

146 Jeong, E.H., Yang, J., Hong, J.H., Kim, T.G., Kim, J.H., and Youk, J.H. (2007) Effective preparation of montmorillonite/polyurethane nanocomposites by introducing cationic groups into the polyurethane main chain. *Eur. Polym. J.*, **43** (6), 2286–2291.

147 Plummer, C.J.G., Rodlert, M., Bucaille, J.-L., Grunbauer, H.J.M., and Manson, J.-A.E. (2005) Correlating the rheological and mechanical response of polyurethane nanocomposites containing hyperbranched polymers. *Polymer*, **46** (17), 6543–6553.

148 Moon, S.-Y., Kim, J.-K., Nah, C., and Lee, Y.-S. (2004) Polyurethane/montmorillonite nanocomposites prepared from crystalline polyols, using 1,4-butanediol and organoclay hybrid as chain extenders. *Eur. Polym. J.*, **40**, 1615–1621.

149 Singh, C. and Balazs, A.C. (2000) Effect of polymer architecture on the miscibility of polymer/clay mixtures. *Polym. Int.*, **49**, 469–471.

150 Shroff Rama, M. and Sivaram, S. (2010) Influence of structure of organic modifiers and polyurethane on the clay dispersion in nanocomposites via *in situ* polymerization. *J. Appl. Polym. Sci.*, **118** (3), 1774–1786.

151 Qian, Y., Liu, W., Park, Y.T., Lindsay, C.I., Camargo, R., Macosko, C.W., and Stein, A. (2012) Modification with tertiary amine catalysts improves vermiculite dispersion in polyurethane via *in situ* intercalative polymerization. *Polymer*, **53**, 5060–5068.

4
Thermal Properties of Formaldehyde-Based Thermoset Nanocomposites

Byung-Dae Park

4.1
Introduction

Polymers depending on their response to temperature may be classified into two main groups: thermoplastics and thermosets. Thermoplastic polymers become a fluid above a certain temperature, while thermosetting polymers are degraded by heating without going through a fluid state. They are heavily cross-linked polymers that are normally rigid and intractable. The cross-linking in thermosetting polymers makes it impossible to melt once it has undergone cross-linking. Thus, they consist of a dense three-dimensional molecular network on the application of heat. Typical examples of thermosetting polymers are formaldehyde-based resins, unsaturated polyesters, and epoxy resins. The formaldehyde-based thermosetting polymers include melamine-formaldehyde (MF) resins, phenol-formaldehyde (PF) resins, urea-formaldehyde (UF) resins, resorcinol-formaldehyde (RF) resins, and melamine-urea-formaldehyde (MUF) resins. PF resins are regarded the first synthetic polymer in polymer science.

Among these formaldehyde-based thermosetting polymers, MF resin is one of the hardest and stiffest existing polymer systems that provide outstanding scratch resistance and surface gloss as well as good performance and appearance. Thus, MF resins are used to improve the mechanical properties, moisture resistance, or fire resistance in many applications. For example, MF resin is impregnated into cellulose paper to fabricate decorative or protective laminates on the surface of wood-based composite products. These laminate papers are hot pressed on composite wood products to provide a hard surface finish. Various factors of manufacturing resin-impregnated paper also influence on the laminate surface quality and resin distribution [1]. The authors found that the level of UF resin treatment strongly influenced the surface quality of the MF coating in laminates, and also reported that defects in MF coatings occur at the unfilled voids by UF resin.

PF resin is one of the most versatile resins, providing an outstanding performance in adhesion, high-temperature resistance, flame resistance, and electric insulation. Thus, PF resins have been used to improve mechanical properties, moisture resistance, or fire resistance in many applications. For example, PF resin is

Thermoset Nanocomposites, First Edition. Edited by Vikas Mittal.
© 2013 Wiley-VCH Verlag GmbH & Co. KGaA. Published 2013 by Wiley-VCH Verlag GmbH & Co. KGaA.

impregnated into cellulose paper to fabricate decorative or protective laminates on the surface of wood-based composite products. These laminate papers are hot-pressed onto the surface of composite wood products to provide a hard surface finish. Amino resins that include UF resins and MUF resins are mainly used as adhesives for manufacturing wood-based composite panels such as plywood, particleboard, and fiberboard. Hence, wood industry is a major consumer of the amino resins.

In recent years, there has been a tremendous increase in the amount of research, and the number of publications on nanocomposite using nanomaterials such as carbon nanotubes [2, 3], nanoclay [4–6], and nanofibers [7, 8]. Polymer-based nanocomposite is commonly defined as the combination of a polymer matrix and additives that have at least one dimension in the nanometer range. Nanomaterial-reinforced polymer composites have its roots in the pioneering work of researchers at Toyota [9, 10]. They first reported on the improvement in the properties of nylon nanocomposites by incorporating nanoclay into the polymer matrix. Subsequently, an enormous amount of research has been done on nanocomposite materials based on a variety of polymers [5]. Although the majority of research and development has focused on thermoplastics, a large volume of work has been done on nano-reinforced thermosetting polymers. In particular, nanocomposites based on layered nanoclay and epoxy resins have been widely studied [4, 11–13]. In these systems, the elastic forces exerted by the cross-linking epoxy molecules inside the clay galleries lead to clay exfoliation, and accelerate gel formation in the extra-gallery regions. Research on cure kinetics of MF resin-based nanocomposites is very limited. For example, Kohl et al. [14] reported curing behaviors of MF resins using differential scanning calorimetry (DSC), resulting in the activation energies of 81.3 kJ/mol for neat MF resin and 67.8–57.6 kJ/mol for catalyzed MF resins.

As mentioned, PF resin has a three-dimensional structure, which makes it tough for the layered silicate gallery to intercalate even if the lamella of the layered silicate can be easily intercalated by linear polymers. Therefore, a limited number of attempts have been made to understand layered silicates in PF resins. Early studies on organoclay-reinforced linear novolac PF resin nanocomposite demonstrated that the structural affinities between the clay modifier and the polymer enhance intercalation [15, 16]. The same research group also reported an ideal silicate concentration of 3% that greatly improved tensile strength, tensile modulus, toughness, and elongation-at-break [17].

The first study on producing nanocomposites from resole PF resins demonstrated that exfoliation was more difficult with resole PF resins than novolac PF resins, since the former possessed more three-dimensional structure even prior to curing [18]. However, the authors reported that 3–5% clay in resole PF resins was completely exfoliated, thereby significantly improving impact strength of the resultant nanocomposite. Employing a suspension condensation polymerization method, Wu et al. [19] showed that both novolac and resole PF resins were suitable for the preparation of PF resin/clay nanocomposite. However, they also reported that the clay exfoliation or intercalation in novolac was easier than that in resole. Kaynak and Tasan [20] also showed that the cure method, clay amount, and clay modification had significant impact on the mechanical properties of the PF resin/clay nanocomposites. The same group found that unmodified pristine Na-montmorillonite clay provided better strength improvement, ~16% in Charpy impact strength, and 66% in fracture toughness values [21].

Since the discovery of carbon nanotubes (CNTs) in 1991 [22], single-walled CNTs (SWCNTs) or multiwalled CNTs (MWCNTs) have been widely used for fabricating thermoset-based nanocomposites. SWNTs consist of a single sheet of graphene rolled seamlessly to form a cylinder with a diameter of order of 1 nm and length of up to few centimeters [23]. In contrast, MWCNTs consist of an array of such cylinders formed concentrically and separated by 0.35 nm, similar to the basal plane separation in graphite [22]. MWCNTs can have diameters from 2 to 100 nm and lengths of tens of microns. Owing to unique properties such as excellent strength, modulus, electrical and thermal conductivities along with a low density, CNTs resulted in a tremendous attraction for research on polymer-based nanocomposite [3, 24]. For example, tensile properties of SWCNTs were measured inside an electron microscopy, and reported tensile moduli between 0.32 and 1.47 TPa and strengths between 10 and 52 GPa [25]. Tensile modulus and strength using bundles of MWCNTs prepared by chemical vapor deposition (CVD) were reported as 0.45 TPa and ~4 GPa, respectively [26]. Xie *et al.* [26] also employed the same method, and obtained modulus values of 0.27–0.95 TPa and tensile strengths in the range 11–63 GPa. However, there is a variation in these properties of CNTs upon fabrication methods such as laser ablation [27, 28] and CVD [29, 30], or decomposition of CO [10]. In order to exploit excellent mechanical properties of CNTs, varieties of polymers matrices were used to provide with reinforcements as reported in a review by Spitalsky *et al.* [31]. A quite number of works have been done on CNTs reinforced nanocomposite based on thermoset polymers even though it was much less than those on thermoplastic polymers [31]. Since the first report [32], epoxy resins have been widely studied as a potential matrix for CNTs-based nanocomposite due to their wide range of industrial uses [33–36].

Excellent mechanical properties of MWCNTs are expected to provide a very effective reinforcement with PF resin that is impregnated into cellulose paper. A very limited work has been done on PF resin as a matrix for CNTs reinforcement because CNTs are extremely difficult to disperse in a polymer matrix due to its agglomeration caused by van der Waals force [37]. For example, Yin *et al.* [38] used CNTs to reinforce novolac PF resin in the powder form for the preparation of PF resin/graphite composites, and found that the surface modification of CNTs by introducing hydroxyl groups improved bending strength and conductivity of the composites. Liu and Ye [39] also modified MWCNTs by the introduction of carboxyl group, benzene ring, or boric acid, and found an enhanced thermal stability of MWCNTs/PF resin nanocomposite. The improved thermal stability was ascribed to a better interfacial interaction between MWCNTs and PF resin matrix.

As can be expected from the above literature review, nanomaterials such as clay, CNTs, or graphene could be used to improve the properties of formaldehyde-based resins/cellulose systems for a possible application of surface laminates for wood-based composite panels. Since properties of formaldehyde-based resin/cellulose nanocomposites are strongly affected by the degree of resin cure, it is essential to understand the influence of their cure kinetics prior to further developments. Therefore, this chapter attempts to report recent progresses on thermal properties of formaldehyde-based resin/cellulose nanocomposites such as MF resin/clay/cellulose nanocomposites, PF resin/clay/cellulose nanocomposites, and PF resin/MWCNT/cellulose nanocomposites.

4.2
Theoretical Background of Thermal Kinetics

4.2.1
Conventional Kinetics of Thermal Cure and Degradation

The DSC technique is used to monitor either the heat evolution or absorption for any reactions that may have taken place during the heating process. The basic assumption underlying the application of DSC to a cure kinetic is that the measured heat flow (dH/dt) is proportional to the reaction rate ($d\alpha/dt$). In practice, it has proven to be a good assumption. All kinetic models start with the basic rate equation that relates the conversion rate of reactant(s) at constant temperature, $d\alpha/dt$, to a function, which depends on kinetic model applied, $f(\alpha)$, through a rate constant, k,

$$\frac{d\alpha}{dt} = kf(\alpha) \tag{4.1}$$

where α is the degree of chemical conversion, or extent of reaction, k is the rate constant, and $f(\alpha)$ is assumed to be independent of temperature. The degree of conversion (α) is given as

$$\alpha = \frac{(\Delta H_p)_t}{\Delta H_o} \tag{4.2}$$

where $(\Delta H_p)_t$ is the heat released up to a time t and ΔH_o is the total reaction heat of a reaction. In contrast, the single-step kinetic equation for thermal degradation is also expressed as Eq. (4.1) [40, 41]. However, the α represents the extent of conversion ($\alpha = 0 - 1$) in this case. The extent of conversion, α, is experimentally determined by the below equation using thermogravimetry analysis (TGA) data

$$\alpha = \frac{m_0 - m(T)}{m_0 - m_\infty} \tag{4.3}$$

where m_0 is the initial mass, m_∞ is the final mass, and $m(T)$ is the mass at a temperature.

In a usual manner, the temperature dependence is assumed to reside in the rate constant through an Arrhenius relationship given by

$$k(T) = A\exp\left(\frac{-E_a}{RT}\right) \tag{4.4}$$

where A is the preexponential factor or Arrhenius frequency factor (s^{-1}), E_a is the activation energy (J/mol), R is the gas constant (8.314 J/mol K), and T is the absolute temperature (K). Combining Eqs. (4.1) and (4.4) gives the complete rate equation:

$$\frac{d\alpha}{dt} = A\exp\left(\frac{-E_a}{RT}\right)f(\alpha) \tag{4.5}$$

For nonisothermal conditions, Eq. (4.5) is converted into Eq. (4.6) when the temperature is raised at a constant heating rate, β:

$$\beta\frac{d\alpha}{dT} = A\exp\left(\frac{-E_a}{RT}\right)f(\alpha) \tag{4.6}$$

Assuming β = constant, and rearrangement in terms of two variables gives

$$\frac{d\alpha}{f(\alpha)} = \frac{A}{\beta}\exp\left(\frac{-E_a}{RT}\right)dT \tag{4.7}$$

Integration of Eq. (4.7) involves solving the temperature integral in

$$g(\alpha) = \int_0^\alpha \frac{d\alpha}{f(\alpha)} = \frac{A}{\beta}\int_{T_0}^{T_\alpha}\exp\left(\frac{-E_a}{RT}\right)dT = \frac{AE_a}{\beta R}p(x) \tag{4.8}$$

where $x = E/RT$. Since the $p(x)$ function, the temperature integral, has no exact analytical solution, it can be solved using either approximations or numerical integration [41]. One of the simplest approximations by Doyle [42] is given as

$$p(x) = 0.0048e^{-1.0516x} \tag{4.9}$$

From Eqs. (4.7) and (4.8), it follows

$$\ln\beta = -5.331 - 1.0516\left(\frac{-E_a}{RT}\right) + \ln\left(\frac{AE_a}{Rg(\alpha)}\right) \tag{4.10}$$

The E_a can be calculated from a plot of $\ln\beta$ versus $1/T$ of Eq. (4.10), which is a simple relationship of the Ozawa method among multiheating rate methods [43].

By differentiating Eq. (4.5) with time t, following equation is obtained:

$$\frac{d^2\alpha}{dt^2} = \frac{d\alpha}{dt}\left(\frac{df(\alpha)}{d\alpha}A\exp\left(\frac{-E_a}{RT}\right) + \frac{\beta E_a}{RT^2}\right) \tag{4.11}$$

Assuming the reaction rate reaches the maximum at T_p where DSC curve displays the peak

$$\left(\frac{d^2\alpha}{dt^2}\right)_{T_p} = 0 \tag{4.12}$$

Therefore, Eq. (4.12) results in

$$\frac{df(\alpha)}{d\alpha}A\exp\left(\frac{-E_a}{RT}\right) + \frac{\beta E_a}{RT^2} = 0 \tag{4.13}$$

By transforming Eq. (4.13) and taking the logarithm of both sides of the equation, the equation of Kissinger method [44] is obtained as

$$\ln\left(\frac{\beta}{T_p^2}\right) = \ln\left(\frac{df(\alpha)}{d\alpha}\frac{AR}{E_a}\right) - \frac{E_a}{RT_p} \qquad (4.14)$$

which is the equation of a straight line between $\ln(\beta/T_p^2)$ and $-1/T_p$. The E can be calculated from the slope and the preexponential factor from the intercept. Thus, the Kissinger method provides a single value of activation energy for the whole reaction process.

The isoconversional kinetic analysis or so-called model-free kinetics method requires performance of a series of experiments at different temperature programs and yields the values of effective activation energy (E_α) as a function of conversion (α). Coat and Redfern [45] suggested the following approximations for the $p(x)$ function:

$$p(x) = x^{-2}e^{-x}\left(1 - \frac{2}{x}\right) \qquad (4.15)$$

In general, the part of $(1 - 2/x)$ of Eq. (4.11) is neglected and an oversimplified approximation leads to Eq. (4.13), which is used in this study [46]

$$p(x) = x^{-2}e^{-x} \qquad (4.16)$$

Submitting Eq. (4.16) into Eq. (4.8) and taking the logarithm results in an isoconversional equation

$$\ln\left(\frac{\beta_i}{T_{\alpha,i}^2}\right) = -\ln\frac{A_\alpha R}{E_\alpha g(\alpha)} - \frac{E_\alpha}{RT_{\alpha,i}} \qquad (4.17)$$

The subscript i denotes different heating rates. Equation. (4.17) provides a straight line between $\ln(\beta/T_{\alpha,i}^2)$ and $1/T_{\alpha,i}$. The E_α can be calculated from the slope and the preexponential factor from the intercept. The above isoconversional method was proposed by Vyazovkin and Sbirrazzuoli [47], which was used in this study. Equation (4.17) that has been employed in this study has similar expression of the Kissinger method expressed by Eq. (4.14). But, the Kissinger method uses the peak temperatures at different heating rates, and provides a single value of activation energy for the whole process. As a result, the obtained activation energy is correct only if there is no variation of E_α with α throughout the process.

4.2.2
Theory of Temperature-Modulated DSC (TMDSC)

To further study the cure kinetics of the resole PF resin/clay nanocomposites, temperature-modulated DSC (TMDSC) was utilized. Unlike conventional DSC, TMDSC permits simultaneous measurement of heat flow and heat capacity by applying the sinusoidal oscillation on a DSC temperature ramp and separately analyzing the sinusoidal and nonsinusoidal response [48, 49]. In particular, this technique has received considerable attention for monitoring the cure kinetics of thermosetting systems [50–52]. In TMDSC, the heat flow equation has the following form:

$$\frac{dH}{dt} = C_p \frac{dT}{dt} + f(t, T) \tag{4.18}$$

where dH/dT is the total heat flow (J/s or W), C_p is the heat capacity (J/K), and dT/dt is the heating rate. The reversing heat flow component "$C_p(dT/dt)$" is dependent on the rate of change of temperature and the heat capacity. The reversing heat flow includes the glass transition, and the melting of slow-melting polymers, or highly impure materials. In contrast, the "nonreversing" heat flow component, $f(t,T)$, contains the heat flow contribution from kinetically controlled events that are dependent on both temperature and time. So, the nonreversing heat flow contains all kinds of kinetic transitions such as crystallization, curing reactions, evaporation, degradation, and the melting of fast-melting polymers and pure crystalline materials. The C_p component is calculated from the response to the oscillation via

$$C_p = K \frac{A_{mhf}}{A_{mhr}} \tag{4.19}$$

where A_{mhf} is the amplitude of the modulated heat flow, A_{mhr} is the amplitude of the heating rate signal, and K is the heat capacity calibration constant. The nonreversing signal is calculated as the difference between the total heat flow response and the reversing signal which is obtained from Eq. (4.18).

4.2.3
Theory of Temperature-Modulated TG (MTG)

In contrast, in the early 1970s, kinetics of modulated TG (MTG) analysis was proposed by Flynn and Wall [53, 54]. In MTG, a periodic temperature modulation is overlaid on the conventional linear heating rate experiments. A general temperature program of MTG is characterized by a sinusoidal modulation with amplitude (A_T) and period (P):

$$T = T_0 + \beta t + A_T \sin(\omega t) \tag{4.20}$$

where T_0 is the initial temperature, β is the average or underlying heating rate, and ω is the period divided by 2π. For an experiment of MTG by a sine wave, Eq. (4.4) may be evaluated as the ratio for adjacent peaks, and valleys, of the periodic rate of reaction [28]. Therefore, the ratio of reaction rates between adjacent peaks (p) and valleys (v) can be written as follows:

$$\frac{(d\alpha_p/dT)}{(d\alpha_v/dT)} = \frac{A f \alpha_p \exp(-E_m/T_p)}{A f \alpha_v \exp(-E_m/T_v)} f(\alpha) \tag{4.21}$$

If the reacted fraction does not change significantly between adjacent half cycles, $f(\alpha_p) \approx f(\alpha_v)$, and their ratio approaches unity, Eq. (4.21) can be simplified and solved for E_m to yield Eq. (4.22).

$$E_m = \frac{RT_p T_v \ln(d\alpha_p/d\alpha_v)}{T_p - T_v} \tag{4.22}$$

In MTG, the oscillatory temperature forcing function is defined by an average temperature T, the temperature amplitude A, and its period or frequency. Because $T_p = T + A$ and $T_v = T - A$, Eq. (4.22) may be rewritten as follows [39]:

$$E_m = \frac{RL(T^2 - A^2)}{2A} \tag{4.23}$$

where $L = \ln(d\alpha_p/d\alpha_v)$, the ratio of the maximum and the minimum conversion rate during a cycle. Thus, Eq. (4.23) allows a continuous calculation of E_m.

4.3
Thermal Properties of Nanocomposites

4.3.1
Cure Kinetics of MF Resin/Clay/Cellulose Nanocomposites

MF resin/clay/cellulose nanocomposites were prepared by mixing a commercial clay lay particles (Cloisite®-Na, Southern Clay Products, Inc., Texas, USA) and then mixed with MF resin by hand followed by ultrasonication for 30 minutes. These mixtures were impregnated into cellulose paper to obtain a target resin loading of 60%, which gave a paper weight of about 200 g/m². The resin-impregnated papers were dried at room temperature for 3 h and then procured at 70 °C in a drying oven for overnight. Figure 4.1 shows typical DSC thermograms of different heating rates for the nanocomposites with 0.5 wt% clay. As the temperature increased according to the heating rate, the heat flow decreased at first because the nanocomposite sample became soften prior to starting curing process. After reaching a minimum, the heat flow started to increase due to the curing of MF resin in the nanocomposites. As expected, the nanocomposites showed an exothermic peak. And the exothermic peak temperatures of the nanocomposite increased as an increase in the heating rate. This result was due to a thermal lag between the programmed heating rate and the real temperature in the furnace, which made it possible to calculate the activation energy by applying different methods such as Ozawa, Kissinger, and isoconversional methods.

Changes of the peak temperature at the maximum rate of cure of the MF resin were shown in Figure 4.2. Regardless of the heating rates, the peak temperature increased to a maximum and thereafter decreased and then increased as the clay content increased. An increase in the peak temperature indicated a slower cure rate of the resin. So, these results suggested that the cure rates of MF resin in the nanocomposites decreased up to 0.5 wt% clay content at both heating rates of 5 and 10 °C/min and up to 1 wt% clay content at the heating rate of 20 °C/min. This result could be due to the nature of clay used in this study. In other words, the clay could be exfoliated at lower levels, like 0.5 wt%, which retarded the cure rate

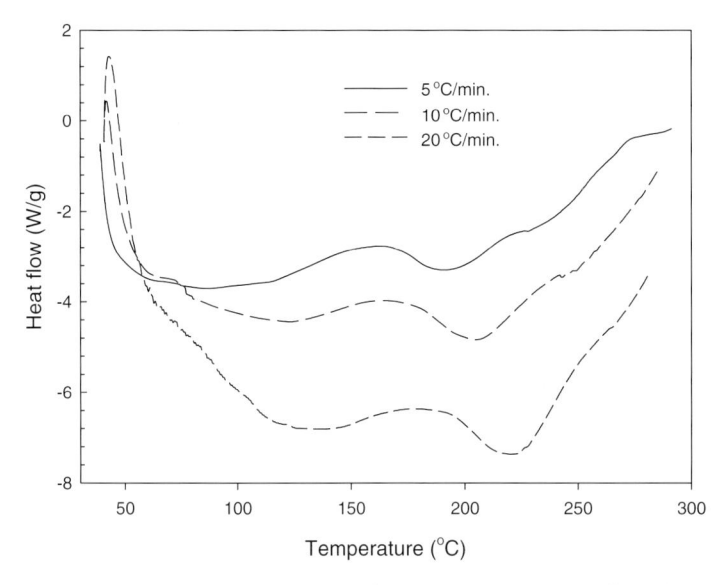

Figure 4.1 Typical DSC thermograms of the nanocomposites at different heating rates (0.5 wt% clay level). Reproduced from Ref. [55] with permission from Elsevier (© Elsevier, 2010).

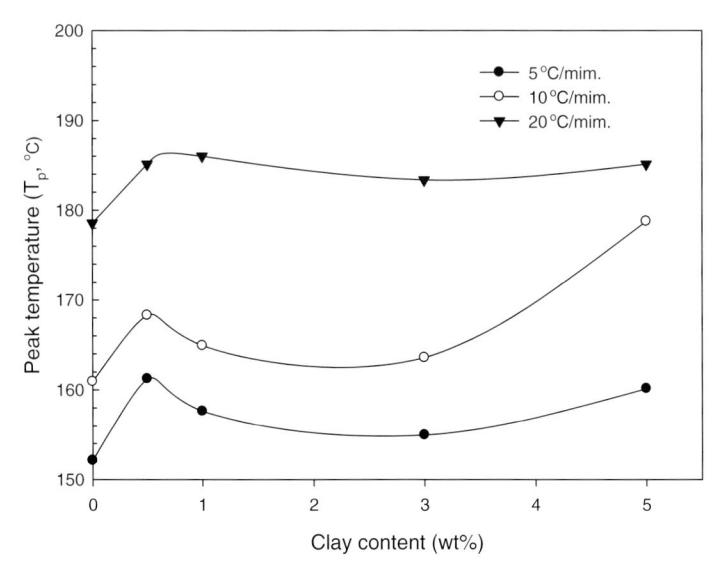

Figure 4.2 Changes of exothermic peak temperatures of the nanocomposites at different heating rates and clay levels. Reproduced from Ref. [55] with permission from Elsevier (© Elsevier, 2010).

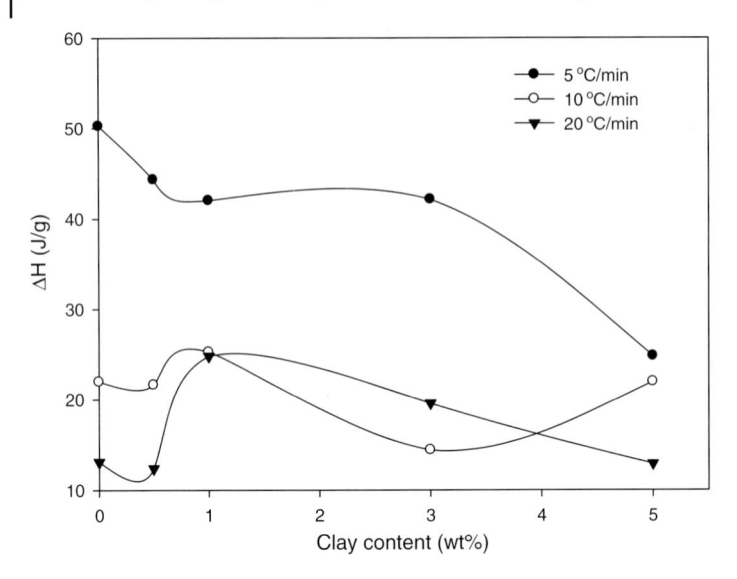

Figure 4.3 Changes of ΔH values of the nanocomposites at different heating rates and clay contents. Reproduced from Ref. [55] with permission from Elsevier (© Elsevier, 2010).

of MF resin by interfering its network formation [12]. After reaching a maximum value, the peak temperature gradually decreased as the clay content increased. The peak temperature of MF resin in the nanocomposites decreased because sodium of the clay accelerated the cure of MF resin.

Changes in the heat of reaction (ΔH) of the nanocomposites depending on the heating rates and clay contents were shown in Figure 4.3. The ΔH that was the area under the exothermic curve of DSC thermogram increased up to 1 wt% clay content at heating rates of both 10 and 20 °C/min except the heating rate of 5 °C/min. These results could be due to an exfoliation of clay, which had retarded the curing of MF resin during their network formation. In other words, the area under the exothermic curve of MF resin increased as the peak temperature increased.

Figure 4.4 illustrates changes of the activation energies of the nanocomposites as a function of clay content. The E_a values were calculated by two different methods, that is, Ozawa and Kissinger methods as expressed by Eqs. (4.10) and (4.14), respectively. As the clay increased, the E_a values reached a maximum at 0.5 wt% clay, and decreased up to 3 wt% clay content. This result could be due to the exfoliation of clay at a lower addition level, that is, 0.5 wt%. In other words, the exfoliation of clay retarded the curing of MF resin in the nanocomposites. The presence of exfoliated nanoclays requires more energy to start the cross-linking process, which increased the activation energy. This result was also compatible with changes of the peak temperatures as shown in Figure 4.2. The minimum E_a

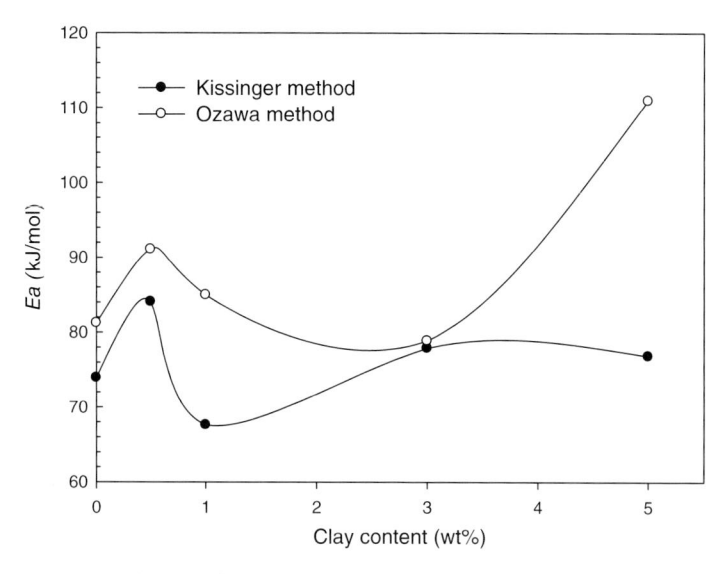

Figure 4.4 Changes of E_a values of curing reaction of the nanocomposites with different clay contents determined by two methods. Reproduced from Ref. [55] with permission from Elsevier (© Elsevier, 2010).

values of the nanocomposites occurred at 1% and 3% levels of the clay for the E_a values obtained from the Ozawa and Kissinger methods, respectively (Figure 4.4). This result could be ascribed to inhomogeneity of the samples, but it requires further work to clearly explain what factors made contribution to this phenomena. At 5 wt% clay content, the E_a values increased again, which could be resulted from the vitrification of MF resin after the gelation.

The E_a values obtained by the Ozawa method were greater than those of the Kissinger method. This result was consistent with the reported results for PF resins [56]. But these methods were more accurate than a single heating rate method, so-called Borchardt–Daniels method.

Figure 4.5 shows changes of the activation energy (E_α) values of the nanocomposite as a function of the degree of conversion (α) at different clay contents. In general, the E_α values of the nanocomposites gradually increased as the degree of the conversion increased. A similar trend was also reported for novolak-type lignophenolic resins [57]. This result could be attributed to that the rate of conversion at the later stage of cross-linking was limited by the mobility of longer polymer chain [47]. However, a decrease in the E_α values was also reported for PF resin [58, 59]. In addition, with an exception of 0 wt% clay content, the E_α value reached a maximum at the 5 wt% clay content and then decreased as it increased. This result also indicated that the clay particles at 5 wt% level were exfoliated, which delayed the cure of MF resin in the nanocomposites [5].

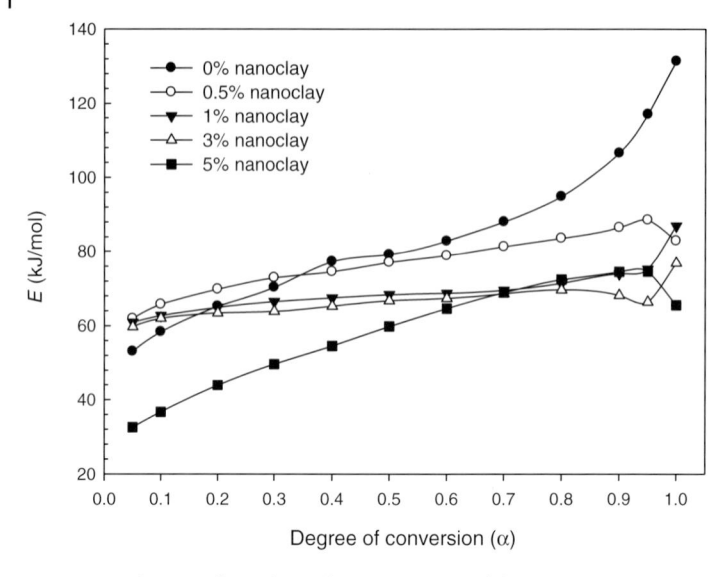

Figure 4.5 Changes of E_α values of curing reaction of the nanocomposites with different clay contents determined by the isoconversion method. Reproduced from Ref. [55] with permission from Elsevier (© Elsevier, 2010).

4.3.2
Cure Kinetics of PF Resin/Clay/Cellulose Nanocomposites

4.3.2.1 Cure Kinetics by the Ozawa and Kissinger Methods

Figure 4.6 shows the typical DSC thermograms for the control (no clay) composites at different heating rates (5–20 °C/min). In all cases, there is an initial decrease in heat flow as temperature is increased. This is likely due to the softening of impregnated resole PF resins prior to curing. After reaching a minimum, the heat flow increases to a maximum peak temperature (T_p) due to the exothermic curing of the resole PF resins.

Figure 4.7 shows the variation in T_p for various nanocomposites with different clay contents and DSC heating rates. As the clay content increased, the T_p decreases up to 3–5% clay, and then increases upon further addition to 10% clay regardless of the heating rate. This result suggests that the addition of 3–5% clay to the resole PF resin accelerates curing of the nanocomposite. The accelerated cure rate could be caused by the Lewis acidity of the clay used in this study [61]. Clay cations like Al^{3+}, Fe^{3+}, or Na^+ act as Lewis acids, which can accept electrons from various aromatic compounds [62, 63]. The oxidizing power of the clay is proportional to its Lewis acidity. However, the addition of more than 10% clay retarded the cure rate, possibly due to agglomeration and, consequently, poor dispersion throughout the nanocomposite. As expected, the T_p of the nanocomposite increased as the heating rate increased. This result suggests that the curing of the resole PF resin is a function of temperature, which makes it possible to calculate the activation

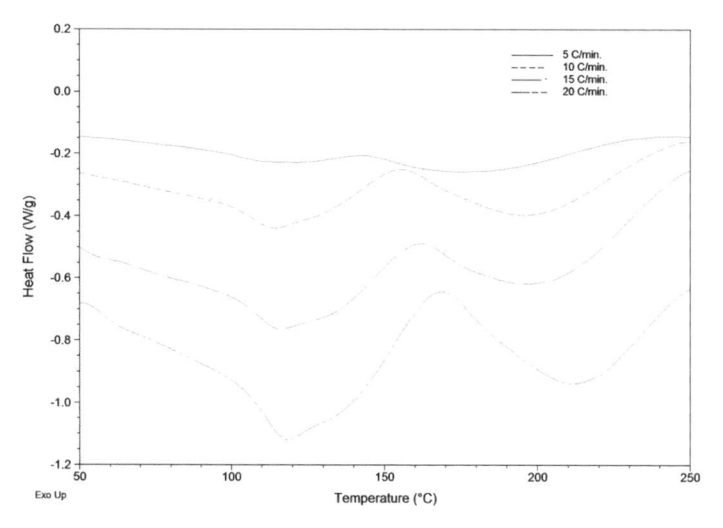

Figure 4.6 Typical DSC thermograms for the control nanocomposites (0% clay) at different heating rates. Reproduced from Ref. [60] with permission from Mokchae Konghak (© Mokchae Konghak, 2010).

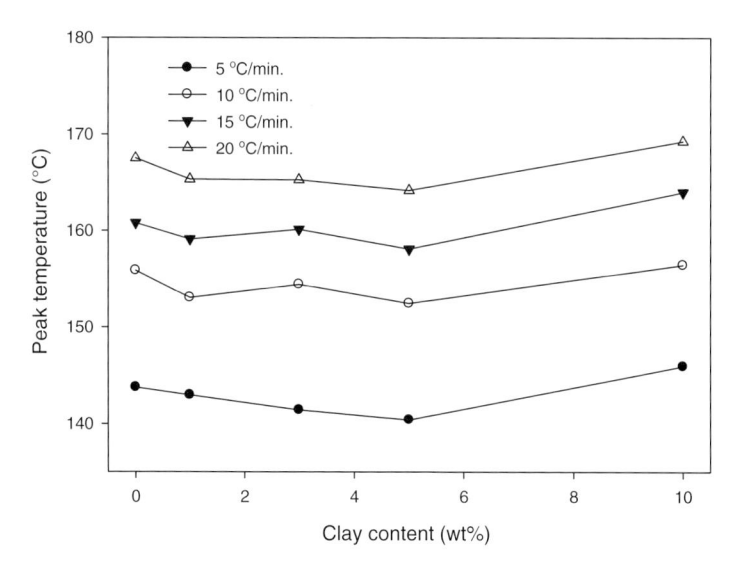

Figure 4.7 Effect of clay addition on the peak temperature (T_p) for resole PF resin/clay/cellulose nanocomposites at different heating rates. Reproduced from Ref. [60] with permission from Mokchae Konghak (© Mokchae Konghak, 2010).

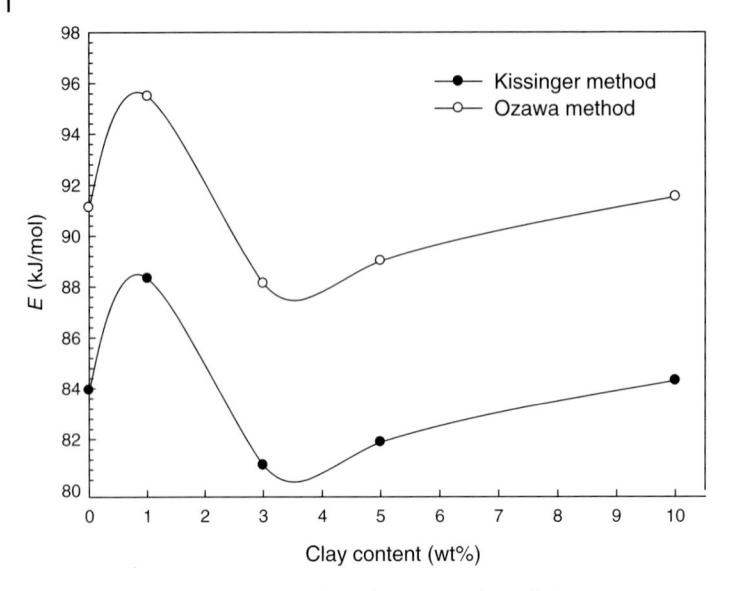

Figure 4.8 Activation energy (E) of resole PF resin/clay/cellulose nanocomposites as a function of the clay content using the Kissinger and Ozawa methods. Reproduced from Ref. [60] with permission from Mokchae Konghak (© Mokchae Konghak, 2010).

energy by applying different methods such as the Ozawa, Kissinger, and isoconversional methods.

Typically, multiheating rate methods such as the Ozawa, Kissinger, and isoconversional methods are utilized. The Ozawa method is based on a simple relationship between the activation energy (E_a) and the heating rate (β) [43]. If the sample temperature is increased by a controlled and constant β, the variation in the degree of conversion can be analyzed as a function of temperature, this temperature being dependent on the time of heating. Figure 4.8 illustrates the changes in the E_a determined by the Ozawa and Kissinger methods using Eqs. (4.10) and (4.14), respectively. The E_a values obtained by the Ozawa method were greater than those obtained by the Kissinger method. However, the E_a values obtained by both methods behave almost identically with increasing clay content.

The higher E_a values obtained using the Ozawa method is consistent with the results previously reported for resole PF resins [56]. As the clay was increased, the E_a values determined using both methods increased to a maximum at 1% clay content and then decreased to a minimum at 3% clay content followed by a slow increase to 10% clay content.

The increase in E_a value at 1% clay could be due to exfoliation of the clay. It has been reported that the presence of exfoliated clays acts as a barrier to network formation during the curing of epoxy resins [12]. Thus, a similar effect may likely be happening here, wherein the exfoliated clay increases the activation energy for the cross-linking process. Another possibility is that at 1% clay loading, there is increased inhomogeneity in the nanocomposite, which acts as a barrier to the curing of the resole PF resin. More work is required to clearly explain what factors contribute to this phenomenon.

4.3.2.2 Cure Kinetics by the Isoconversional Method

A third kinetic method is the isoconversional kinetic analysis or so-called model-free kinetics method, and has been used to characterize cure kinetics of both neat and modified PF resins [18, 41, 44, 56–59, 62, 63]. This method requires a series of experiments performed at different temperature programs and yields the effective activation energy (E_a) as a function of conversion (α) as Eq. (4.14).

Figure 4.9 illustrates the relationships between the degree of conversion and temperature profiles for various resole PF/clay nanocomposites. Comparing the various profiles, the clear impact of clay content on curing rate is evident; the curing reaction was accelerated as the clay content increased from 0% to 3% and then retarded with further increases to 10% clay content. As anticipated, the cure temperature decreases with increasing clay loading up to 3% clay, and then increased with further clay addition. At the same degree of conversion ($\alpha = 0.4$) and heating rate (5 °C/min), the cure temperatures for the resole PF resins were 141 °C, 139 °C 139 °C, and 164 °C for 0%, 1%, 3%, and 10% clay levels, respectively.

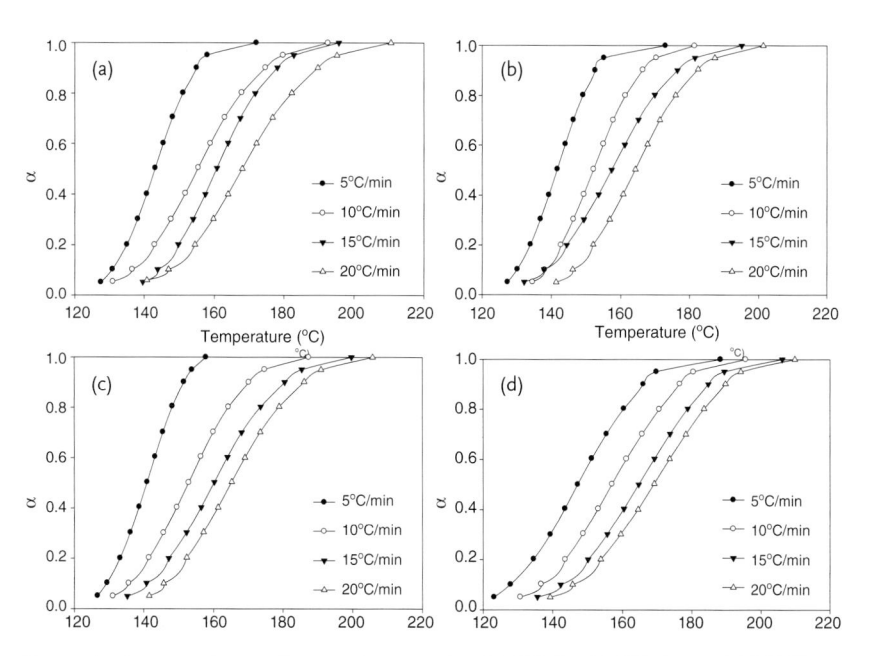

Figure 4.9 Apparent degree of conversion versus temperature profiles for the resole PF resin/clay/cellulose nanocomposites at different clay levels and heating rates: (a) 0% clay, (b) 1% clay (c) 3% clay, and (d) 10% clay. Reproduced from Ref. [60] with permission from Mokchae Konghak (© Mokchae Konghak, 2010).

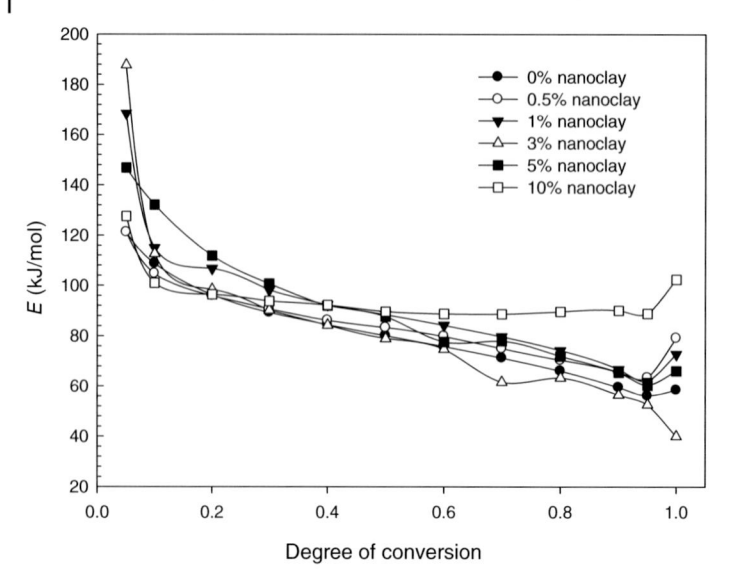

Figure 4.10 Relationship between E_α and degree of conversion for the resole PF resin/clay/cellulose nanocomposite at different clay contents. Reproduced from Ref. [60] with permission from Mokchae Konghak (© Mokchae Konghak, 2010).

Using the isoconversional method, the E_α as a function of α was calculated and is presented in Figure 4.10. All of the resole PF resin/clay nanocomposites showed similar behavior in calculated E_α values as a function of degree of conversion (α). For all clay contents, there was an initial sharp decrease in E_α at α up to 0.1, followed by a slow gradual decrease with increasing degree of conversion α. At $\alpha >$ 0.9, all of the nanocomposites showed an increase in E_α except for the 3% clay nanocomposites with sharply decreased. The E_α values ranged from 40 to 187 kJ/mol, with those obtained for α between 0.1 and 0.9 falling within the reported range between 60 and 120 kJ/mol for the cure of resole PF resins [18, 61, 64]. The reduction in E_α values with increasing α between 0.1 and 0.9 could be ascribed to a cure acceleration of resole PF resins by the clay. In fact, a similar trend in E_α value has been reported for resole PF resins [58, 59]. An interesting point is that the E_α of the nanocomposite with 10% clay does not seem to decrease with increasing α as per the other clay loadings, which may suggest a different curing behavior. This may be due to agglomeration of clay particles in the resole PF resins, which hinder resin network formation as a result of the increased E_α. As well this may be a result of transition from a chemical-controlled curing reaction regime to one controlled by diffusion [57].

4.3.2.3 Cure Kinetics by TMDSC

Figure 4.11 shows the typical dynamic TMDSC thermograms for the control nanocomposite without clay, and the nanocomposite with 3% clay; included are

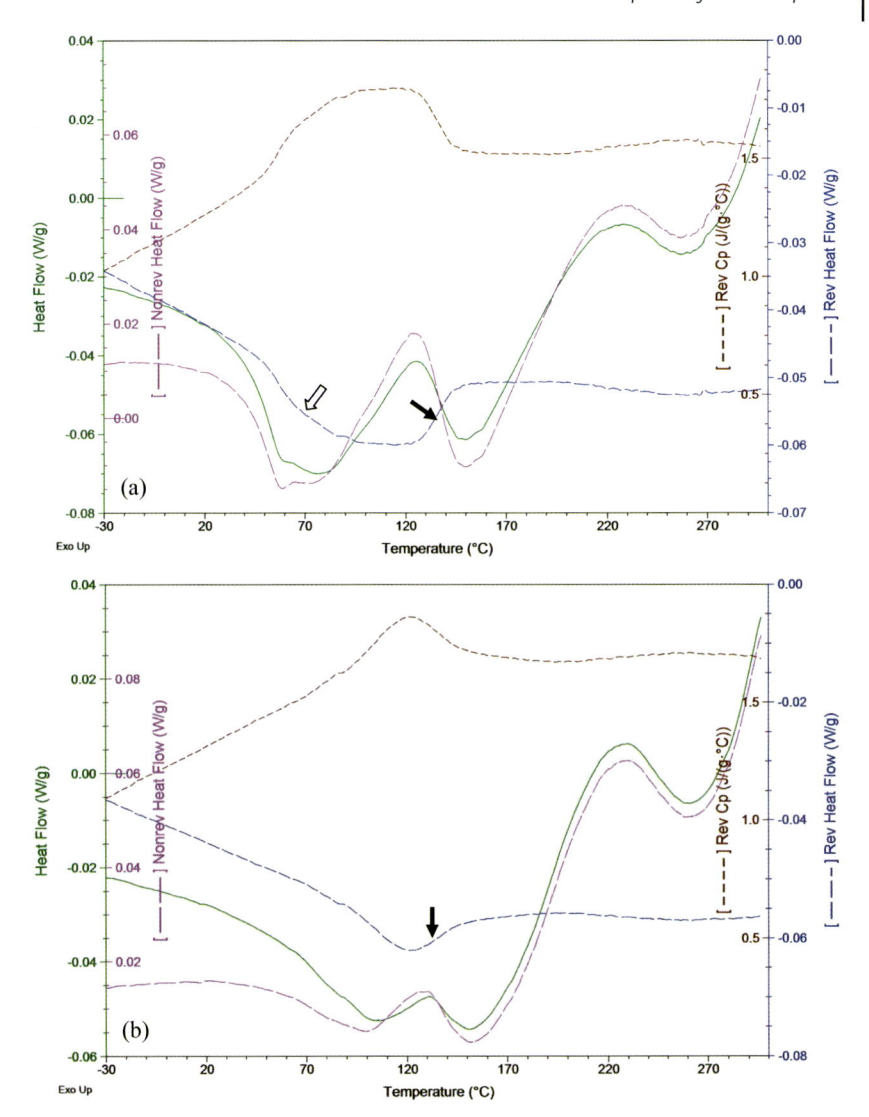

Figure 4.11 Typical TMDSC thermograms for the resole PF resin/clay/cellulose nanocomposites at (a) 0% clay and (b) 3% clay contents. Included are the total heat flow (solid line), reversing heat flow (medium dash line), nonreversing heat flow (long dash line), and reversing heat capacity (short dash line) curves. Reproduced from Ref. [60] with permission from Mokchae Konghak (© Mokchae Konghak, 2010).

the total, reversing, nonreversing heat flows, and the reversing heat capacity curves. The total and nonreversing heat flow curves were very close to each other and displayed an exothermic reaction due to the curing of resole PF resin (Figure 4.11a and b). This phenomenon was pretty much the same for the nanocomposite with clay contents greater than 3%. This suggests that the curing reaction of resole PF resin becomes a predominant event in the thermal behavior of the nanocomposite. However, the reversing heat flow curve of the control nanocomposite shows a rapid decrease around 57.8 °C (empty arrow in Figure 4.11a) and then an increase around 136.7 °C (filled arrow in Figure 4.11a). As temperature increases, the resole PF resin softens to a molten state at first, and then starts to cure by cross-linking, leading to a network formation. After reaching the maximum cure rate at a peak temperature, vitrification begins. Vitrification is a transition from a liquid-rubbery state with high molecular mobility to a glassy state with much lower molecular mobility. The reversing heat capacity of the nanocomposite was closely related to the exothermic cure reaction of the resole PF resin, gradually decreasing from a maximum to lower values as the resole PF resin changed from a gel state to a glassy state. The dynamic vitrification temperature (T_{dv}) was defined as the midpoint of the reversible heat flow curve right after the endothermic peak (filled arrow in Figure 4.11a and b).

As for the conventional DSC, the TMDSC thermograms also provided T_p values for the resole PF resin cure in the nanocomposite. Figure 4.12 illustrates the effect of clay content on the T_p values. The T_p values from both total heat flow and nonreversing heat flows of the clay nanocomposites decreased with increasing clay

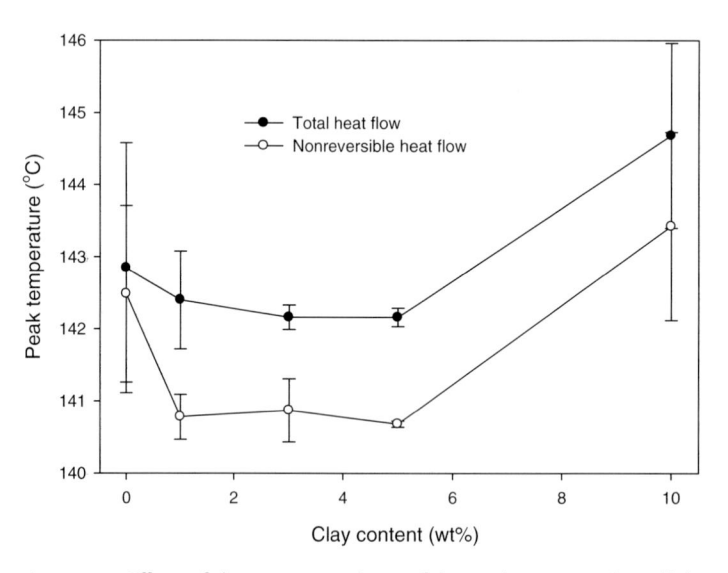

Figure 4.12 Effects of clay content on the T_p of the resole PF resin/clay/cellulose nanocomposite as determined by TMDSC. Reproduced from Ref. [60] with permission from Mokchae Konghak (© Mokchae Konghak, 2010).

content up to 5%, and then rapidly increased at 10% clay content. This is again consistent with a cure acceleration of resole PF resin by adding clay as shown in Figure 4.7. In fact, the T_p range of the total heat flow from the TMDSC was quite similar to that from conventional DSC at a heating rate of 5 °C/min. However, the T_p values obtained from the nonreversing heat flow were lower than those from the total heat flow curve. This is because the nonreversing heat flow is equal to the difference between the total heat flow and the reversing heat flow for the nanocomposites. That is, the T_p of the nonreversing heat flow from TMDSC represents only the cure of resole PF resin in the nanocomposite.

However, the curing reaction heat (ΔH) values from the total and nonreversing heat flow curves were greater than those obtained from the total heat flow of the nanocomposite. Figure 4.13 shows that the ΔH values for the total and nonreversing heat flows decreased with increasing clay content up to ~3% and then increased as clay content increased. Again, this could be due to the cure acceleration effect of the added clay; the addition of clay accelerates the cure of resole PF resin by lowering the T_p, which results in a decrease in the area under the exothermic curve (ΔH).

Figure 4.14 shows the relationship between T_{dv} and clay content for the resole PF resin/clay nanocomposites. As the clay content increased, the T_{dv} increased reaching a maximum at 3% clay content. Increasing the clay content beyond 3% led to a sharp decrease in T_{dv}. This indicates that the vitrification of the resole PF resin is delayed up to a clay content of 3%. Thus, the addition of clay restricted

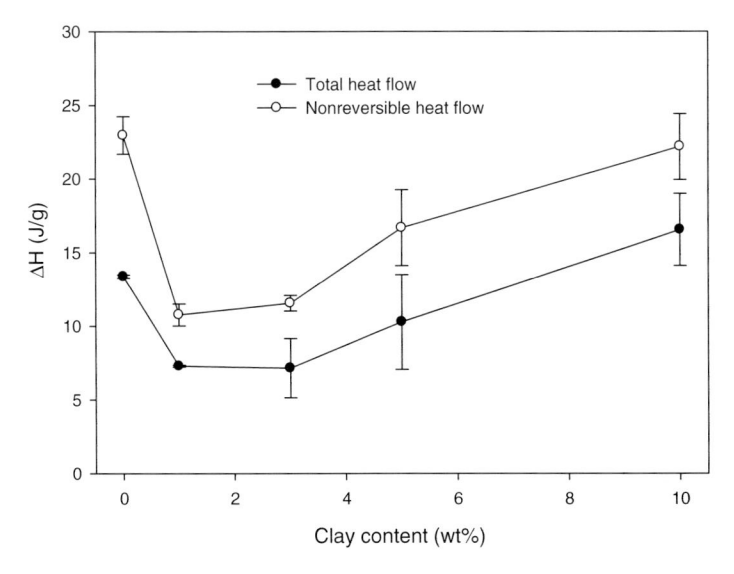

Figure 4.13 Effect of clay content on the curing reaction heat (ΔH) of the resole PF resin/clay/cellulose nanocomposite as determined by TMDSC. Reproduced from Ref. [60] with permission from Mokchae Konghak (© Mokchae Konghak, 2010).

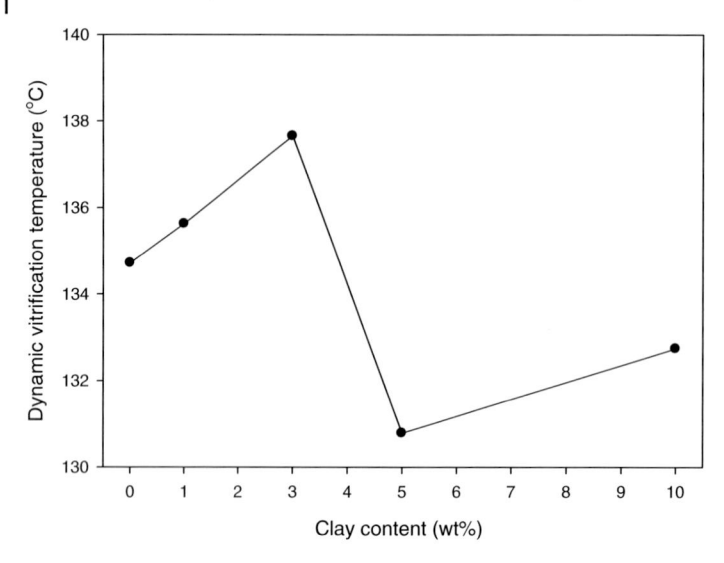

Figure 4.14 Effect of clay content on the dynamic vitrification temperature (T_{dv}) of the resole PF resin/clay/cellulose nanocomposite as determined by the reversing heat flow of the TMDSC thermograms. Reproduced from Ref. [60] with permission from Mokchae Konghak (© Mokchae Konghak, 2010).

the vitrification process even though it accelerated the initial cure reaction in the nanocomposites. The lower T_{dv} values at clay contents of 5% and 10% could also be due to an aggregation of the clay particles, resulting in an increased viscosity of resole PF resin as per finding for epoxy resin curing [65].

4.3.3
Thermal Degradation Kinetics of PF Resin/MWCNT/Cellulose Nanocomposites

4.3.3.1 FT-IR Analysis of Surface-Modified MWCNTs

FT-IR spectra of pristine and surface-modified MWCNTs were obtained to understand chemical changes on the surface of MWCNTs after either oxidation or silanization after the oxidation treatments. As shown in Figure 4.15, two bands appeared at 3740 and 1630 cm^{-1} corresponding to free OH group due to atmospheric moisture present on pristine MWCNTs [67, 68]. The band appeared at 2000–1900 cm^{-1} due to various cumulated C=C bonds [69]. When MWCNTs were oxidized, two new bands appeared at 1789 and 1325 cm^{-1} corresponding to C=O from COOH group and C–O from COOH, respectively [67, 70]. In addition to common bands of pristine and oxidized MWCNTs, a new band occurred at 986 cm^{-1} belongs to Si–O of Si–OH when MWCNTs were treated with 3-aminopropyltrimethoxysilane (APTMS) after their oxidation [71].

On the basis of FT-IR spectra, a schematic diagram of chemical reactions occurred for treated MWCNTs is presented in Figure 4.16. When MWCNTs were

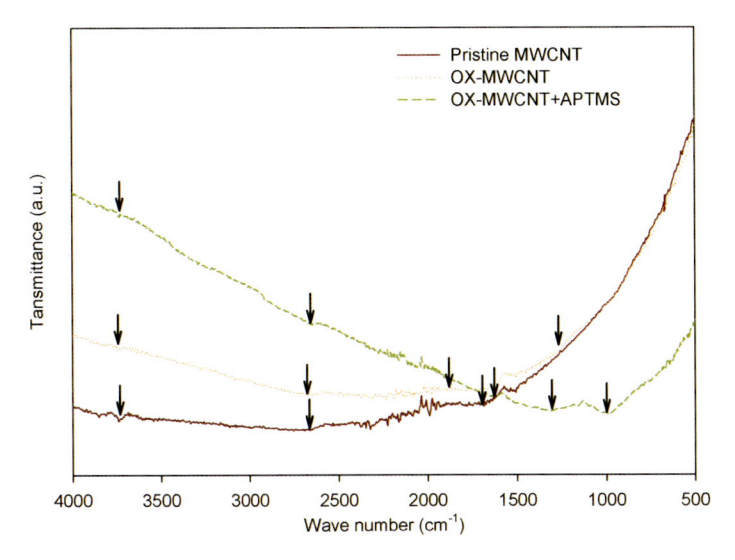

Figure 4.15 FT-IR spectra of pristine and surface-modified MWCNTs. Reproduced from Ref. [66] with permission from Elsevier (© Elsevier, 2012).

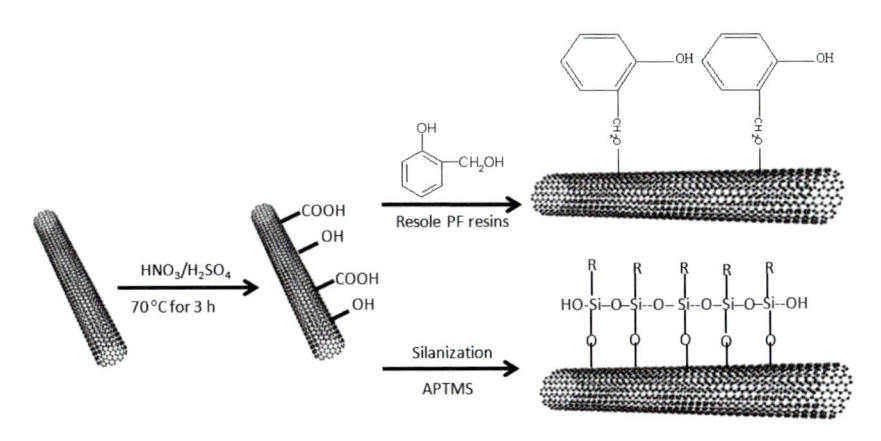

Figure 4.16 Schematic diagram of chemical reactions for surface-modified MWCNTs. Reproduced from Ref. [66] with permission from Elsevier (© Elsevier, 2012).

oxidized, hydroxyl or carboxyl groups were formed on their surface. These functional groups react with hydroxylmethyl group (CH_2OH) of resole PF resins, leading to the attachment of phenolic molecules to the MWCNT. When the oxidized MWCNTs were treated with APTMS, then functional groups formed on their surface also react with hydrolyzed silane, leading to the attachment of silane to the MWCNT surface [71].

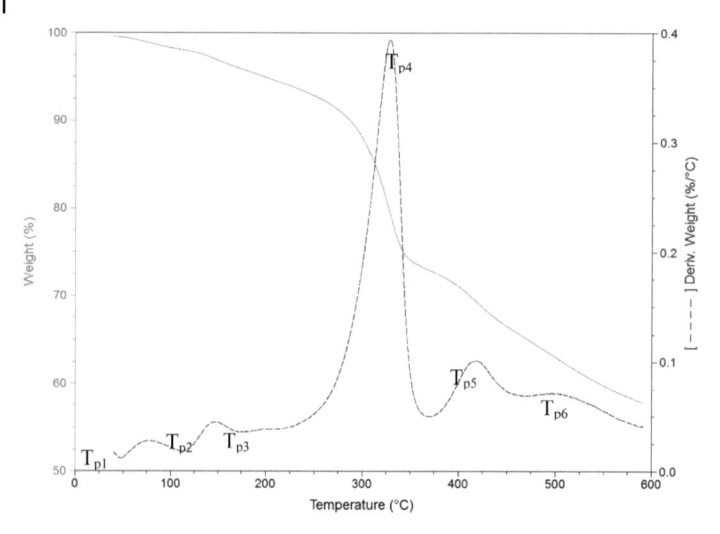

Figure 4.17 A typical CTG thermogram of the nanocomposite with 0.5% MWCNTs from 5 °C/min scan. Reproduced from Ref. [66] with permission from Elsevier (© Elsevier, 2012).

4.3.3.2 Thermal Degradation Kinetics Analysis of Conventional Thermogravimetry

Figure 4.17 shows a typical thermogram of conventional thermogravimetry (CTG) at 5 °C/min for the nanocomposite with 0.5% MWCNTs. Five derivative TG (DTG) peaks corresponding to five mass loss steps were appeared during thermal degradation of the nanocomposite. The first DTG peak (T_{p1}) around 75.7 °C could be due to either the evaporation of water in the sample or thermal softening, or gelation of PF resins. The second DTG peak (T_{p2}) around 145.4 °C could be resulted from the evaporation of water produced by the condensation reaction, or curing of UF resins. And the T_{p3} could be caused by either from the degradation of small terminal groups (i.e., hydroxymethyl groups) or melting of the surfactant in the sample. In fact, the melting temperature of the surfactant (i.e., sodium dodecyl sulfate [SDS]) is 204–207 °C. The most dominant DTG peak, T_{p4}, is attributed to the decomposition of cellulose that is a main component of the nanocomposite. After decomposition of cellulose, the T_{p5} could be due to the degradation of methylene linkages in cured PF resins, while the T_{p6} could be from the decomposition of aromatic phenol rings of PF resins. However, the presence of a DTG peak around 200 °C due to the surfactant, SDS, was inconsistent that was possibly due to a relatively small amount of the SDS compared to PF resins and MWCNTs.

CTG thermograms of the nanocomposite as a function of the MWCNT content are shown in Figure 4.18. As expected, the major decomposition peak occurred by cellulose at around 342 °C. When MWCNTs were added into the nanocomposite, the mass loss (%) at around 200 °C was greater than that of the control sample. This became greater after the decomposition of cellulose, indicating a lower thermal stability for the nanocomposite with 0.1% and 0.5% MWCNTs, which could be due to the decomposition of the surfactant. However, the addition of 1% MWCNTs much improved the thermal stability of the nanocomposite.

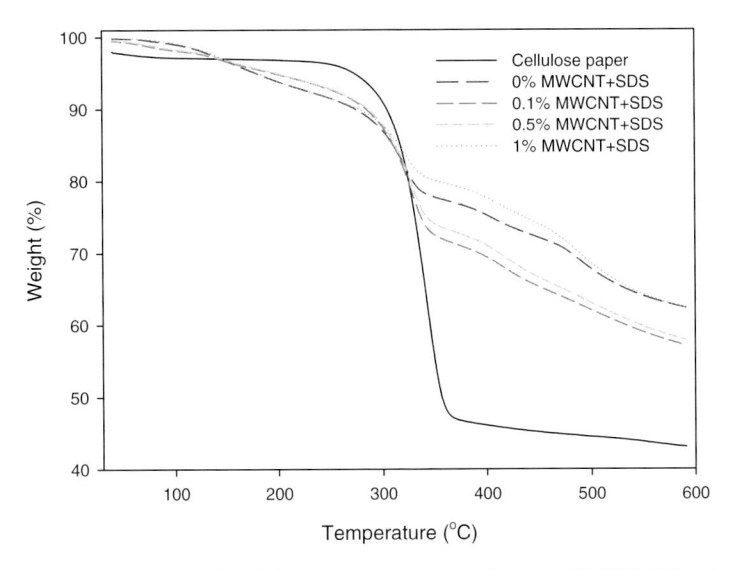

Figure 4.18 CTG results of the nanocomposite as a function of MWCNTs level at a heating rate of 5 °C/min. Reproduced from Ref. [66] with permission from Elsevier (© Elsevier, 2012).

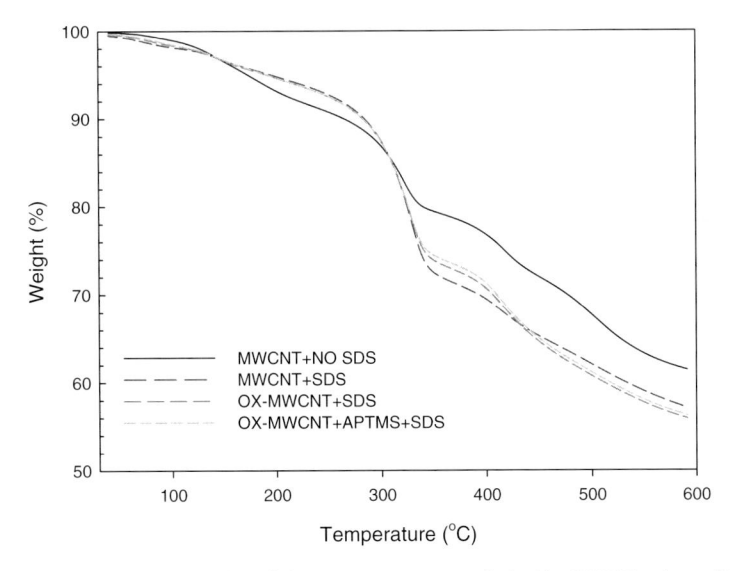

Figure 4.19 CTG results of the nanocomposite with 0.1% MWCNTs, depending on various surface treatments (heating rate: 5 °C/min). Reproduced from Ref. [66] with permission from Elsevier (© Elsevier, 2012).

Figure 4.19 shows CTG thermograms of the nanocomposite with different surface treatments of the MWCNT (0.1 wt%). Thermal stability of the nanocomposite without the surfactant was much improved above 328 °C, which was the main decomposition temperature of cellulose. When the surfactant was used to disperse MWCNTs, the nanocomposite's thermal stability significantly decreased above 328 °C. This could be ascribed to high alkalinity of the surfactant (pH = 7.2) and PF resins (pH = 10.1). This high alkalinity facilitates to decompose cellulose component in the nanocomposite at high temperature, which subsequently reduce resistance to thermal degradation.

When the OX-MWCNTs were dispersed in resole PF resin, thermal resistance of the resultant nanocomposite was slightly weakened below 300 °C and above 400 °C. It is not clear whether an interaction between the oxidized MWCNTs and surfactant accelerates thermal degradation of the nanocomposite, which requires further research work. However, thermal resistance of the nanocomposite was improved when the MWCNTs were treated with APTMS. This could be due to a chemical reaction between MWCNTs and APTMS as presented in Figure 4.16.

In order to evaluate thermal degradation kinetics of each component of the nanocomposite, four heating rates were employed, and a typical effect of heating rate is shown in Figure 4.20. As expected, the DTG peaks moved to higher temperature as the heating rate increased from 2.5 to 20 °C/min. So, these DTG peak temperatures were used to obtain the E_a values for thermal degradation of each component using Eq. (4.14) of the Kissinger method.

The calculated E_a values for pure cellulose paper and three DTG peaks using Eq. (4.14) are shown in Figure 4.21. The E_a value of the thermal decomposition of pure cellulose paper was calculated as 207.9 kJ/mol in the present work, and close to the reported value (145–200 kJ/mol) by Capart et al. [72]. although the E_a values of cellulose thermal decomposition are in the range of 200–270 kJ/mol [72–75]. However, the E_a values of the nanocomposite were lower than that of the pure

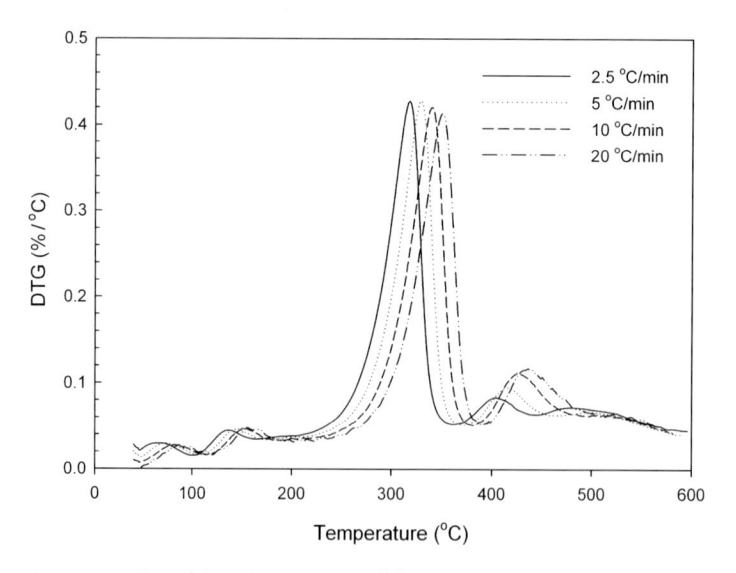

Figure 4.20 Typical CTG thermograms of the nanocomposite with 0.1% MWCNTs as a function of heating rates. Reproduced from Ref. [66] with permission from Elsevier (© Elsevier, 2012).

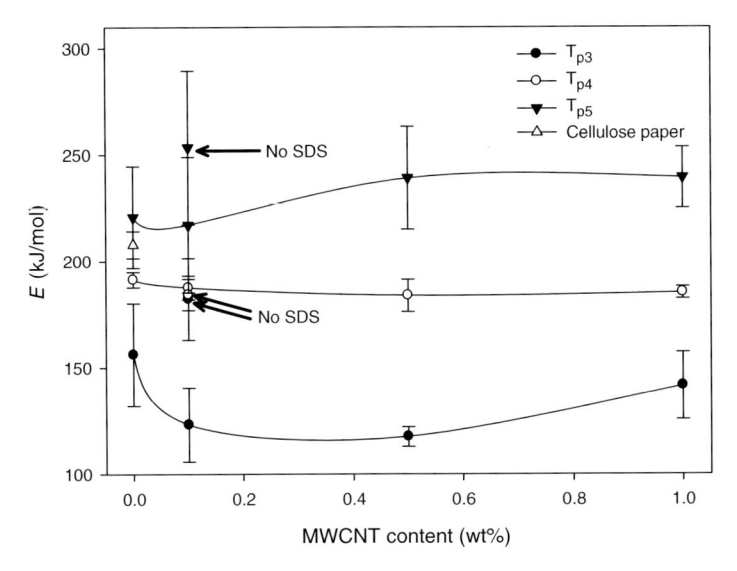

Figure 4.21 Change of the E values of three DTG peaks (T_{p3}, T_{p4}, and T_{p5}) as a function of MWCNT content as well as cellulose paper. Reproduced from Ref. [66] with permission from Elsevier (© Elsevier, 2012).

cellulose paper, which could be due to the presence of PF resins. The addition of MWCNTs did not significantly affect the E_a values of cellulose decomposition of the nanocomposite. And, the addition of SDS also did not much affect the E_a values of cellulose decomposition of the nanocomposite with 0.1% MWCNTs. These results indicate that thermal decomposition of cellulose in the nanocomposite was not much affected by the presence of MWCNTs and SDS.

However, these two components affected the E_a values of T_{p3} and T_{p5}. As shown in Figure 4.21, overall E values of the three DTG peaks are in the increasing order of T_{p3}, T_{p4}, and T_{p5}, which indicates different amounts of thermal energies for the decomposition of each event. Figure 4.21 also shows a significant influence of the surfactant (i.e., SDS) to the thermal degradation of the nanocomposite. In fact, when no SDS was used, the E values of both T_{p3} and T_{p5} for the nanocomposite were much greater than those of the counterparts at 0.1% MWCNTs content. This result suggests that the dispersion (or agglomeration) of MWCNTs in PF resin changes thermal degradation behavior of the nanocomposite; the addition of SDS made MWCNTs dispersed well in PF resin, which makes thermal degradation of the composites easy, but the agglomeration of MWCNTs without SDS requires much more energy for thermal degradation.

As the MWCNTs content increased, the E_a value of the T_{p3} decreased up to 0.5% MWCNTs followed by a slight increase while that of the T_{p4} increased above 0.5% MWCNTs. We believe that these results are also related to the state of MWCNT's dispersion in PF resin. A greater amount of MWCNTs (i.e., 1%) causes their agglomerations, which requires more thermal energy to be decomposed.

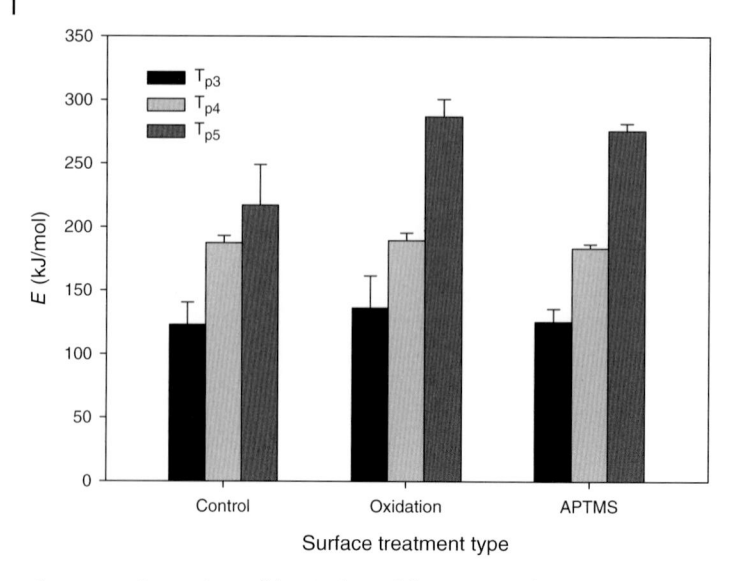

Figure 4.22 Comparison of the E values of three DTG peaks (T_{p3}, T_{p4}, and T_{p5}), depending on the MWCNT's surface treatments for the nanocomposite (0.1% MWCNTs). Reproduced from Ref. [66] with permission from Elsevier (© Elsevier, 2012).

Figure 4.22 also shows E values of the nanocomposite, depending on the surface treatments of MWCNTs. The E values of the nanocomposite increased as the degradation temperature rose, regardless of surface treatments. The E_a values also increased when MWCNTs' surface was modified by either oxidation or silanization. This could be due to a chemical reaction between MWCNTs and PF resins, or APTMS, which improved interfacial adhesion. However, the E value of the nanocomposite prepared with silane-treated MWCNTs slightly decreased when compared to those of the oxidized MWCNTs. This could be ascribed to thermal degradation of APTMS at lower temperature.

4.3.3.3 Isoconversional Analysis of CTG

An isoconversional method provides the effective activation energy (E_α) as a function of conversion (α), which could lead to understand thermal degradation kinetics of the nanocomposite. Figure 4.23 shows changes of E_α values of the nanocomposite at different levels of MWCNTs content. When 0.1% MWCNT was added, the E_α values increased for all degrees of conversions from the one of the control samples (0% MWCNTs). However, further addition of MWCNTs diminished E_α at the degree of conversion up to 0.65. For all MWCNTs' levels, E_α values rapidly increased above the $\alpha = 0.65$. A large deviation of E_α value versus α suggests that multiple thermal degradation steps is dominant above the $\alpha = 0.65$. This is quite consistent with multiple degradation steps as shown in Figure 4.22. This result could also be due to the fact that much more energy was required to ther-

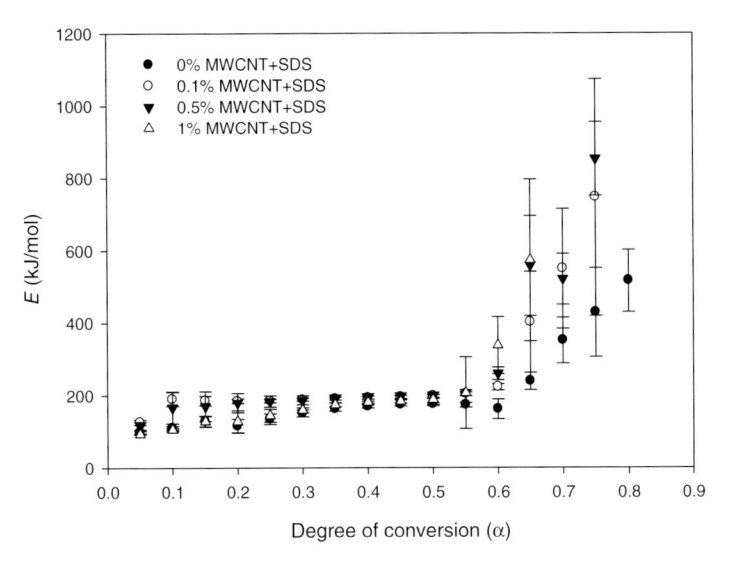

Figure 4.23 Changes of E_α values of resole PF resin/MWCNT/cellulose nanocomposite as the degree of conversion. Reproduced from Ref. [66] with permission from Elsevier (© Elsevier, 2012).

mally degrade methylene linkages and aromatic rings of PF resins, leading to greater E_α values.

The isoconversional method was also employed to understand thermal degradation kinetics of the nanocomposite, depending on the type of the surface modification of MWCNTs, and the result was presented in Figure 4.24. When the surfactant (SDS) was not added, the E_α value between 0.1 and 0.5 degree of conversion of the nanocomposite (MWCNTs + no SDS) was dropped from those of the control ones (MWCNTs + SDS). However, the nanocomposite prepared by the silane-treated MWCNTs showed moderate increase in the E_α value between 0.1 and 0.5 degree of conversion. The E_α values were not obtained above 0.5 degree of conversion for the nanocomposite prepared with the silane-treated MWCNTs. However, the E_α value rapidly rises above the $\alpha = 0.65$ for all samples, which is compatible with those of the nanocomposites with different MWCNT contents. These results also indicate that multiple thermal degradation steps occur above the $\alpha = 0.65$. As observed for the addition levels of MWCNTs, this result also suggests that more energy is required to thermally degrade methylene linkages, or benzene rings of PF resins.

4.3.3.4 Thermal Kinetic Analysis of MTG

In order to compare thermal degradation kinetics of the nanocomposite using CTG technique, MTG was employed to obtain a continuous measurement of activation energy (E_m). Figure 4.25 displays a typical MTG thermogram of the nanocomposite with 0.1% MWCNTs and SDS. As shown in Figure 4.17, the

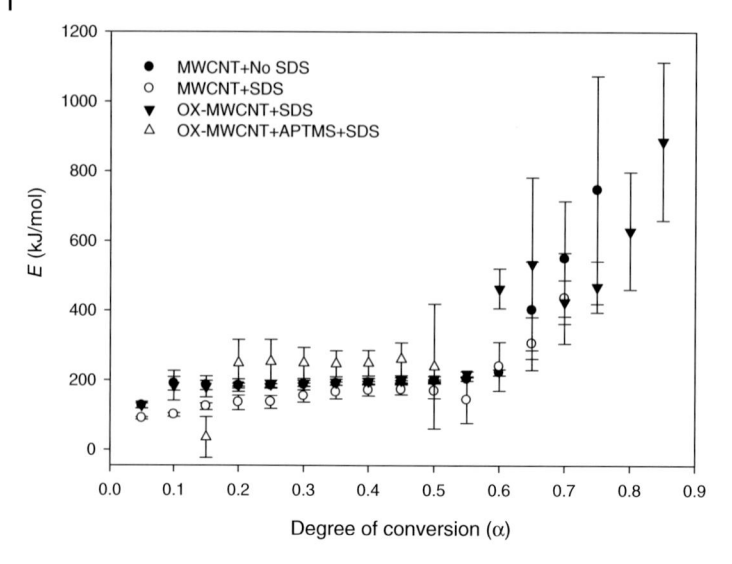

Figure 4.24 Changes of E_a values of resole PF resin/MWCNT/cellulose nanocomposite, depending on surface modifications at 0.1% MWCNT. Reproduced from Ref. [66] with permission from Elsevier (© Elsevier, 2012).

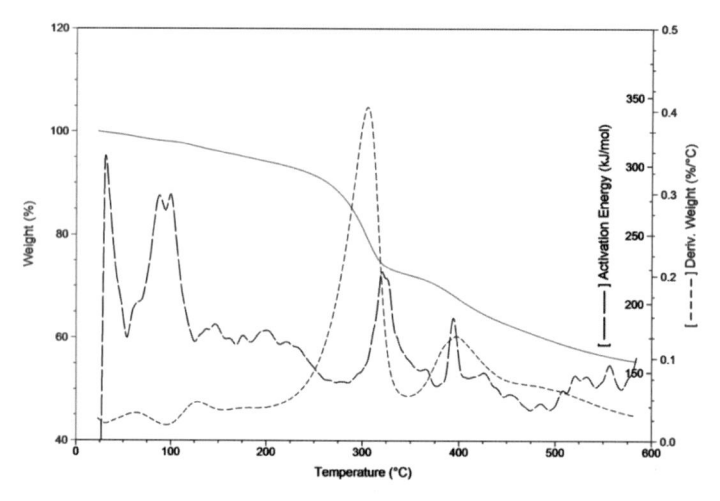

Figure 4.25 A typical MTG thermogram of the nanocomposite at 0.1% MWCNT content. Reproduced from Ref. [66] with permission from Elsevier (© Elsevier, 2012).

weight loss curve of the nanocomposite was quite similar to the counterpart of CTG. Figure 4.25 also shows a continuous change of the E_m value as thermal degradation temperature rose. Thermal degradation of the nanocomposite includes multistep reactions corresponding to DTG peaks such as T_{p3}, T_{p4}, or T_{p5}. As expected, the E_m values are approximately corresponding to each reaction steps

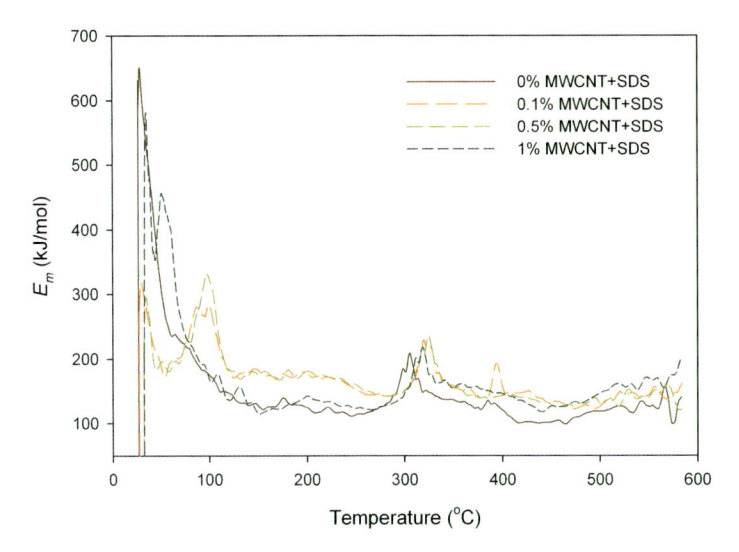

Figure 4.26 Changes of E_m values of resole PF resin/MWCNT/cellulose nanocomposite, depending on the content of MWCNTs. Reproduced from Ref. [66] with permission from Elsevier (© Elsevier, 2012).

even though there is a temperature difference between the DTG peaks and E_m peaks. For example, the DTG peak occurs at 305.2 °C while the E_m peak corresponding to this temperature appears at 319.5 °C.

Figure 4.26 illustrates changes of the E_m value of the nanocomposite at different levels of the MWCNTs added. When 0.1% and 0.5% MWCNTs were added, the E_m at low temperature (i.e., around 100 °C) was much greater than that when 1% MWCNTs were added. However, the E_m value of T_{p4} (i.e., cellulose degradation step) slightly increased as the content of MWCNTs rose. This result is quite consistent with the E value obtained by Kissinger method as shown in Figure 4.19. The E_m value obtained by MTG is in the range of the E_a value calculated by Kissinger method for the thermal degradation of cellulose that is a main component of PF resin/MWCNT/cellulose nanocomposite. These results indicate that both methods deliver similar activation energy for cellulose thermal decomposition of the nanocomposite, and are compatible with the result of polytetrafluoroethylene [76].

Figure 4.27 displays changes of the E_m value of the nanocomposite at different types of surface modifications for MWCNTs. In most of cases, the E_m values of the nanocomposite prepared by OX-MWCNTs without the surfactant were lower than those of the nanocomposite prepared with OX-MWCNTs and surfactant as well as silane-treated MWCNTs and surfactant. However, the E_m values of cellulose pyrolysis were greater than those of the nanocomposite prepared by OX-MWCNTs without the surfactant. These results could be ascribed to chemical reactions between MWCNTs and PF resins or APTMS in the nanocomposite as presented in Figure 4.22.

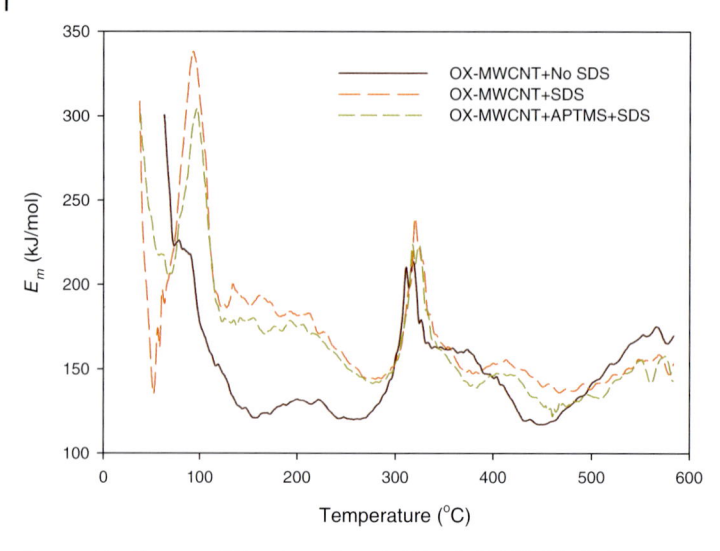

Figure 4.27 Changes of E_m values of resole PF resin/MWCNT/cellulose nanocomposite, depending on the surface modifications of MWCNTs (0.1%). Reproduced from Ref. [66] with by permission from Elsevier (© Elsevier, 2012).

4.3.4
Dynamic Mechanical Analysis of PF Resin/MWCNT/Cellulose Nanocomposites

Figure 4.28 shows a relationship between the degree of cure (β) and cure temperature for the resole PF resin/MWCNT/cellulose nanocomposites with the surfactant, SDS. The degree of cure determined by dynamic mechanical analysis (DMA) is obtained by the equation

$$\beta = \frac{E'_T - E'_{min}}{E'_{max} - E'_{min}} \quad (4.24)$$

where E'_T, E'_{min}, and E'_{max} are storage moduli at a certain temperature, minimum, and maximum of the DMA thermograms. In general, as the temperature increases, the β value tends to increases. The incorporation of 0.1% and 0.5% MWCNT into the nanocomposites speed up the β value compared to the control one, which indicates an acceleration of the degree of cure of the nanocomposites. However, the degree of cure decreases when 1% MWCNT is added. This results shows that the degree of cure could be different for different measuring techniques such as DSC, DMA, etc. It has been reported that the degree of cure measured by DMA is retarded when compared with those by DSC measurements [77].

Figure 4.29 illustrates changes of the E' value of the nanocomposites as a function of the surface treatment. E' value was highest for the nanocomposites with 0.1% MWCNT, and then decreased as the MWCNT content increased. This result is closely related to the degree of cure development in the nanocomposites as shown in Figure 4.28. In other words, the maximum E' value could be obtained

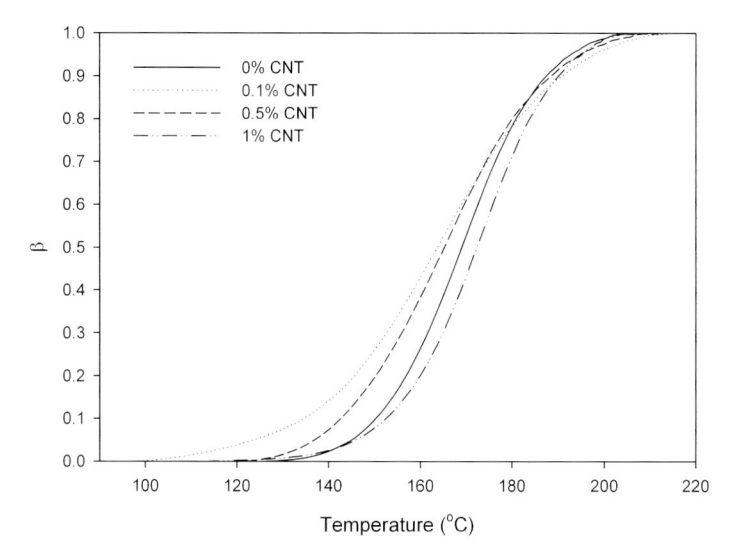

Figure 4.28 Relationship between degree of conversion (β) and cure temperature for the resole PF resin/MWCNT/cellulose nanocomposites with SDS.

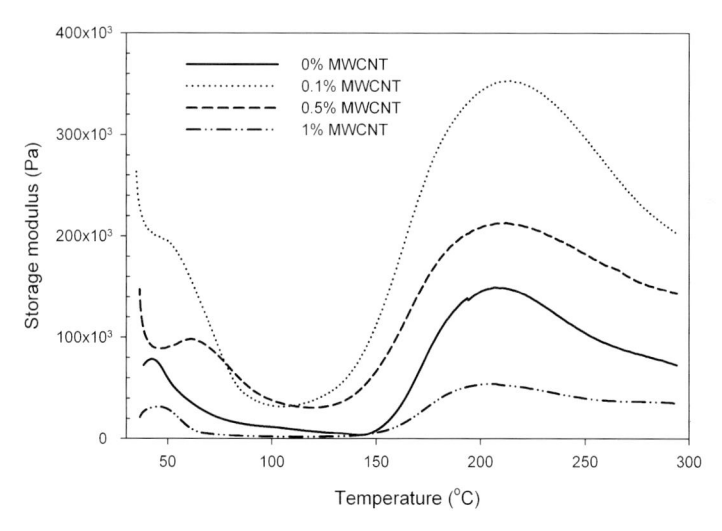

Figure 4.29 Change of E' of the resole PF resin/MWCNT/cellulose nanocomposites with SDS as a function of MWCNT levels and temperatures.

by tridimensional structure of cured PF resins reached at highest in the degree of cure. This result is also compatible with thermal stability of the nanocomposites as measured by CTG as shown in Figure 4.18.

The surface modification of MWCNTs also influenced E' values of the nanocomposites as depicted in Figure 4.30. The E' values of the nanocomposites with

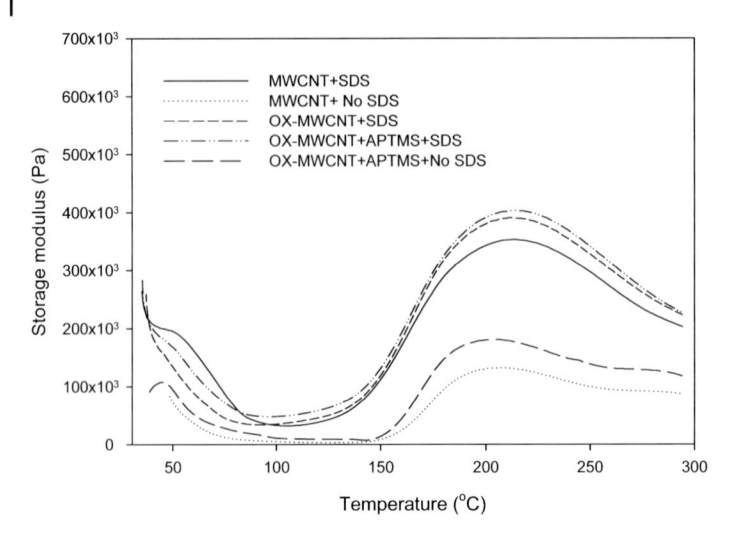

Figure 4.30 Change of E' of the resole PF resin/MWCNT/cellulose nanocomposites with SDS as a function of the surface treatments of MWCNT (0.1 wt%).

control and no surfactant SDS-treated MWCNTs are greater than for those with oxidation of MWCNTs. The highest E' values among the nanocomposites occurred for the nanocomposites prepared by MWCNTs with surfactant, silane, and oxidation treatments. These results could be ascribed to a better dispersion of MWCNTs by SDS and interfacial adhesion between MWCNTs and PF resins as shown in Figure 4.16. In addition, the surface oxidation treatment of MWCNTs also positively influenced the maximum E' value of the nanocomposites.

4.4
Mechanical Properties of the Nanocomposites

Tensile strength and elongation of MF resin/clay/cellulose nanocomposites are shown in Figures 4.31 and 4.32. Tensile strength of the nanocomposite increased to a maximum at 0.1% clay content and then gradually decreased thereafter (Figure 4.31). This result could be ascribed to exfoliated nanoclay particles at 0.1% level, while nanoclay above 0.1% level might be difficult to be exfoliated or to have a proper dispersion in MF resins, which impaired tensile strength of the nanocomposites. The elongation of the nanocomposites did not change very much up to 1% clay level followed by a gradual decrease with an increase in the clay content (Figure 4.32). This indicates that the nanocomposite increasingly became brittle as the clay content increased. As a result, the elongation decreases as well as the tensile strength of the nanocomposites.

The effect of clay addition on the mechanical properties of the resole PF resin/clay/cellulose nanocomposites is presented in Table 4.1. When compared to the

4.4 Mechanical Properties of the Nanocomposites | 101

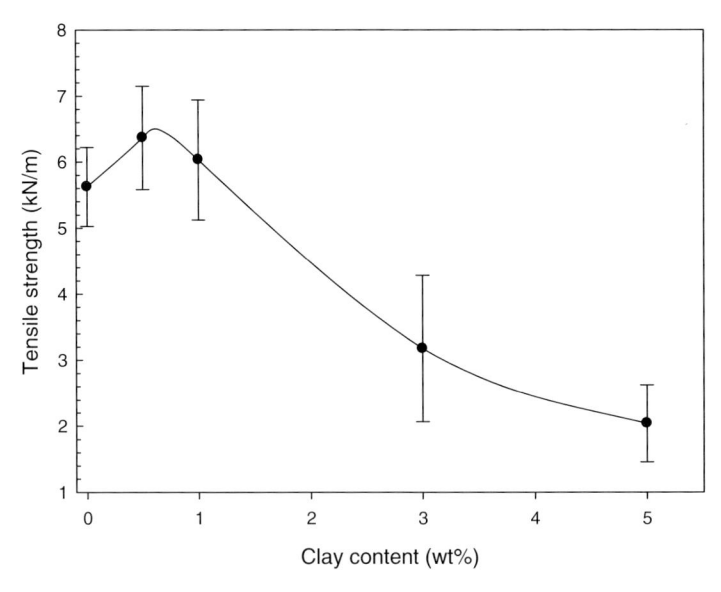

Figure 4.31 Tensile strength of MF resin/clay/cellulose nanocomposites as a function of clay content.

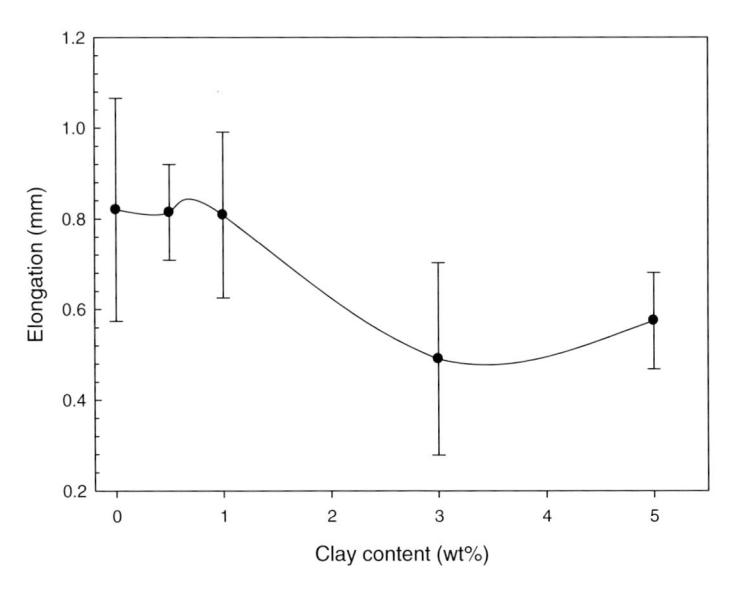

Figure 4.32 Elongation of MF resin/clay/cellulose nanocomposites as a function of clay content.

Table 4.1 Tensile strength and elongation of the resole PF resin/clay/cellulose nanocomposites as a function of the clay content.

Clay content (wt%)	Tensile strength (kN/m)[a]	Elongation (%)[a]
0	4.74 ± 1.608	0.96 ± 0.288
1	7.52 ± 1.829	0.88 ± 0.196
3	6.84 ± 1.631	0.79 ± 0.160
5	6.98 ± 2.065	0.77 ± 0.213
10	6.70 ± 1.923	0.83 ± 0.243

a) The values are presented as average and standard deviation.

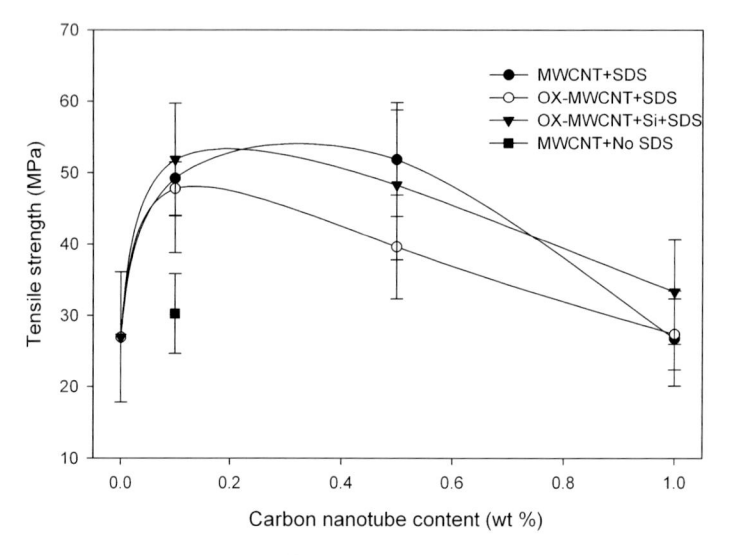

Figure 4.33 Tensile strength of PF resin/MWCNT/cellulose nanocomposites as a function of MWCNT content.

control nanocomposite (i.e., the ones lacking clay), tensile strength of the nanocomposite was improved by the addition of 1% clay; tensile strength decreased at higher clay contents. These results show a reinforcement effect of the added clay in the nanocomposite through either exfoliation or intercalation. As shown in Table 4.1, the elongation at break was reduced up to 3% clay and then leveled off afterward. Since the added clay reinforced resole PF resin matrix in the nanocomposite, the resultant nanocomposite consequently became stiffer, which reduced the elongation at break.

Tensile strength and modulus of PF resin/MWCNT/cellulose nanocomposites are shown in Figures 4.33 and 4.34. As shown in Figure 4.33, the tensile strength

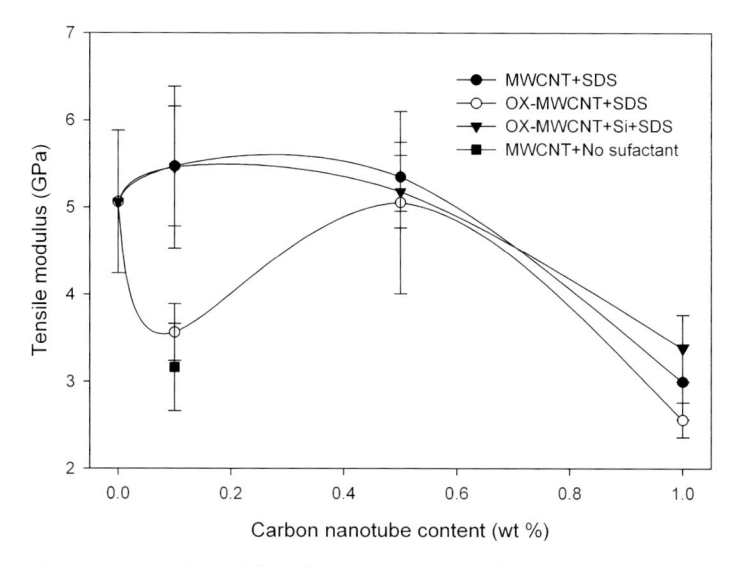

Figure 4.34 Tensile modulus of PF resin/MWCNT/cellulose nanocomposites as a function of MWCNT content.

of the nanocomposites increased to a maximum at 0.1% MWCNTs regardless of types of the surface treatments and then decreased afterward. The greatest tensile strength occurs for nanocomposites prepared by MWCNTs with oxidation, silane-treated, and surfactant addition. This could be resulted from a better dispersion of MWCNTs and interfacial adhesion between MWCNTs and PF resins. This result is quite consistent with the outcomes of thermal stability, degree of cure, and maximum E' values of the nanocomposites with 0.1 wt% MWCNTs. As shown in Figure 4.33, tensile strength value is quite low at 0.1% MWCNTs when no surfactant is used, which could be due to a poor dispersion of MWCNTs in PF resins. Similar trends were observed for tensile modulus of the nanocomposites as illustrated in Figure 4.34. As expected, tensile modulus reached a maximum at 0.1% MWCNTs level and decreased thereafter. A poor tensile modulus was also found when no surfactant was used for the nanocomposites.

4.5 Summary

This chapter attempts to review recent progress of thermal properties of nanocomposites such as MF resin/clay/cellulose, PF resin/clay/cellulose, or PF resin/MWCNT/cellulose systems by focusing on cure kinetics, thermal degradation kinetics, or dynamic mechanical analysis, using DSC, TGA, or DMA. Three different methods such as Ozawa, Kissinger, and isoconversional methods of DSC were used to characterize cure kinetics of the nanocomposites by demining their activation energies. TMDSC was also employed to study cure kinetics of PF resin/clay/cellulose nanocomposites. Both Ozawa and Kissinger methods showed that the E_a of the MF resin/clay/cellulose nanocomposite at 0.5 wt% clay content reached a maximum value and then decreased thereafter. But, the Ozawa method provided greater E_a values than those of the Kissinger method. The E_α values obtained from the isoconversional method increased as the degree of conversion increased, while the influence of clay levels followed similar trends of the overall E_a values from both the Ozawa and Kissinger methods.

As the clay content increased, T_p values of the PF resin/clay/cellulose nanocomposite curing reaction decreased up to 3–5% clay content, showing the cure acceleration effect of the clay. Both the Ozawa and Kissinger methods showed the lowest E at 3% clay content. The isoconversional method provided E_α as a function of the degree of conversion, wherein the E_α showed a gradual decrease with increasing the degree of conversion, the lowest value at 3% clay loading. TMDSC further confirmed the cure acceleration effect of clay addition to the resole PF resin system, as indicated by both T_p and ΔH measurements. Interestingly, the reversing heat flow and heat capacity values showed that the addition of clay up to 3% delayed the vitrification process of PF resin in the nanocomposite, indicating an inhibition effect of the clay in the final stage of the curing reaction.

Study of thermal degradation kinetics of MWCNTs-reinforced resole PF resin/cellulose nanocomposite, using TG analysis as a function of the content or surface treatment of MWCNTs with or without a surfactant resulted in a few important outcomes. FT-IR spectroscopy showed that the oxidation provided hydroxyl or carboxyl groups with MWCNT's surface, while the silanization resulted in the silane attachment to MWCNT's surface. The E_a values of the PF resin/MWCNT/cellulose nanocomposite slightly increased as the MWCNT content increased, while the one of cellulose degradation was independent of the MWCNT content. Both the oxidation and silanization of MWCNTs' surface resulted in an increase of the E_a values compared to that of the control sample. The E_α values gradually increased up to $\alpha = 0.5$, and then were rapidly elevated to fluctuations at $\alpha > 0.5$. The E_m value obtained by the MTG was within the range of the E value calculated by the Kissinger method for cellulose thermal degradation that is a main component of the nanocomposite. These results show that MTG method provides similar activation energy to that of CTG method for thermal degradation of the nanocomposite, and indicate that MTG method could be efficiently used to obtain activation energy without many scans from multiple heating rates.

The DMA results show that the degree of cure and maximum E' value are closely dependent on each other, affecting the thermal curing and mechanical properties. All tensile strengths of MF resin/clay/cellulose, PF resin/clay/cellulose, and PF resin/MWCNT/ cellulose nanocomposites were slightly improved by the addition of 0.1% layered clay, showing a reinforcing effect, while the elongation at break slightly decreased as the clay content increased. These results demonstrate that an optimum level of clay addition to either MF resin or PF resin/clay/cellulose nanocomposites is a balance between the impact on cure kinetics and tensile strength.

Acknowledgment

This research was supported by Kyungpook National University Research Fund, 2012.

References

1 Roberts, R.J. and Evans, P.D. (2005) Effects of manufacturing variables on surface quality and distribution of melamine formaldehyde resin in paper laminates. *Compos. Part A: Appl. Sci. Manuf.*, **36**, 95–104.
2 Shen, J., Huang, W., Wu, L., Hu, Y., and Ye, M. (2007) Thermo-physical properties of epoxy nanocomposite reinforced with amino-functionalized multi-walled carbon nanotubes. *Compos. Part A: Appl. Sci. Manuf.*, **38**, 1331–1336.
3 Coleman, J.N., Khan, U., Blau, W.J., and Gun'ko, Y.K. (2006) Small but strong: a review of the mechanical properties of carbon nanotube–polymer composites. *Carbon*, **44**, 1624–1652.
4 Tjong, S.C. (2006) Synthesis and structure–property characteristics of clay–polymer nanocomposite, in *Nanocrystalline Materials* (ed. S.C. Tjong), Elsevier, the Netherlands, Chapter 10, pp. 311–348.
5 Ray, S.S. and Okamoto, M. (2003) Polymer/layered silicate nanocomposite: a review from preparation to processing. *Prog. Polym. Sci.*, **28**, 1539–1641.
6 Pavlidou, S. and Papaspyrides, C.D. (2008) A review on polymer-layered silicate nanocomposite. *Prog. Polym. Sci.*, **33**, 1119–1198.
7 Cheng, Q., Wang, S., and Rials, T.G. (2009) Poly(vinyl alcohol) nanocomposite reinforced with cellulose fibrils isolated by high intensity ultrasonication. *Compos. Part A: Appl. Sci. Manuf.*, **40**, 218–224.
8 Prolongo, S.G., Campo, M., Gude, M.R., Chaos-Moran, R., and Urena, A. (2009) Thermo-physical characterization of epoxy resin reinforced by amino-functionalized carbon naonofibers. *Compos. Sci. Technol.*, **69**, 349–357.
9 Kojima, Y., Usuki, A., Kawasumi, M., Okada, A., Kurauchi, T., and Kamigaito, O. (1993) Synthesis of nylon 6–clay hybrid by montmorillonite intercalated with ε-caprolactam. *J. Polym. Sci. Part A Polym. Chem.*, **31**, 983–986.
10 Usuki, A., Kojima, Y., Kawasumi, M., Okada, A., Fukushima, Y., Kurauchi, T., and Kamigaito, O. (1993) Synthesis of nylon 6–clay hybrid. *J. Mater. Res.*, **8**, 1179–1184.
11 Ingram, S., Rhoney, I., Liggat, J.J., Hudson, N.E., and Pethrick, R.A. (2007) Some factors influencing exfoliation and physical property enhancement in clay epoxy resins based on diglycidyl ethers of bisphenol A and F. *J. Appl. Polym. Sci.*, **106**, 5–19.
12 McIntyre, S., Kaltzakorta, I., Liggat, J.J., Pethrick, R.A., and Rhoney, I. (2005) Influence of the epoxy structure on the physical properties of epoxy resin nanocomposites. *Ind. Eng. Chem. Res.*, **44**, 8573–8579.
13 Park, J.H. and Jana, S.C. (2003) Mechanism of exfoliation of nanoclay particles in epoxy–clay nanocomposite. *Macromolecules*, **36**, 2758–2768.
14 Kohl, W.S., Frei, J., and Tretheway, B.R. (1996) Characterization of the cure process in melamine–formaldehyde laminating resins using high-pressure differential scanning calorimetry. *Tappi J.*, **79** (9), 199–205.
15 Choi, M.H., Chung, I.J., and Lee, J.D. (2000) Morphology and curing behaviors of phenolic resin-layered silicate nanocomposite prepared by melt intercalation. *Chem. Mater.*, **12**, 2977–2983.
16 Choi, M.H. and Chung, I.J. (2003) Mechanical and thermal properties of phenolic resin-layered silicate nanocomposite synthesized by melt intercalation. *J. Appl. Polym. Sci.*, **90**, 2316–2321.

17 Byun, H.Y., Choi, M.H., and Chung, I.J. (2001) Synthesis and characterization of resol type phenolic resin/layered silicate nanocomposite. *Chem. Mater.*, **13**, 4221–4226.

18 Wang, J., Laborie, M.-P.G., and Wolcott, M.P. (2005) Comparison of model-free kinetic methods for modeling the cure kinetics of commercial phenol–formaldehyde resins. *Thermochim. Acta*, **439**, 68–73.

19 Wu, Z., Zhou, C., and Qi, R. (2002) The preparation of phenolic resin/montmorillonite nanocomposite by suspension condensation polymerization and their morphology. *Polym. Compos.*, **23**, 634–646.

20 Kaynak, C. and Tasan, C.C. (2006) Effects of production parameters on the structure of resol type phenolic resin/layered silicate nanocomposite. *Eur. Polym. J.*, **42**, 1908–1921.

21 Tasan, C.C. and Kaynak, C.C. (2009) Mechanical performance of resol type phenolic resin/layered silicate nanocomposite. *Polym. Compos.*, **30**, 343–350.

22 Iijima, S. (1991) Helical microtubules of graphitic carbon. *Nature*, **354**, 56–58.

23 Iijima, S. and Ichihashi, T. (1993) Single-shell carbon nanotubes of 1-nm diameter. *Nature*, **363**, 603–605.

24 Chou, T.-W., Gao, L., Thostenson, E.T., Zhang, Z., and Byun, J.-H. (2010) An assessment of the science and technology of carbon nanotube-based fibers and composites. *Compos. Sci. Technol.*, **70**, 1–19.

25 Yu, M., Lourie, O., Dyer, M.J., Kelly, T.F., and Ruoff, R.S. (2000) Strength and breaking mechanism of multiwalled carbon nanotubes under tensile load. *Science*, **287**, 637–640.

26 Xie, S., Li, W., Pan, Z., Chang, B., and Sun, L. (2000) Mechanical and physical properties on carbon nanotube. *J. Phys. Chem. Solids*, **61**, 1153–1158.

27 Puretzky, A.A., Geohegan, D.B., Fan, X., and Pennycook, S.J. (2000) *In situ* imaging and spectroscopy of single-wall carbon nanotube synthesis by laser vaporization. *Appl. Phys. Lett.*, **76**, 182–184.

28 Hernadi, K., Fonseca, A., Nagy, J.B., Bernaerts, D., Fudala, A., and Lucas, A.A. (1996) Catalytic synthesis of carbon nanotubes using zeolite support. *Zeolites*, **17**, 416–423.

29 Che, G., Lakshmi, B.B., Martin, C.R., Fisher, E.R., and Ruoff, R.S. (1998) Chemical vapor deposition based synthesis of carbon nanotubes and nanofibers using a template method. *Chem. Mater.*, **10**, 260–267.

30 Zhou, W., Ooi, Y.H., Russo, R., Papanek, P., Luzzi, D.E., and Fischer, J.E. (2001) Structural characterization and diameter-dependent oxidative stability of single wall carbon nanotubes synthesized by the catalytic decomposition of CO. *Chem. Phys. Lett.*, **350**, 6–14.

31 Spitalsky, Z., Tasis, D., Papagelis, K., and Galiotis, C. (2010) Carbon nanotube–polymer composites: chemistry, processing, mechanical and electrical properties. *Prog. Polym. Sci.*, **35**, 357–401.

32 Ajayan, P.M., Stephan, O., Colliex, C., and Trauth, D. (1994) Aligned carbon nanotube arrays formed by cutting a polymer resin–nanotube composite. *Science*, **265**, 1212–1214.

33 Allaoui, A., Bai, S., Cheng, H.M., and Bai, J.B. (2002) Mechanical and electrical properties of a MWNT/epoxy composite. *Compos. Sci. Technol.*, **62**, 1993–1998.

34 Bai, J. (2003) Evidence of the reinforcement role of chemical vapour deposition multi-walled carbon nanotubes in a polymer matrix. *Carbon*, **41**, 1325–1328.

35 Breton, Y., Desarmot, G., Salvetat, J.P., Delpeux, S., Sinturel, C., and Beguin, F. (2004) Mechanical properties of multiwall carbon nanotubes/epoxy composites: influence of network morphology. *Carbon*, **42**, 1027–1030.

36 Li, X.D., Gao, H.S., Scrivens, W.A., Fei, D.L., Xu, X.Y., and Sutton, M.A. (2004) Nanomechanical characterization of single-walled carbon nanotube reinforced epoxy composites. *Nanotechnology*, **15**, 1416–1423.

37 Sahoo, N.G., Rana, S., Cho, J.W., Li, L., and Chan, S.H. (2010) Polymer nanocomposites based on functionalized carbon nanotubes. *Prog. Polym. Sci.*, **35**, 837–867.

38 Yin, Q.K., Sun, A., Li, L., Shao, S., Liu, S., and Sun, C. (2008) Study on carbon nanotube reinforced phenol–formaldehyde resin/graphite composite for bipolar plate. *Power Sources*, **175**, 861–865.

39 Liu, L., and Ye, Z. (2009) Effects of modified multi-walled carbon nanotubes on the curing behavior and thermal stability of boron phenolic resin. *Polym. Degrad. Stab.*, **94**, 1972–1978.

40 Peterson, J.D., Vyazovkin, S., and Wight, C.A. (2001) Kinetics of the thermal and thermo-oxidative degradation of polystyrene, polyethylene and polypropylene. *Macromol. Chem. Phys.*, **202**, 775–784.

41 Vyazovkin, S. and Sbirrazzuoli, N. (2006) Isoconversional kinetic analysis of thermally stimulated processes in polymers. *Macromol. Rapid Commun.*, **27**, 1515–1532.

42 Doyle, C.D. (1962) Estimating isothermal life from thermogravimetric. *J. Appl. Polym. Sci.*, **6**, 639–642.

43 Ozawa, O.A. (1965) A new method of analyzing thermogravimetric data. *Bull. Chem. Soc. Jpn.*, **38**, 1881–1886.

44 Kissinger, H.E. (1957) Reaction kinetics in differential thermal analysis. *Anal. Chem.*, **29**, 1702–1706.

45 Coat, A.W. and Redfern, J.P. (1964) Kinetic parameters from thermogravimetric data. *Nature*, **201**, 68–69.

46 Vyazovkin, S. and Dollimore, D. (1996) Linear and nonlinear procedures in isoconversional computations of the activation energy of nonisothermal reactions in solids. *J. Chem. Inf. Comput. Sci.*, **36**, 42–45.

47 Vyazovkin, S. and Sbirrazzuoli, N. (1997) Confidence intervals for the activation energy estimated by few experiments. *Anal. Chim. Acta*, **355**, 175–180.

48 Reading, M., Elliott, D., and Hill, V.L. (1992) Some aspects of the theory and practice of modulated differential scanning calorimetry. In: Proceedings of the 21st NATAS Conference, pp. 145–150.

49 Reading, M., Luget, A., and Wilson, R. (1994) Modulated differential scanning calorimetry. *Thermochim. Acta*, **238**, 295–307.

50 Van Assche, G., Van Hemelrijck, A., Rahier, H., and Van Mele, B. (1995) Modulated differential scanning calorimetry: isothermal cure and vitrification of thermosetting systems. *Thermochim. Acta*, **268**, 121–142.

51 Van Assche, G., Van Hemelrijck, A., Rahier, H., and Van Mele, B. (1996) Modulated differential scanning calorimetry: non-isothermal cure, vitrification, and devitrification of thermosetting systems. *Thermochim. Acta*, **286**, 209–224.

52 Van Assche, G., Van Hemelrijck, A., Rahier, H., and Van Mele, B. (1997) Modulated temperature differential scanning calorimetry; consideration for a quantitative study of thermosetting systems. *J. Therm. Anal.*, **49**, 443–447.

53 Flynn, J.H. and Wall, L.A. (1966) A general treatment of the thermogravimetry of polymers. *J. Res. Nat. Bur. Stand.*, **70**, 487–523.

54 Flynn, J.H. and Wall, L.A. (1996) A quick direct method for determination of activation energy from thermogravimetric data. *J. Polym. Sci. Part C: Polym. Lett.*, **4**, 323–328.

55 Park, B.-D. and Jeong, H.-W. (2010) Cure kinetics of melamine–formaldehyde resin/clay/cellulose nanocomposites. *J. Ind. Eng. Chem.*, **16** (3), 375–379.

56 Alonso, M.V., Oliet, M., Perez, J.M., Rodriguez, F., and Echeverria, J. (2004) Determination of curing kinetic parameters of lignin-phenol-formaldehyde resol resins by several dynamic differential scanning calorimetry methods. *Thermochim. Acta*, **419**, 161–167.

57 Tejado, A., Kortaberria, G., Echeverria, J.M., and Mondragon, I. (2008) Isoconversional kinetic analysis of novolac-type lignophenolic resin cure. *Thermochim. Acta*, **471**, 80–85.

58 He, G.B., Riedl, B., and Ait-Kadi, A. (2003) Model-free kinetics: curing behavior of phenol formaldehyde resins by differential scanning calorimetry. *J. Appl. Polym. Sci.*, **87** (3), 433–440.

59 He, G.B. and Riedl, B. (2004) Curing kinetics of phenol formaldehyde resin and wood–resin interactions in the

60 Park, B.-D. and Kadla, J.F. (2012) Thermal curing behavior and tensile properties of resole phenol–formaldehyde resin/clay/cellulose nanocomposites. *Mokchae Konghak*, **40** (2), 110–122.

61 Yong, R.N., Desjardins, S., Farant, J.P., and Simon, P. (1997) Influence of pH and exchangeable cation on oxidation of methylphenols by a montmorillonite clay. *Appl. Clay Sci.*, **12**, 93–110.

62 López, M., Blanco, M., Vazquez, A., Gabilondo, N., Arbelaiz, A., Echeverría, J.M., and Mondragon, I. (2008) Curing characteristics of resol-layered silicate nano-composites. *Thermochim. Acta*, **467**, 73–79.

63 López, M., Blanco, M., Vazquez, A., Ramos, J.A., Arbelaiz, A., Gabilondo, N., Echeverría, J.M., and Mondragon, I. (2009) Isoconversional kinetic analysis of resol–clay nanocomposite. *J. Therm. Anal. Calorim.*, **96**, 567–573.

64 Park, B.D., Riedl, B., Kim, Y.S., and So, W.T. (2002) Effect of synthesis parameters on thermal behavior of phenol-formaldehyde resole resin. *J. Appl. Polym. Sci.*, **83**, 1415–1424.

65 Fraga, I., Montserrat, S., and Hutchison, J.M. (2008) Vitrification during the isothermal cure of thermosets. Part I. An investigation using TOPEM, a new temperature modulated technique. *J. Therm. Anal. Calorim.*, **91**, 687–695.

66 Park, B.-D. and Kadla, J.F. (2012) Thermal degradation kinetics of resole phenol–formaldehyde resin/multi-walled carbon nanotube/cellulose nanocomposite. *Thermochim. Acta*, **540**, 107–115.

67 Kathi, J., Rhee, K.-Y., and Lee, J.H. (2009) Effect of chemical functionalization of multi-walled carbon nanotubes with 3-aminopropyltriethoxysilane on mechanical and morphological properties of epoxy nanocomposites. *Compos. Part A: Appl. Sci. Manuf.*, **40**, 800–809.

68 Ramanathan, T., Fisher, F.T., Ruoff, R.S., and Brinson, L.C. (2005) Amino-functionalized carbon nanotubes for binding to polymers and biological systems. *Chem. Mater.*, **17**, 1290–1295.

69 Silverstein, R.M. and Webster, F.X. (1998) *Spectrometric Identification of Organic Compounds*, 6th edn, John Wiley & Sons Inc., New York, p. 71.

70 Yuen, S.M., Ma, C.C.M., Chiang, C.L., Chang, J.A., Huang, S.W., Chen, S.C., Chuang, C.Y., Yang, C.C., and Wei, M.H. (2007) Silane-modified MWCNT/PMMA composites – preparation, electrical resistivity, thermal conductivity and thermal stability. *Compos. Part A: Appl. Sci. Manuf.*, **38** (2007), 2527–2535.

71 Ma, P.C., Kim, J.K., and Tang, B.Z. (2006) Functionalization of carbon nanotubes using a silane coupling agent. *Carbon*, **44**, 3232–3238.

72 Capart, R., Khezami, L., and Burnham, A.K. (2004) Assessment of various kinetic models for the pyrolysis of a microgranular cellulose. *Thermochim. Acta*, **417**, 79–89.

73 Antal, M.J., Mok, W.S.L., Varhegyi, G., and Szekely, T. (1990) Review of methods for improving the yield of charcoal from biomass. *Energy Fuels*, **4**, 221–225.

74 Conesa, J.A., Caballero, J.A., Marcilla, A., and Font, R. (1995) Analysis of different kinetic model, in the dynamic pyrolysis of cellulose. *Thermochim. Acta*, **254**, 175–192.

75 Burnham, A.K., Braun, R.L., Coburn, T.T., Sandvik, E.I., and Curry, D.J. (1996) An appropriate kinetic model for well-preserved algal kerogens. *Energy Fuels*, **10**, 49–59.

76 Schubnell, M. (2000) Comparison of activation energies obtained from modulated and conventional non-modulated TG. *J. Therm. Anal. Calorim.*, **61**, 1005–1011.

77 Follensbee, R.A., Koutsky, J.A., Christiansen, A.W., Myers, G.E., and Geimer, R.L. (1993) Development of dynamic mechanical methods to characterize the cure state of phenolic resole resins. *J. Appl. Polym. Sci.*, **47** (8), 1481–1496.

5
Mechanical Performance of Thermoset Clay Nanocomposites

Nourredine Aït Hocine, Said Seghar, Hanaya Hassan, and Saïd Azem

5.1
Introduction

In the late 1980s, polymer nanocomposites were widely developed in both academic laboratories and the commercial organizations. The term "Nanocomposites" was used for the first time in 1984 by Roy Komarneni and it referred to hybrid organic materials, in which mixing of the filler phase is less than 100 nm. In such materials, the nanophase material is often reinforcing particles and the matrix is a thermoset resin, thermoplastic, or elastomer. Various kinds of reinforcing nanofillers are used in polymer nanocomposites: mineral (clay, silica) [1–3], vegetal (cellulose whiskers) [4, 5], organic (rigid or flexible plastics) [6], metallic [7] and so on. Nowadays, the nanocomposite describes a very large family of materials, whose properties are of interest due to the nanosize of their structures. Since then, several big industrial companies have started commercializing various nanocomposites, and subsequently have begun expanding the field of research on those materials. However, most commercial interests have been focused on thermoplastic nanocomposites and less on the thermosetting ones.

The properties of polymers are sensibly improved by adding to a matrix only a small fraction of reinforcing nanofillers compared with microfillers, which extend the use of such materials to various applications [8–15]. Recent reviews draw up an inventory of the most prominent studies on preparation, processing, and characterization of polymer-layered silicate nanocomposites [16–18], and an overview of the numerous polymeric matrices used to design those nanostructured materials [19]. As far as the nanocharges are concerned, organically modified montmorillonites are the most commonly used layered silicates for the preparation of polymer nanocomposites since their introduction by a Toshiba firm in the 1990s [20–22].

Polymer-layered silicate nanocomposites have been shown to exhibit various structural states, as evidenced by X-ray diffraction experiments, and scanning or transmission electron micrographic observation. Their structure may be either ordered or disordered, intercalated or exfoliated or, in most cases, partly intercalated/exfoliated [23]. Better enhancement of mechanical properties is

Thermoset Nanocomposites, First Edition. Edited by Vikas Mittal.
© 2013 Wiley-VCH Verlag GmbH & Co. KGaA. Published 2013 by Wiley-VCH Verlag GmbH & Co. KGaA.

expected to occur with the best dispersed structure, that is, the exfoliated one since it favors the interaction between the matrix and the reinforcing fillers.

Several studies have shown that the design of the mixing device and the melt processing conditions play a key role in achieving good final properties. Using either extruder or internal mixer, the average processing shear rate [24–26], the residence time [24, 27, 28], and mixing temperature [26, 29] have been shown to be key parameters to optimize the mechanical properties of nanocomposites. However, even under good processing conditions, the nanoclays are usually only partially exfoliated, as manifested by the presence of small stacks of a few layers [30].

Some other works done on polymer/nanoclay composites have pointed out the effect of each component on the final morphology of the microstructure and on the mechanical properties of the material [31–33]. One of the main parameters playing a key role in the modification of the microstructure of a thermoset matrix is the kind and the shape of the added reinforcing fillers, as proved by the work of Lan et al. [34] on an epoxy. These authors underlined that the alkyl ammonium ions used for modifying the clay layers favor the improvement of the mechanical properties of epoxy/elastomer nanocomposites. Such properties seem to be better as the chains of the alkyl ammonium ions are longer. In fact, they observed that alkyl ammonium ions with chain length larger than eight carbon atoms help in the synthesis of delaminated nanocomposites whereas alkyl ammonium ions with shorter chains lead to the formation of intercalated nanocomposites. In subsequent studies, they showed that the catalytic effect of the acidic primary alkyl ammonium ions at ion-exchange sites of the clay enhanced the exfoliation of the silicate nanolayers in the glassy epoxy network during polymerization with diamine curing agents compared with nonacidic quaternary onium ions, which formed mainly intercalated clay nanocomposites. The exfoliation degree of the clay depends not only on the reactivity of the epoxy system but also on the rate of diffusion of the organic molecules between the clay layers. Indeed, if this rate is too slow, the polymerization rate outside the layers (extragallery polymerization) will be faster than that between the layers (intragallery polymerization) and only intercalated nanocomposites will be formed. However, this interpretation assumes that the polymerization reaction is slower inside of the galleries than outside [35].

The polymerization conditions of thermosetting matrices represent another parameter having an effect on the mechanical properties of thermosetting nanocomposites. For instance, the temperature of such polymerization should be carefully selected [36–38].

Kornmann et al. [39] explain the exfoliation of nanoclay layers in an unsaturated polyester (UP) matrix by the delamination of these silicate layers caused by polymer molecules during the cross-linking reaction of the UP. During the cross-linking of UP, the monomer of this matrix diffuses into the galleries between the silicate layers. The final morphology depends on the degree of penetration of the monomer into the organolayered silicate structure.

Suh et al. [40] highlighted a relationship between the variation of the glass transition temperature of thermosetting nanocomposites and the order in which components are introduced during the mixing process. In fact, prior intercalation of matrix macromolecules between the clay layers and the increase of the mixing time allow a better dispersion of monomers in whole reactive field.

It is also highlighted that the clay dispersion was enhanced when a high-amplitude sonication dispersion technique were combined with a rapid heating rate during cure [41]. XRD and SAXS spectra showed that optimizing the processing conditions resulted in a twofold increase in clay gallery distances. Clay dispersion is thought to be improved as a result of the low viscosity achieved in the nanocomposites. An increase in the heating rate from 3 to 10 °C/min during cure reduced the viscosity by 60%, facilitating the penetration of polymer chains into clay galleries.

In short, the exfoliation of the clay layers depending on several parameters plays a key role in the microstructure of thermosetting nanocomposites and, thus, controls the final properties of the material [42]. Several other factors have also a direct effect on these properties: the interaction matrix/fillers and fillers/fillers, the surface modification of the clays and so on [42]. The interaction forces between clay layers, even small, can lead to the formation of aggregates and, in the best case, to the formation of a percolation network playing a role of a frame reinforcing the nanocomposite [3, 43]. A percolation network represents a three-dimensional structure of interconnected long silicate layers, strengthening the material.

This chapter represents a succinct literature review on mechanical performance of thermoset clay nanocomposites. We will analyze the parameters playing a key role in the improvement of mechanical properties of such nanocomposites. We particularly expose the effects of nanoclay reinforcing fillers on the viscoelasticity behavior, rigidity, fracture strain, fracture stress, and toughness of these nanocomposites. It could be noticed that thermosetting polymers are the base of a large domain of engineering applications such as computer chip insulation, protective coating, adhesives, and advanced aerospace composites. This is due to the easy processing of those polymers, a good affinity to heterogeneous materials, considerable solvent and creep resistance, higher operating temperature, and their great strength.

5.2
Viscoelasticity Analysis: Dynamical Mechanical Thermal Analysis (DMTA)

In the 2000s, Agag and Takeichi [44] examined the effect of the clay nanolayers on the thermomechanical properties of polyimide films. A small decrease in the storage modulus was observed around glass transition temperature T_g, and this critical temperature increased with the increase of the clay content: T_g values were 321, 342, 330, 336, and 343 °C at a clay content of 0, 1, 2, 4, and 6 wt%, 6 wt%, respectively. This increase can be attributed to maximizing of the adhesion between the polymer and layered silicate surfaces because of the nanometer size, restricting segmental motion near the organic inorganic interface, which is a typical effect of the clay inclusions on polymer systems [44].

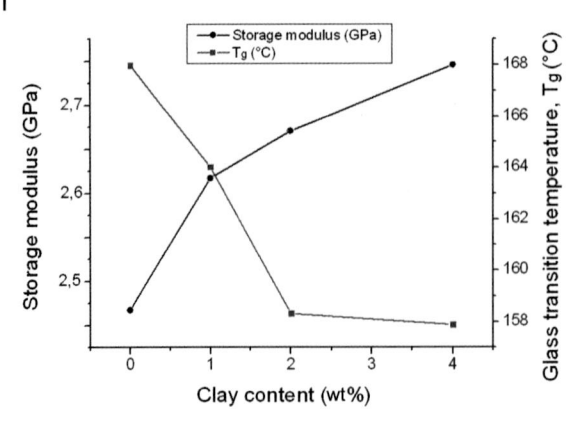

Figure 5.1 Variation of storage modulus (at 50 °C) and glass transition temperature (T_g) of epoxy/clay nanocomposites as a function of clay content.

In 2004, Xiong et al. [45] analyzed the mechanical properties of polyurethane (PU)/organic-montmorillonite (MMT) nanocomposites. It was found that the tan δ peaks of these nanocomposites are narrower and lower than those of the PU, indicating that the addition of the organic-MMT lessens the damping ability of the PU. This result is in agreement with that reported by Mishra et al. [46].

Liu et al. [47] studied the effect of the organoclay nanoparticles on the properties of epoxy/organoclay nanocomposites. As shown in Figure 5.1, dynamical mechanical analysis (DMA) achieved at 50 °C indicated a steady increase in storage modulus and a gradual decrease in high glass transition temperature T_g as the clay loading increased. In fact, compared with the T_g of neat epoxy resin, that of the nanocomposite decreases by about 10 °C (from 168 to 158 °C) with increasing clay concentration up to only 4 wt%. This decrease in T_g implies that the molecular motion becomes less restrictive, probably because of a lower degree of cross-linking. Figure 5.2 illustrates the DMA plots of storage modulus (E') versus temperature for four clay concentrations. It can be seen that the storage modulus steadily increases as the clay content increases because of an enhancement effect of the addition of rigid inorganic clay particles. Only a moderate increase in storage modulus (by 10%, from 2.49 to 2.75 GPa) was observed as a result of incorporation of 4 wt% clay into the epoxy matrix. However, in a rubbery epoxy matrix, the improvement in modulus was reported to be much higher (250%) by adding organically modified clays [36]. This clearly indicates that organically modified clay is more effective in reinforcing a soft (rubbery) matrix compared to that in a hard (glassy) matrix. The similar observations have been reported in the literature [36, 48–52].

In 2005, Dean et al. [53] examined the effect of the organoclay nanoparticles on the properties for epoxy/organoclay nanocomposites. They showed the utility of rheological analyses in understanding the effect of nanoparticles on the overall curing process in thermoset polymer nanocomposites. They found that the improvement of the storage modulus is higher in the rubbery region, due to the

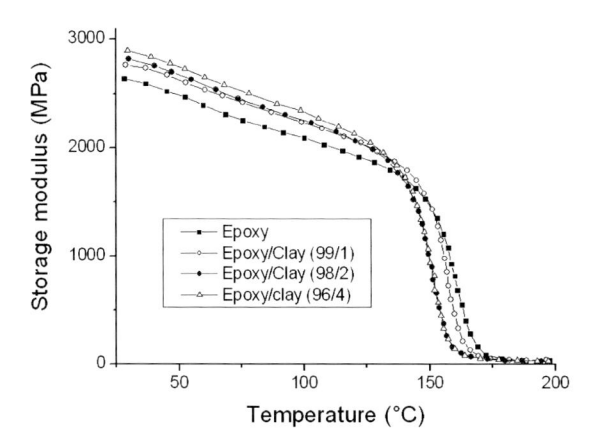

Figure 5.2 DMA curves: storage modulus of epoxy/clay nanocomposites versus temperature.

constraining effects of the clays [13, 54, 55]. As the clay loading increases, the modulus increases by 35%, 15%, and 30% for the 2 wt%, 4 wt%, and 6 wt% of reinforcing particles, respectively.

Liu et al. [56] studied the mechanical properties of epoxidized soybean oil/clay nanocomposites. DMA study showed the ESO/clay nanocomposites with 5–10 wt% clay content possess storage modulus ranging from 2×10^6 to 2.70×10^6 Pa at 30 °C. The glass transition temperature T_g of the ESO/clay nanocomposite was obtained from the peak of the loss factor curve. The matrix showed the lower T_g, which is 11.8 °C, and T_g of the nanocomposite with 5 wt% clay content is 20.7 °C. As the clay loading increases, the resulting nanocomposites show a higher value of T_g.

In 2006, Yasmin et al. [57] analyzed the mechanical behavior of clay/epoxy nanocomposites. They used two kinds of clay: the Nanomer I.28E and the Cloisite 30B. It was found that the addition of clay particles improves significantly the elastic modulus of pure epoxy, as shown in Figure 5.3. In fact, the higher the clay content, the higher the elastic modulus of the nanocomposite. They also analyzed the variation of storage modulus with temperature. At room temperature, it was found that the Cloisite 30B/epoxy shows much higher storage modulus than the Nanomer I.28E/epoxy for the same clay content. As the temperature increases, both nanocomposites show a gradual drop in storage modulus up to T_g. The nanocomposites with both types of particles show decreasing T_g with increasing clay content. However, the Cloisite 30B/epoxy shows much lower T_g above 3 wt% clay content than the Nanomer I.28E/epoxy nanocomposite. Agag et al. [58] reported higher T_g values for polyimide/clay nanocomposites with higher clay contents. This was mainly attributed to the good adhesion between the polymer and the particles so that the nanometer-sized particles can restrict the segmental motion of cross-links near the organic–inorganic interface [58]. However, a lower T_g with higher clay content was reported by many other investigators [59–65].

Wang et al. [66] studied the mechanical properties of epoxy/silane-modified clay (SMC) nanocomposite by using DMA. They showed that in the glassy region, the

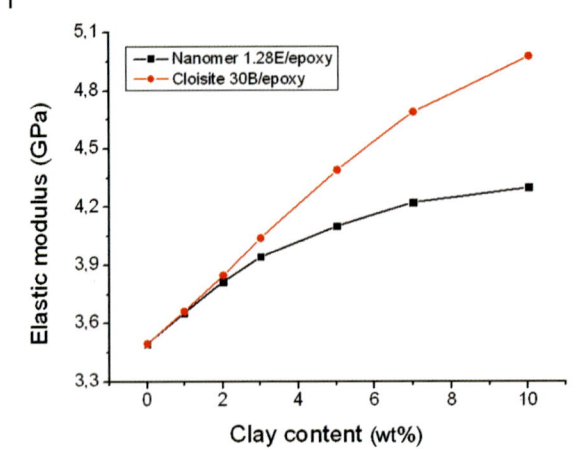

Figure 5.3 Effect of clay content on the elastic modulus of Nanomer I.28E/ and Cloisite 30B/epoxy nanocomposites.

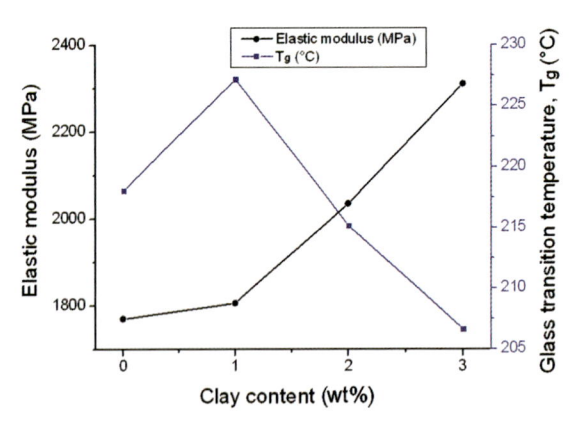

Figure 5.4 Elastic modulus and glass transition temperature of epoxy/silane-modified clay nanocomposites, versus clay concentration.

storage modulus increases with the clay mass fraction. However, T_g of the nanocomposites identified from tan δ increased with incorporation of 1 wt% of clay but dropped with further addition. Those results can be seen clearly in Figure 5.4.

5.3
Rigidity – Young's Modulus

In thermoset/nanoclay composites, the reinforcement mechanism of the matrix is mainly controlled by the properties of the nanofillers: size, shape factor, specific surface, rigidity, and so on. In fact, when rigid filler is added to a pure polymer matrix, it will carry the major portion of the applied load to the polymer matrix [67]. Thus, the interaction particle/matrix and, therefore, the contact-specific surface area particle/matrix, play a key role in the reinforcement of the matrix.

It has been well demonstrated that the enhancement in the interfacial adhesion properties between clay and polymer leads to the increasing in the mechanical properties of nanocomposite. This can be explained by the best transfer of the applied stress through the whole structure of the composite. The adhesion between the reinforcing particles and the matrix is commonly improved by surface modification of polymer chains using a suitable polar compatibilizer or by organically modifying the surface of the particle [68]. However, it should be noticed that using high quantity of compatibilizers may cause plasticization due to their lower molecular weight, which negatively affects the modulus of the nanocomposite [69].

The use of organically modified clay layers favors the intercalated/exfoliated structures, which increases the rigidity of the nanocomposite. Better the exfoliation of the clay layers in the matrix, greater is the platelet aspect ratio. Because of the rigid structure of clay layers and their high aspect ratio, they have proven to be very effective in the increasing of rigidity, modulus, and stiffness of the polymer matrix in well-dispersed and exfoliated condition. This is observed in the case of montmorillonite and hectorite clays, contrary to the laponite clays that have a smaller platelet aspect ratio [50, 54]. It has been observed that the clay nanolayers are more effective in improving mechanical properties, particularly Young's modulus as shown in Figure 5.2 [48, 49], when the polymer is in its rubbery state versus the glassy state [54, 68]. Lan et al. [38] explain this result by the fact that, under deformation, clay layers align in the direction of applied strain in the case of the polymer rubbery state, which provide higher stiffness of the nanocomposite. A disordered morphology obtained in a glassy state of the polymer may have the opposite effect. Figure 5.5 represents schematically ordered and disordered morphology of nanocomposites.

The role of the curing agent is also prominent in the improvement of nanocomposite mechanical properties. Such a role was highlighted in the work of Kronmann et al. [35]. These authors used three kinds of cured agents in an epoxy matrix filled with organically modified montmorillonite, an aliphatic diamine curing agent, Jeffamine D-230 and two cycloaliphatic polyamines: 3,3-dimethylmethylenedi–cyclohexylamine (3DCM) and Amicure bisparaaminocyclohexylmethane (PACM). The flexural modulus is 2.95 and 2.61 GPa at zero clay content for the matrix, respectively, cured with Jeffamine D-230 and 3DCM. For both systems, these moduli increase substantially with only adding a small amount of clay. In fact, for instance, for the nanocomposite cured with Jeffamine D-230, the modulus is increased by 43% with the addition of 4.2% of the clay mass fraction. This result seems to be closely related to the variation of the microstructure depending on the kind of the curing agents used in the polymerization process. In fact, X-ray diffraction analyses were achieved on three kinds of nanocomposites studied, and the diffractogram obtained is shown in Figure 5.6. No peak reflection is detectable at small angles for the organoclay swollen DGEBA resin cured by an

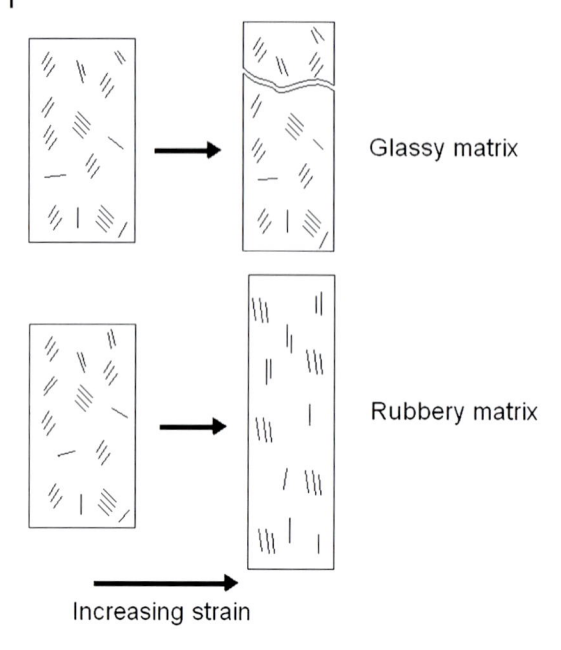

Figure 5.5 Deformation mechanism of a thermosetting nanocomposite: glassy matrix and rubbery matrix.

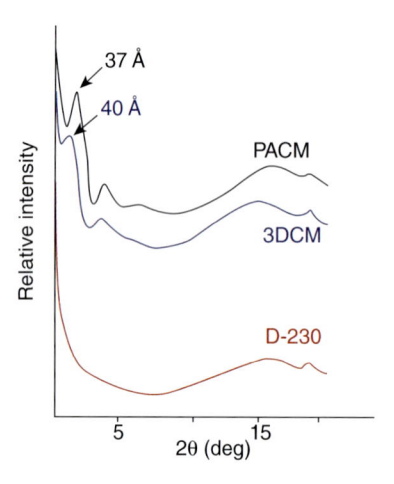

Figure 5.6 X-ray diffraction of nanocomposites elaborated with PACM, 3DCM and D-230 curing agents.

aliphatic diamine, Jeffamine D-230. This is certainly related to the fact that the clay is exfoliated in the matrix and the distance between the clay layers is too large to be detected by X-ray analysis. Reflections can be observed at small angles in the diagram of nanocomposite obtained with 3DCM as a curing agent, which corresponds to an intercalated microstructure. The average spacing between the clay layers is 40 Å. Finally, the (001) and (002) reflections are clearly highlighted at low angles by the X-ray diffractogram of nanocomposite obtained using PACM as a curing agent. The interlamellar spacing in this intercalated nanocomposite is of about 37 Å. One may conclude that the nature of the curing agent is also an important parameter controlling the final microstructure of epoxy/clay nanocomposites and, thus, the final properties of such nanocomposites.

Xidas et al. [68] studied the effect of various montmorillonites on the mechanical properties of the glassy and rubbery epoxy. The epoxy resin used was a diglycidyl ether of bisphenol A (DGEBA) (EPON 828RS). The curing agent was an aliphatic polyoxypropylene diamine (Jeffamine D-230) for the glassy epoxy system, and (Jeffamine D-2000) for the rubbery epoxy system. The clays used were an inorganic Na+-PGW montmorillonite, Nanomer I.30E, Nanomer I.28E, Cloisite 10A, and Cloisite 15A. The types of the alkyl ammonium ion used to modify the different organoclays are Primary Octadecylammonium for I.30E clay, Quaternary Trimethyl stearyl ammonium for I.28 E, Quaternary Dimethyl benzyl hydrogenated tallow ammonium for C10A, and Quaternary Dimethyl dihydrogenated tallow ammonium C15A clays. More details about the characteristics of these clays are given in [68]. Figure 5.7 shows, as an example, the relative Young's modulus of epoxy–clay nanocomposites prepared by Nanomer I.30E organoclay, as a function of loadings, for glassy and rubbery epoxy nanocomposites. As it can be seen, adding the organoclay continuously improves Young's modulus of the nanocomposite. This effect is more pronounced in the case of a rubbery matrix.

The study of Liu et al. [56] on the mechanical properties of epoxidized soybean oil/clay nanocomposites showed that Young's modulus increases from 1.20 to 3.64 MPa with increasing clay content from 0 to 8 wt%. Slight decrease in this modulus was observed when the clay content increases beyond 8 wt%. Similar results were

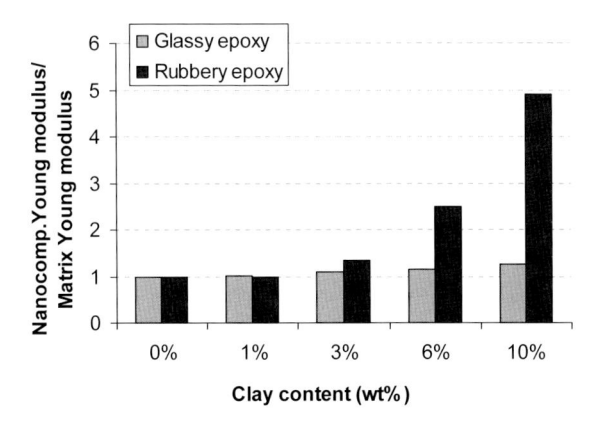

Figure 5.7 Relative Young's modulus of epoxy–clay nanocomposites versus clay content for glassy epoxy (EPON 828RS + D-230 Jeffamine) and rubbery epoxy (EPON 828RS + D-2000 Jeffamine).

reported by Liu et al. [70] on the polypropylene/clay nanocomposites prepared by grafting–melt compounding method.

Yasmin et al. [57] found that the addition of clay particles to epoxy leads to significant improvement of both the elastic modulus and storage modulus of the pure matrix. Two types of clay nanoparticles were used as reinforcement, one of them, Nanomer I.28E, came from Nanocor, Inc., and the other, Cloisite 30B, came from Southern Clay Products. It is found that the modulus of the nanocomposites increases monotonically with increasing clay content. Both Nanomer I.28E/ and Cloisite 30B/epoxy nanocomposites show similar values up to 2 wt% clay content. However, at higher clay contents, the epoxy/ Cloisite 30B shows a faster increase in modulus than the Nanomer epoxy/I.28E nanocomposite. In both cases, the rate of modulus increase decreases and levels off at higher clay contents.

In the study by Wang et al. [66] on the mechanical properties of epoxy/SMC nanocomposites, it was found that there is linear increase in tensile modulus with respect to SMC content. This result is typical of polymeric systems in which nanoscale clay dispersion is prevalent [50, 71, 72].

5.4
Strain at Break

Important parameters favoring the improvement in the deformation at break of thermosetting clay–nanocomposites are often related to the reinforcing particles: geometry, specific surface, pretreatment, added volume fraction, degree of exfoliation, interactions of matrix/particle and particle/particle, and so on.

Wang et al. [48] and Zilg et al. [60] pointed out that the increase in the ultimate deformation of elastomeric epoxy and polyurethane nanocomposites is only possible when the reinforcing clay particles are enough exfoliated in the matrix. Polyurethane matrix exhibited a twofold enhancement in elongation at break, with 10 wt% organoclay. These tensile properties increase with increase in the fraction of silicate nanolayers. The improved elasticity was attributed in part to the plasticizing effect of gallery onium ions, which contribute to dangling chain formation in the matrix, as well as to conformational effects on the polymer at the clay–matrix interface. Another parameter influencing this improvement is probably the alignment of clay platelets in the direction of loading. This ordered morphology is present in rubber matrix and not in glassy matrix.

The main published works dealing with nanocomposite of glassy thermoset showed that the strain at break often decreases with adding modified or unmodified clay particles. This decline can be related to the presence of small aggregates forming stress concentrators acting as failure initiators [73].

The glassy thermoset/clay elasticity can be also improved by adding adequate adjuvant during the mixing process. Balakrishnan et al. [74] prepared rubber-modified epoxy nanocomposites by adding octadecyl ammonium ion-exchanged clay to a dispersion of preformed acrylic rubber particles in liquid epoxy. The detailed synthetic steps for dispersing acrylic rubber (poly(2-ethylhexyl acrylate-*co*-glycidyl methacrylate)) in epoxy can be found elsewhere [75, 76].

It was observed that the elongation at break of epoxy resin is improved by adding acrylic rubber dispersants. This improvement is better for the rubber-modified epoxy resin than for an unmodified epoxy resin with no dispersants. This is expected because it is well known that the rubber particles increase the elongation at break of polymer matrix [74]. Moreover, ductility of epoxy resin is enhanced without considerable reduction in the modulus and strength when organically clay particles and rubber dispersants are added to the matrix. By examining the fracture behavior of rubber-dispersed epoxy and clay-filled rubber-dispersed epoxy, it has been inferred that a combination of the mechanisms, including rubber particle cavitation, yielding and plastic deformation of matrix initiated by rubber particles, crack diversion by clay platelets, and energy dissipation to create additional roughened area, may contribute to the improvement in the ductility of the three-phase nanocomposite system.

In the work of Xidas *et al.* [68], the tensile properties of the epoxy–clay nanocomposites were determined by stress–strain measurements. The relative critical strain of epoxy–3 wt% clay nanocomposites is shown in Figure 5.8 for glassy epoxy (EPON 828RS + D-230 Jeffamine) and rubbery epoxy (EPON 828RS + D-2000 Jeffamine). Figure 5.9 shows the relative critical strain of both nanocomposites versus clay content. A decrease in the stain at break was observed for the nanocomposites compared with the pristine glassy epoxy, whichever the kind of organoclay used. The opposite effect is pointed out when the rubbery matrix is concerned (Figure 5.8). Those results are confirmed for all mass fractions of Nanomer I.30E organoclay, as illustrated in Figure 5.9.

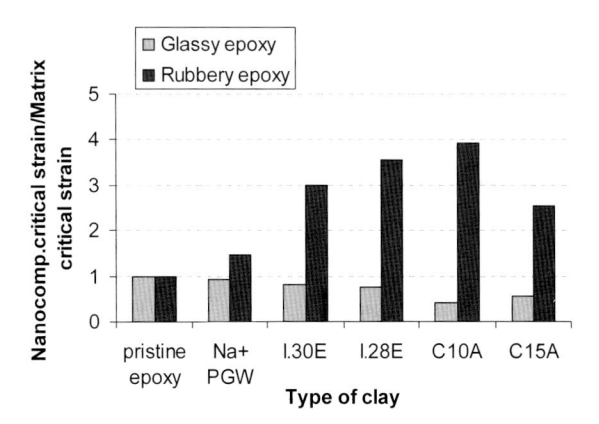

Figure 5.8 Relative critical strain of epoxy–3 wt% clay nanocomposites for glassy epoxy (EPON 828RS + D-230 Jeffamine) and rubbery epoxy (EPON 828RS + D-2000 Jeffamine), and for various type of clay.

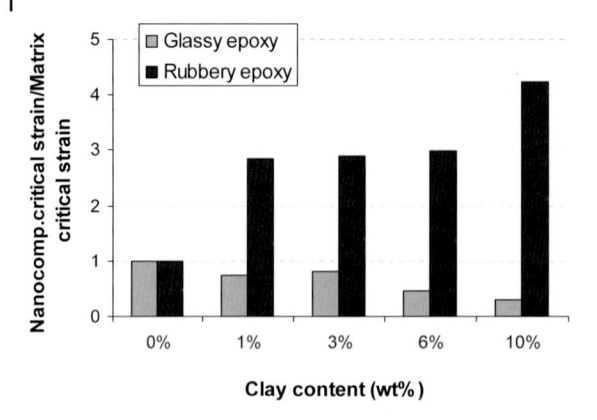

Figure 5.9 Relative critical strain of epoxy–Nanomer I.30E organoclay nanocomposites for glassy epoxy (EPON 828RS + D-230 Jeffamine) and rubbery epoxy (EPON 828RS + D-2000 Jeffamine) versus clay content.

5.5
Stress at Break – Fracture Toughness

Tensile strength of polymeric matrix is generally significantly increased by adding only a small amount of nanoclays [34, 50]. Polymer with more polarity such as nylon and rubber have more strong interfacial interactions with polar clays and favors high degree of exfoliation of these fillers in the matrix, which leads to a good stress at break of their nanocomposites compared with that of the unfilled matrix. However, this property decreases or slightly increases when the matrix is nonpolar because there are no attractive interactions between the silicate surface and the chains of such a kind of matrix [16, 48, 49].

The alignment of clay platelets in the direction of the applied strain provides high stress at break to the material since this one resists to the crack propagation perpendicularly to the loading axis [50]. This kind of morphology is commonly observed in rubbery matrix. A disordered morphology, observed in glassy matrix, may have the opposite effect.

The tensile strength of the polymer/nanoclay composites is better when the added reinforcing particles are previously chemically modified (e.g., silane-treated nanoclay). Such treatment seems have no effect on the stress at break of the final nanocomposite above the glass transition temperature. Ha *et al.* [77] showed that the mechanical properties of an epoxy/Cloisite samples are greater than those of unmodified samples at several temperatures. However, beyond the glass transition temperature, the tensile properties decreased sharply, and were similar for all samples.

As mentioned earlier, the role of the curing agent seems to be major in the improvement of nanocomposite mechanical properties since this parameter controls the exfoliation degree of the clays in the matrix [35, 68].

Akesson et al. [78] analyzed the effects of nanoclay on three different biobased thermoset resins: an acrylate epoxidized soybean oil (AESO), a synthesized methacrylic anhydride modified soybean oil (MMSO), and a semicommercial resin based on star-shaped oligomers of lactic acid named Pollit. These resins were cross-linked by UV-radiation and these can all undergo free radical polymerization reactions and can therefore be cross-linked into rigid materials. Clay is a montmorillonite organically modified. No or small improvement in stress at break of all the studied matrix has been observed, independently of the clay volume fraction. The nonreinforcing effect of the nanoclay is probably caused by low adhesion between the filler and the resin and likely the chosen clay was not optimal for the three resins. However, it could be noticed that this kind of clay has previously been used to reinforce epoxy resins with good result [79–81].

It was also found that the organic-MMT content has a remarkable effect on the strength of the polyurethane (PU)/organic-MMT nanocomposites [45]. The ultimate strength increased dramatically with increasing organic-MMT content and reached maximum at 5 wt% of organic-MMT content.

Liu et al. [47] studied the effect of the organoclay nanoparticles on the properties of epoxy/organoclay nanocomposites. The fracture toughness of the nanocomposites significantly increased with increasing clay concentration, suggesting a toughening effect of the clay particles.

The study of Liu et al. [56] on the mechanical properties of epoxidized soybean oil/clay nanocomposites showed that the tensile strength of nanocomposites increases rapidly from 1.27 to 4.54 MPa with increasing clay content from 0 to 8 wt%. Similar results were already reported by Liu et al. [70] on the polypropylene/clay nanocomposites prepared by grafting–melt compounding method.

Wang et al. [66] showed in his study on epoxy/SMC nanocomposites that the tensile strength increased by 25% with addition of 2 wt% of SMC, but dropped with further increasing SMC content. Similar phenomena have been observed in their previous study [82].

The effects of silane treatment on the fracture behavior of epoxy/clay nanocomposites were analyzed by Ha et al. [77] by comparing the critical fracture load of silane-treated nanocomposites with those of untreated nanocomposites. It appeared that the fracture behavior of the nanocomposites was significantly affected by silane treatment. In fact, the critical fracture loads of epoxy/silane-treated clay nanocomposites were greater than those of untreated samples. Moreover, mainly due to the improvement in interfacial bonding between epoxy and silane-treated clay, the critical fracture load, that is, the load at which crack propagation occurs, increased when the clay treated with 3-aminotriethoxysilane was used. The intercalated microstructure of epoxy/silane-treated clay nanocomposites favors the dissipation of the energy release rate, which prevents the crack propagation.

Xidas et al. [68] also analyzed the effect of various montmorillonites on the stress at break of the glassy and rubbery epoxy. More details about the different constituents used in their study are given in Section 5.3. The relative critical stress of epoxy–3 wt% clay nanocomposites is shown in Figure 5.10 for the glassy epoxy (EPON 828RS + D-230 Jeffamine) and rubbery epoxy (EPON 828RS + D-2000

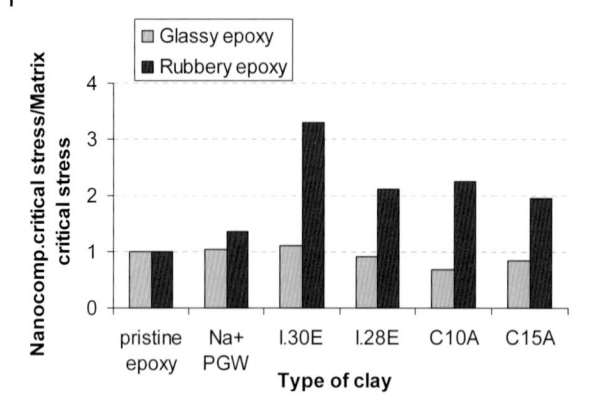

Figure 5.10 Relative critical stress of epoxy–3 wt% clay nanocomposites for glassy epoxy (EPON 828RS + D-230 Jeffamine) and rubbery epoxy (EPON 828RS + D-2000 Jeffamine), and for various type of clay.

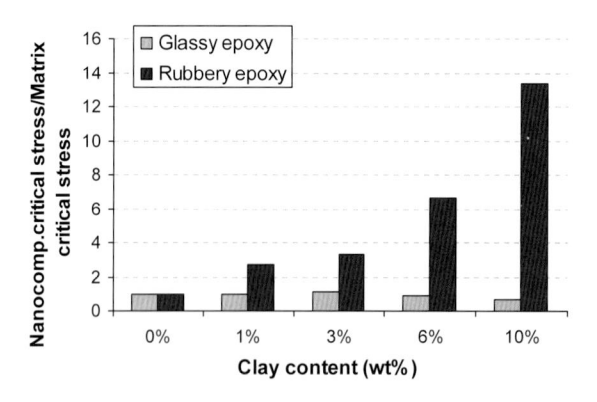

Figure 5.11 Relative critical stress of epoxy–Nanomer I.30E organoclay nanocomposites for glassy epoxy (EPON 828RS + D-230 Jeffamine) and rubbery epoxy (EPON 828RS + D-2000 Jeffamine) versus clay content.

Jeffamine). Figure 5.11 shows the relative critical stress of both nanocomposites versus clay content. In the case of the glassy epoxy nanocomposites prepared by the various organoclays, it can be seen that the stress at break of the pristine epoxy are globally reduced with the addition of any clay kind except the I.30E organoclay (Figure 5.10). It should be noticed that the glassy epoxy resins are very stiff materials and exhibit high strength, which cannot be easily improved. With regard to the rubbery epoxy nanocomposites, the improvement induced in the stress at break by the clay nanofillers is mainly due to the larger differences in toughness and stiffness between the rubbery network and the inorganic fillers. It can also be suggested that improvement derived from the interaction of the epoxy matrix with the external surfaces of the reinforcing particles. The effect of clay loading on the stress at break of glassy epoxy and rubbery epoxy nanocomposites was analyzed with the I.30E organoclay, which provided further improvements. As it can be seen in Figure 5.11, low organoclay concentration, that is, ~1 wt% has a minor effect on the tensile strength, while further increase of the organoclay loading up to 10 wt% leads to concomitant decrease of the tensile strength of glassy epoxy (~70% decrease) and to the greater improvement of this property in the case of rubbery epoxy. An explanation of these trends is given in [68].

A stress intensity factor K_I is a common indicator of the fracture properties of brittle materials as glassy thermoset–clay nanocomposites. Its critical value K_{IC}, named "tenacity," is used as a fracture criterion and it evaluates the ability of a material to prevent crack growth.

Zilg et al. [59] highly improved the fracture toughness of epoxy/anhydride matrix by adding nanoclay layers. This improvement is not attributed to exfoliated morphology of the structure but to the nondelaminated clay aggregates that might act as stress concentrators and induce plastic deformation around them, permitting therefore an enhancement of the fracture toughness. In fact, the intercalated nanocomposite obtained by adding 10 wt% of bentonite exhibits a value of 1.3 MPa$\sqrt{m}$ for the K_{IC} parameter, while the exfoliated nanocomposite obtained by adding 10 wt% of hectorite leads to a value of about 0.8 MPa$\sqrt{m}$ for this parameter.

Interestingly, preaging an intercalated epoxy/layered silicate polymer mixture before curing has been shown to result in exfoliated nanocomposites and in enhancement of toughness. Zilg et al. [59] showed, for various anhydride-cured, DGEBA-layered silicate nanocomposites, that well-dispersed, intercalated epoxy nanocomposites primarily improve the toughness, while the exfoliated state more likely reinforces the stiffness of the materials.

Becker et al. [65] obtained 100% improvement in the critical stress intensity factor in the case of highly cross-linked epoxy networks by adding 10 wt% of organophilic montmorillonite.

Zerda et al. [83] analyzed nanocomposites of modified montmorillonite clays in a glassy epoxy prepared by cross-linking with commercially available aliphatic diamine curing agents. The fracture of these materials was investigated and improvements in toughness values of 100% over unmodified resin were observed. This result is related to the intercalated microstructures obtained and highlighted using scanning electron and tapping-mode atomic force microscopy. These authors also investigated the fracture properties of intercalated diglycidylether of bisphenol A (DGEBA)/Jeffamine D230-layered silicate nanocomposites. The stress intensity factor of these materials showed significant improvements above 3.5 wt% of layered silicate contents. Zerda et al. [83] confirmed that the improvement in the tenacity of the nanocomposite is not related to the exfoliation clays in the matrix but is attributed to the dissipation of the energy release rate in the intercalated clay aggregates.

5.6
Conclusion

The effects of nanoclay particles on the mechanical properties of thermoplastic and thermosetting matrices are widely analyzed in the literature during the past two decades. Reinforcing these matrices by nanoclay fillers successfully led to the elaboration of a new class of materials, the polymer nanocomposites. Nowadays, nanocomposites have been commercialized by several industrial companies.

Nanoclay particles provide to thermosets interesting mechanical properties when the constituents and the processing parameters are adequately selected. These properties are highly improved in the case of rubbery matrix than in the case of glassy matrix. They seem to be mainly controlled by the adhesion polymer/filler and the final morphology of the nanocomposite: dispersed, exfoliated, or intercalated microstructure. The magnitude of those controlling parameters depends on the kinds of the mixed nanoclays and matrix, the clay pretreatment, the elobaration process of the nanocomposite, the polymerization conditions of the polymer, etc.

References

1 Motomatsu, M., Takahashi, T., Nie, H.Y., Mizutani, W.H., and Tokumoto, H. (1997) Microstructure study of acrylic polymer silica nanocomposite surface by scanning force microscopy. *Polymer*, **38**, 177–182.

2 Wen, J. and Mark, J.E. (1994) Precipitation of silica-titania mixed-oxide fillers into poly(dimethylsiloxane) networks. *Rubber Chem. Technol.*, **67**, 806–819.

3 Pu, Z., Mark, J.E., Jethmalani, J.M., and Ford, W.T. (1996) Mechanical properties of a poly(methyl acrylate) nanocomposite containing regularly-arranged silica particles. *Polym. Bull.*, **37**, 545–551.

4 Helbert, W., Cavaillé, J.Y., and Dufresne, A. (1996) Thermoplastic nanocomposites filled with wheat straw cellulose whiskers. Part I: processing and mechanical behaviour. *Polym. Compos.*, **17**, 604–611.

5 Hajji, P., Cavaillé, J.Y., Favier, V., Gauthier, C., and Vigier, G. (1996) Tensile behaviour of nanocomposites from latex and cellulose whiskers. *Polym. Compos.*, **17**, 612–619.

6 Ruckenstein, E. and Yuan, Y. (1997) Nanocomposites of rigid polyamide dispersed in flexible vinyl polymer. *Polymer*, **38**, 3855–3860.

7 Zabihi, O., Hooshafza, A., Moztarzadeh, F., Payravand, H., Afshar, A., and Alizadeh, R. (2012) Isothermal curing behavior and thermo-physical properties of epoxy-based thermoset nanocomposites reinforced with Fe_2O_3 nanoparticles. *Thermochim. Acta*, **527**, 190–198.

8 Okada, A., Kawasumi, M., Usuki, A., Kojima, Y., Kurauchi, T., and Kamigaito, O. (1990) Nylon 6–clay hybrid. *Mater. Res. Soc. Proc.*, **171**, 45–50.

9 Ma, H., Xu, Z., Tong, L., Gu, A., and Fang, Z. (2006) Studies of ABS-graft-maleic anhydride/clay nanocomposites: morphologies, thermal stability and flammability properties. *Polym. Degrad. Stab.*, **91**, 2951–2959.

10 Kojima, Y., Usuki, A., Kawasumi, M., Okada, A., Fukushima, Y., Kurauchi, T., and Kamigaito, O. (1993) Mechanical properties of nylon 6–clay hybrid. *J. Mater. Res.*, **8**, 1185–1189.

11 Kojima, Y., Usuki, A., Kawasumi, M., Okada, A., Kurauchi, T., and Kamigaito, O. (1993) Sorption of water in nylon 6–

clay hybrid. *J. Appl. Polym. Sci.*, **49**, 1259–1264.

12 Gilman, J.W., Kashiwagi, T., and Lichtenhan, J.D. (1997) Nanocomposites: a revolutionary new flame retardant approach. *SAMPE J.*, **33**, 40–46.

13 Abdalla, M.O., Dean, D., and Campbell, C. (2002) Viscoelastic and mechanical properties of thermoset polyimide–clay nanocomposites. *Polymer*, **43**, 5887–5893.

14 Suguna Lakshmi, M., Narmadha, B., and Reddy, B.S.R. (2008) Enhanced thermal stability and structural characteristics of different MMT-clay/epoxy–nanocomposite materials. *Polym. Degrad. Staby*, **93**, 201–213.

15 Aït Hocine, N., Médéric, P., and Aubry, T. (2008) Mechanical properties of polyamide-12 layered silicate nanocomposites and their relations with structure. *Polym. Test.*, **27** (3), 330–339.

16 Alexandre, M. and Dubois, P. (2000) Polymer-layered silicate nanocomposites: preparation, properties and uses of a new class of materials. *Mater. Sci. Eng. R Rep.*, **28**, 1–63.

17 Ray, S.S. and Okamoto, M. (2003) Polymer/layered silicate nanocomposites: a review from preparation to processing. *Prog. Polym. Sci.*, **28**, 1539–1641.

18 Theng, B.K.G. (1979) *Formation and Properties of Clay–Polymer Complexes*, Elsevier, Amsterdam, p. 133.

19 Sur, G.S., Sun, H.L., Lyu, S.G., and Mark, J.E. (2001) Synthesis, structure, mechanical properties, and thermal stability of some polysulfone/organoclay nanocomposites. *Polymer*, **42**, 9783–9789.

20 Fukushima, Y. and Inagaki, S. (1987) Synthesis of an intercalated compound of montmorillonite and 6-polyamide. *J. Incl. Phenom.*, **5**, 473–482.

21 Usuki, A., Kojima, Y., Kawasumi, M., Okada, A., Fukushima, Y., Kurauchi, T., and Kamigaito, O. (1993) Synthesis of nylon 6–clay hybrid. *J. Mater. Res.*, **8**, 1179–1184.

22 Kojima, Y., Usuki, A., Kawasumi, M., Okada, A., Kurauchi, T., Kamigaito, O., and Kaji, K. (1994) Fine structure of nylon-6–clay hybrid. *J. Polym. Sci. B*, **32** (4), 625–630.

23 Vaia, R.A. and Giannelis, E.P. (1997) Polymer melt intercalation in organically-modified layered silicates: model predictions and experiment. *Macromolecules*, **30**, 8000–8009.

24 Zhu, L. and Xanthos, M. (2004) Effects of process conditions and mixing protocols on structure of extruded polypropylene nanocomposites. *J. Appl. Polym. Sci.*, **93** (4), 1891–1899.

25 Médéric, P., Razafinimaro, T., Aubry, T., Moan, M., and Klopffer, M.H. (2005) Rheological and structural investigation of layered silicate nanocomposites based on polyamide or polyethylene: influence of processing conditions and volume fraction effects. *Macromol. Symp.*, **221**, 75–84.

26 Lertwimolnun, W. and Vergnes, B. (2005) Influence of compatibilizer and processing conditions on the dispersion of nanoclay in a polypropylene matrix. *Polymer*, **46**, 3462–3471.

27 Dennis, H.R., Hunter, L.D., Chang, D., Kim, S., White, J.L., Cho, J.W., and Paul, D.R. (2001) Effect of melt processing conditions on the extent of exfoliation in organoclay-based nanocomposites. *Polymer*, **42**, 9513–9522.

28 Cho, J.W. and Paul, D.R. (2001) Nylon 6 nanocomposites by melt compounding. *Polymer*, **42**, 1083–1094.

29 Di, Y., Iannace, S., Di Maio, E., and Nicolais, L. (2003) Nanocomposites by melt intercalation based on polycaprolactone and organoclay. *J. Polym. Sci. B Polym. Phys.*, **41** (7), 670–678.

30 Huang, X., Lewis, S., Brittain, W.J., and Vaia, R.A. (2000) Synthesis of polycarbonate-layered silicate nanocomposites via cyclic oligomers. *Macromolecules*, **33** (6), 2000–2004.

31 Zheng, H., Zhang, Y., Peng, Z., and Zhang, Y. (2004) Influence of clay modification on the structure and mechanical properties of EPDM/montmorillonite nanocomposites. *Polym. Test.*, **23** (2), 217–223.

32 Jawahar, P., Gnanamoorthy, R., and Balasubramanian, M. (2006) Tribological behaviour of clay–thermoset polyester nanocomposites. *Wear*, **261** (7–8), 835–840.

33 Shen, L., Wang, L., Liu, T., and He, C. (2006) Nanoindentation and morphological studies of epoxy

nanocomposites. *Macromol. Mater. Eng.*, **291** (11), 1358–1366.

34 Lan, T., Karivatna, P.D. and Pinnavaia, T.J. (1995) Mechanism of clay tactoid exfoliation in epoxy–clay nanocomposites. *Chem. Mater.*, **7**, 2144–2150.

35 Kronmann, X., Lindberg, H., and Berglund, L.A. (2001) Synthesis of epoxy–clay nanocomposites. Influence of the nature of curing agent on structure. *Polymer*, **45**, 4493–4499.

36 Wang, M.S. and Pinnavaia, T.J. (1994) Clay polymer nanocomposites formed from acidic derivatives of montmorillonite and an epoxy resin. *Chem. Mater.*, **6**, 468–474.

37 Kaviratna, P.D., Lan, T., and Pinnavaia, T.J. (1994) Synthesis of polyether–clay nanocomposites: kinetics of epoxy self-polymerization in acidic smectic clay. *Polym. Prepr.*, **35** (1), 788–789.

38 Lan, T., Karivatna, P.D., and Pinnavaia, T.J. (1996) Epoxy self-polymerizations in smectite clays. *J. Phys. Chem. Solids*, **57**, 1005–1010.

39 Kornmann, X., Berglund, L.A., Sterte, J., and Giannelis, E.P. (1998) Nanocomposites based on montmorillonite and unsaturated polyester. *Polym. Eng. Sci.*, **38** (8), 1351–1358.

40 Suh, D.J., Lim, Y.T., and Park, O.O. (2000) The property and formation mechanism of unsaturated polyester-layered silicate nanocomposites depending on the fabrication method. *Polymer*, **41**, 8557–8563.

41 Nuhiji, B., Attard, D., Thorogood, G., Hanley, T., Magniez, K., and Fox, B. (2011) The effect of alternate heating rates during cure on the structure–property relationships of epoxy/MMT clay nanocomposites. *Compos. Sci. Technol.*, **71**, 1761–1768.

42 Ha, S.R. and Rhee, K.Y. (2008) Effect of surface-modification of clay using 3-aminopropyltriethoxysilane on the wear behavior of clay/epoxy nanocomposites. *Colloids Surf. A Physicochem. Eng. Asp.*, **322**, 1–5.

43 Pu, Z., Mark, J.E., Jethmalani, J.M., and Ford, W.T. (1997) Effects of dispersion and aggregation of silica in the reinforcement of poly(methyl acrylate) elastomers. *Chem. Mater.*, **9**, 2442–2447.

44 Agag, A., and Takeichi, T. (2000) Polybenzoxazine–montmorillonite hybrid nanocomposites: synthesis and characterization. *Polymer*, **41**, 7083–7090.

45 Xiong, J., Liu, Y., Yang, X., and Wang, X. (2004) Thermal and mechanical properties of polyurethane/montmorillonite nanocomposites based on a novel reactive modifier. *Polym. Degrad. Staby*, **86** (3), 549–555.

46 Mishra, J.K., Kim, I., and Ha, C.S. (2003) New millable polyurethane/organoclay nanocomposite: preparation, characterization and properties. *Macromol. Rapid Commun.*, **24**, 671–675.

47 Liu, T., Tjiu, W.C., Tong, Y., He, C., Goh, S.S., and Chung, T.S. (2004) Morphology and fracture behavior of intercalated epoxy/clay nanocomposites. *Macromolecules*, **94** (3), 1236–1244.

48 Wang, Z. and Pinnavaia, T.J. (1998) Hybrid organic-inorganic nanocomposites: exfoliation of magadiite nanolayers in an elastomeric epoxy polymer. *Chem. Mater.*, **10**, 1820–1826.

49 Wang, S.J., Long, C.F., Wang, X.Y., Li, Q., and Qi, Z.N. (1998) Synthesis and properties of silicon rubber organo-montmorillonite hybrid nanocomposites. *J. Appl. Polym. Sci.*, **69**, 1557–1561.

50 Lan, T. and Pinnavaia, T.J. (1994) Clay-reinforced epoxy nanocomposites. *Chem. Mater.*, **6**, 2216–2219.

51 Lan, T., Karivatna, P.D., and Pinnavaia, T.J. (1994) Synthesis, characterization and mechanical properties of epoxy–clay nanocomposites. *ACS Polym. Mater. Sci. Eng.*, **71**, 527–528.

52 Ratna, D., Manoj, N.R., Varley, R., Singh Raman, R.K., and Simon, G.P. (2003) Clay-reinforced epoxy nanocomposites. *Polym. Int.*, **52**, 1403–1407.

53 Dean, D., Walker, R., Theodore, M., Hampton, E., and Nyairo, E. (2005) Chemorheology and properties of epoxy/layered silicate nanocomposites. *Polymer*, **46** (9), 3014–3021.

54 Messersmith, P.B. and Giannelis, E.P. (1994) Synthesis and characterization of layered silicate–epoxy nanocomposites. *Chem. Mater.*, **6**, 1719–1725.

55 Ganguli, S., Dean, D., Jordan, K., Price, G., and Vaia, R. (2003) Mechanical properties of intercalated cyanate ester-layered silicate nanocomposites. *Polymer*, **44** (4), 1315–1319.

56 Liu, Z., Erhan, S.Z., and Xu, J. (2005) Preparation, characterization and mechanical properties of epoxidized soybean oil/clay nanocomposites. *Polymer*, **46**, 10119–11012.

57 Yasmin, A., Luo, J.J., Abot, J.L., and Daniel, I.M. (2006) Mechanical and thermal behavior of clay/epoxy nanocomposites. *Compos. Sci. Technol.*, **66** (14), 2415–2422.

58 Agag, T., Koga, T., and Takeichi, T. (2001) Studies on thermal and mechanical properties of polyimide–clay nanocomposites. *Polymer*, **42**, 3399–3408.

59 Zilg, C., Muelhaupt, R., and Finter, J. (1999) Morphology and toughness/stiffness of nanocomposites based upon anhydride cured epoxy resins and layered silicates. *Macromol. Chem. Phys.*, **200**, 661–670.

60 Zilg, C., Thomann, R., Muelhaupt, R., and Finter, J. (1999) Polyurethane nanocomposites containing anisotropic nanoparticles derived from organophilic layered silicates. *Adv. Mater.*, **11**, 49–52.

61 Gârea, S.A., Iovu, H., and Bulearca, A. (2008) New organophilic agents of montmorillonite used as reinforcing agent in epoxy nanocomposites. *Polym. Test.*, **27**, 100–113.

62 Kornmann, X., Berglund, L.A., and Lindberg, H. (2000) Stiffness improvements and molecular mobility in epoxy–clay nanocomposites. MRS Proceedings, 628.

63 Lee, A. and Lichtenhan, J.D. (1996) Thermal and viscoelastic property of epoxy clay and hybrid inorganic–organic epoxy nanocomposites. *J. Appl. Polym. Sci.*, **73**, 1993–2001.

64 Chen, J.S., Poliks, M.D., Ober, C.K., Zhang, Y., Wiesner, U., and Giannelis, E. (2002) Study of the interlayer expansion mechanism and thermal-mechanical properties of surface-initiated epoxy nanocomposites. *Polymer*, **43**, 4895–4904.

65 Becker, O., Varley, R., and Simon, G. (2002) Morphology, thermal relaxation and mechanical properties of layered silicate nanocomposites based upon high-functionality epoxy resins. *Polymer*, **43**, 4365–4373.

66 Wang, L., Wang, K., Chen, L., Zhang, Y., and He, C. (2006) Preparation, morphology and thermal/mechanical properties of epoxy/nanoclay composite. *Compos. Part A Appl. Sci. Manuf.*, **37** (11), 1890–1896.

67 Gorrasi, G., Tortora, M., Vittoria, V., Pollet, E., Lepoittevin, B., and Alexandre, M. (2003) Vapor barrier properties of polycaprolactone montmorillonite nanocomposites: effect of clay dispersion. *Polymer*, **44**, 2271–2279.

68 Xidas, P.I. and Triantafyllidis, K.S. (2010) Effect of the type of alkylammonium ion clay modifier on the structure and thermal/mechanical properties of glassy and rubbery epoxy–clay nanocomposites. *Eur. Polym. J.*, **46**, 404–417.

69 Mittal, V. (2008) Mechanical and gas permeation properties of compatibilized polypropylene-layered silicate nanocomposites. *J. Appl. Polym. Sci.*, **107**, 1350–1361.

70 Liu, X. and Wu, Q. (2001) PP/clay nanocomposites prepared by grafting-melt intercalation. *Polymer*, **42**, 10013–10019.

71 Delozier, D.M., Orwoll, R.A., Cahoon, J.F., Johnston, N.J., Smith, J.G., Jr., and Connell, J.W. (2002) Preparation and characterization of polyimide/organoclay nanocomposites. *Polymer*, **43**, 813–822.

72 Liu, T.X., Liu, Z.L., Ma, K.X., Shen, L., Zeng, K.Y., and He, C.B. (2003) Morphology, thermal and mechanical behavior of polyamide 6/layered-silicate nanocomposites. *Compos. Sci. Technol.*, **63**, 3–4, 331–337.

73 Seghar, S., Azem, S., and Ait Hocine, N. (2011) Effects of clay nanoparticles on the mechanical and physical properties of unsaturated polyester. *Adv. Sci. Lett.*, **4**, 3424–3430.

74 Balakrishnan, S., Start, P.R., Raghavan, D., and Hudson, S.D. (2005) The influence of clay and elastomer concentration on the morphology and fracture energy of preformed acrylic rubber dispersed clay filled epoxy

nanocomposites. *Polymer*, **46**, 11255–11262.
75 Hoffman, D.K., Kolb, G.C., Arends, C.B., and Stevens, M.G. (1985) *Polym. Prepr. (Am. Chem. Soc., Div. Polym. Chem.)*, **26**, 232.
76 Hoffman, D.K. and Arends, C.B. (1987) US Patent 4,708,996.
77 Ha, S.R., Rhee, K.Y., Park, S.J., and Lee, J.H. (2010) Temperature effects on the fracture behavior and tensile properties of silane-trated clay/epoxy nanocomposites. *Composites B*, **41**, 602–607.
78 Akesson, D., Skrifvars, M., Lv, S., Shi, W., and Adekunle, K. (2010) Preparation of nanocomposites from biobased thermoset resins bye UV-curing. *Prog. Org. Coat.*, **67**, 281–286.
79 Lam, C.K., Lau, K.T., Cheung, H.Y., and Ling, H.Y. (2005) Effect of ultrasound sonication in nanoclay clusters of nanoclay/epoxy composites. *Mater. Lett.*, **59**, 1369–1372.
80 Lam, C.K., Cheung, H.Y., Lau, K.T., Zhou, L.M., Ho, M.W., and Hui, D. (2005) Cluster size effect in hardness of nanoclay/epoxy composites. *Compos. Eng.*, **36**, 263–269.
81 Lam, C.K. and Lau, K.T. (2007) Tribological behavior of nanoclay/epoxy composites. *Mater. Lett.*, **61**, 3863–3866.
82 Wang, K., Chen, L., Wu, J.S., Toh, M.L., He, C.B., and Yee, A.F. (2005) Epoxy nanocomposites with highly exfoliated clay: mechanical properties and fracture mechanisms. *Macromolecules*, **38**, 788–800.
83 Zerda, A.S. and Leser, A.J. (2001) Intercalated clay nanocomposites: morphology, mechanics, and fracture behaviour. *J. Polym. Sci. B Polym. Phys.*, **39**, 1137–1146.

6
Unsaturated Polyester Resin Clay Hybrid Nanocomposites

Kanniyan Dinakaran, Subramani Deveraju, and Muthukaruppan Alagar

6.1
Introduction

Unsaturated polyester (UP) resin is one of the most versatile thermoset polymer systems, commercially produced first in 1940. Unsaturated polyester resin is a thermoset matrix resin most widely used in the coatings and composites industry, and constitutes about three-fourths of the total resins used. It should be noted that UP resin is different from polyester resins, which have no double bonds in polymer backbone, whereas polyester resins (two-dimensional linear saturated) are best known for their use as fibers in textiles and clothing, films, and molding powders. Unsaturated polyesters are versatile because of their capacity to be modified or tailored during the building of the polymer chains. They have been found to have an almost unlimited utility in all segments of the adhesives, coatings, and composites industry. The principal advantage of these resins is a balance of properties (including mechanical, chemical, electrical), dimensional stability, low cost, and ease of handling or processing.

Unsaturated polyester resins (UPRs) are produced by the polycondensation of dibasic acids, such as isophthalic acid, phthalic anhydride, unsaturated aliphatic diacid (i.e., maleic acid and fumaric acid), maleic anhydride, and glycols (such as ethylene glycol, propylene glycol, di-ethylene glycol, and mono-ethylene glycol) at 210–230 °C. The unsaturation is introduced through the unsaturated aliphatic dibasic acid or its anhydride, which contains a double bond (e.g., maleic acid, and fumaric acid). Unsaturated polyester resins form highly durable structures and coatings when they are cross-linked with a vinylic reactive monomer, the most commonly used being styrene. This reactive diluent reduces the viscosity of the polyester resin, so that it can be easily processed.

UPR can be cured by a free-radical mechanism. The free radical can be generated by addition of a small amount of the appropriate free-radical initiator under different conditions, such as heat, UV light, or visible light. The initiators disassociate to form radicals, which are needed to begin the cross-linking reaction; this is the first step of the initiation of free radical polymerization. However, the initiators used for unsaturated polyester (UP) resins are not capable of producing

Thermoset Nanocomposites, First Edition. Edited by Vikas Mittal.
© 2013 Wiley-VCH Verlag GmbH & Co. KGaA. Published 2013 by Wiley-VCH Verlag GmbH & Co. KGaA.

radicals at room temperature. Therefore, accelerators are used to promote curing at room temperature by activating the initiator, which provides a room temperature curing ability to UPR. The styrene randomly branches across from the double bonds of polymer chains. At the same time, there is a possibility of homopolymerization of the reactive diluent during the propagation step, because it also contains a reactive double bond. This will affect the length and cross-link density of the cured part. The control of the cross-linking process is extremely important in the commercial processing of unsaturated polyester composites, both before and after the gel point. The mechanical properties of the cross-linked product depend on the average number of cross-links between the polymer chains (cross-link density) and the average length of the cross-links (branches). Too slow or too rapid cross-linking will determine the properties of a desired product. The different monomers possess different chemical structures, which impart their unique characteristics to the polymer chain when they are polymerized together. The properties of the unsaturated polyester resin are determined by the types and amounts of monomers used, as well as the reaction conditions adopted to polymerize them.

6.2
Reinforced Unsaturated Polyester Composites

Thermoplastics are used in certain applications, but constitute a relatively small sector of the structural composites market. Matrices used for structural composites are mainly thermosetting plastics, such as polyester resins, epoxy resins, phenolic resins, and vinyl ester resins. Polyester resins are the most widely used resin systems, particularly in the marine industry. The majority of dinghies, yachts, and work-boats are built of unsaturated polyester resin composites. Thermosetting plastic systems generally consist of liquid mixtures of relatively low molar mass reactants, such as monomers and/or prepolymers, which polymerize upon heating to form highly cross-linked, network polymers. On their own, cross-linked neat unsaturated polyester resins have limited structural integrity, so they are often combined with fiber glass or mineral fillers before cross-linking to enhance their mechanical strength. Unsaturated polyester resins reinforced with fiber glass are lightweight and durable and are used primarily in construction, marine, and land transportation industries, although they are used in a variety of other applications. The resin matrix forms a significant volume fraction of polymer composites and it has a number of critical functions; it binds the reinforcements together, maintains the shape of a component, and transfers the applied load to the reinforcing fibers. The resin protects the reinforcing fibers from degradation, due to abrasion or environmental attack. The polymer resin contributes significantly to the mechanical properties of the structural polymer composites, acting to resist delamination between plies of reinforcements, and to inhibit fiber buckling during compression.

Recently, there has been a focus on the development of nanosized particle-dispersed resin composites, called nanocomposites; they exhibit superior physical, mechanical, thermal, electrical properties, as well as flame retardancy, and gas barrier [1–3]. Polymer nanocomposites represent a new alternative to conventionally filled polymers, because nanoscaled, filler-dispersed nanocomposites exhibit markedly improved properties when compared with those of neat polymers, and have the added advantages of lower density and ease of processability. Systems in which the inorganic particles are the individual layers of a lamellar compound, most typically clay layers, exhibit drastically enhanced physicochemical properties, relative to the neat unreinforced polymer. Due to the nanometer length scale, which minimizes the scattering of light, nanocomposites usually exhibit transparent behavior. Many efforts have been made for the preparation of intercalated and exfoliated polymer/clay nanocomposites with improved properties. A variety of polymer characteristics including polarity, molecular weight, hydrophobicity, reactive groups, as well as clay characteristics, such as charge density and its modified structure and polarity, are influenced by the addition of clay into polymer matrix. Therefore, different synthetic approaches have been used for the preparation of polymer/clay nanocomposites. In general, there are four methods available to prepare polymer/clay nanocomposites: *in situ* template synthesis, solution intercalation, *in situ* intercalative polymerization, and melt intercalation.

6.3
Clay Minerals

6.3.1
Layered Structure

Clay minerals used in polymer nanocomposites can be classified into three groups, such as 2:1 type, 1:1 type, and layered silicic acids. Their structures are depicted in Figure 6.1 [4].

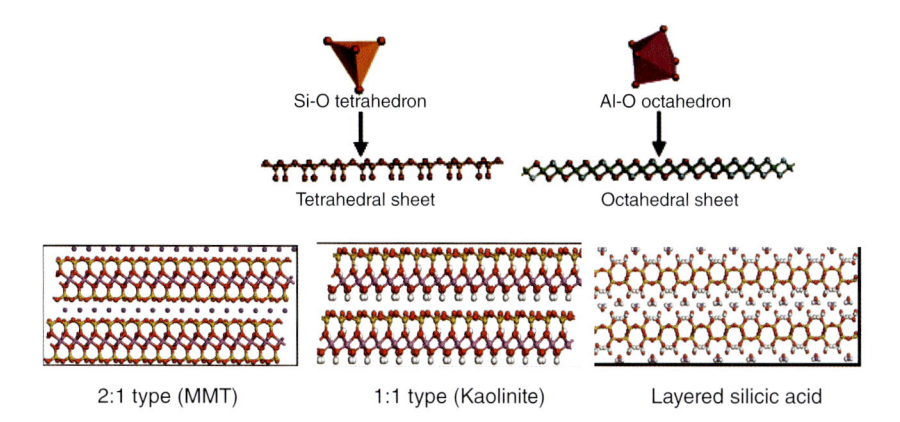

Figure 6.1 Structure of clay minerals.

2:1 type: The clays belong to the smectite family, with the crystal structure consisting of nanometer-thick layers (platelets) of an aluminum octahedron sheet sandwiched between two silicon tetrahedron sheets. Stacking of the layers leads to the existence of van der Waal's force between the layers. The isomorphic substitution of Al with Mg, Fe, Li in the octahedron sheets and/or Si with Al in the tetrahedron sheets gives each three-sheet layer an overall negative charge, which is counterbalanced by the exchangeable metal cations residing in the interlayer space, such as Na, Ca, Mg, Fe, and Li.

1:1 type: The clay consists of layers made of one aluminum octahedron sheet and one silicon tetrahedron sheet. Each layer bears no charge, due to the absence of isomorphic substitution in either the octahedron or the tetrahedron sheet. Thus, except for the water molecules, neither cations nor anions occupy the space between the layers, and the layers are held together by the hydrogen bonding between the hydroxyl groups in the octahedral sheets and oxygen in the tetrahedral sheets of the adjacent layers.

Layered Silicic Acids: The clay minerals mainly consist of silicon tetrahedron sheets with different layer thicknesses. Their basic structures are composed of layered silicate networks and interlayer hydrated alkali metal cations. The silanol groups in the interlayer regions favor the organic modification, by grafting the organic functional groups in the interlayer regions. Layered silicic acids are potential candidates for the preparation of polymer nanocomposites, because they exhibit similar intercalation chemistry as the smectite clays. Besides, they possess high purity and structural properties that are complementary to those of the smectite clays.

6.3.2
General Characteristics

The important characteristics pertinent to application of the clay minerals in polymer nanocomposites are their rich intercalation chemistry, high strength and stiffness, and high aspect ratio of the individual platelets, abundant availability in nature, and low cost. First, their unique layered structure and high intercalation capabilities allow them to be chemically modified and to be compatible with polymers, which make them particularly attractive in the development of clay-based polymer nanocomposites. In addition, the lower charge of the clay layers related to the existence of a weak force between the adjacent layers, making the interlayer cations exchangeable. Therefore, the intercalations of the inorganic and organic cations and molecules into the interlayer space are facile, which is an important aspect of their use in polymer nanocomposites manufacturing. Among the smectite clays, MMT and hectorite are the most commonly used reinforcements, while others are sometimes useful, depending on the specific applications. Although smectite clays are naturally not nanoparticles, they can be exfoliated or delaminated into nanometer platelets with a thickness of about 1 nm and a surface area of 700–800 m²/g. Each platelet has very high strength and stiffness, and can be regarded as a rigid inorganic polymer whose molecular weight (ca. 1.3×10^8) is much greater than that of neat polymer [4]. Thus, a very low concentration of clay is required to achieve properties equivalent to those of conventional composites.

6.3.3
Surface Modification

The main drawback of clay is the incompatibility between the hydrophilic clay and the hydrophobic polymer, which often causes the agglomeration of the clay mineral in the polymer matrix. Therefore, surface modification of clay minerals is necessary to ensure a perfect compatibility or chemical bonding between clay and polymers to achieve polymer nanocomposites. On being subjected to organic treatment, the clay becomes hydrophobic, and hence compatible with the specific polymers (thermoplastics, thermosets, or elastomers). Such modified clays are commonly referred to as organoclays. The most popular modification of clay is to exchange the interlayer of inorganic cations (e.g., Na^+ and Ca^{2+}) with organic ammonium cations. The key aspect of the surface modification is to swell the interlayer space up to a certain extent (normally over 20 Å), and hence reduce the layer–layer attraction, which allows a favorable diffusion and accommodation of the polymer or precursor into the interlayer space (Figure 6.2).

The compatibilizing agents used initially in the synthesis of nanocomposites were amino acids. However, the most popular are alkyl ammonium ions, because they can be exchanged easily with the inorganic ions situated between the clay layers. Some of the common compatibilizing agents used are amino acids, alkylamines, adducts, polyetheramines, dihydroimidazolines, and silanes.

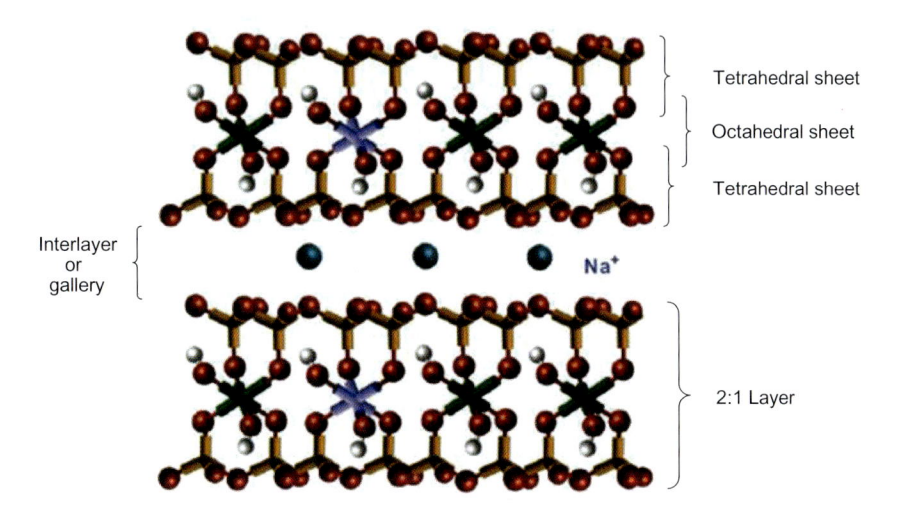

Figure 6.2 Structure of sodium montmorillonite.

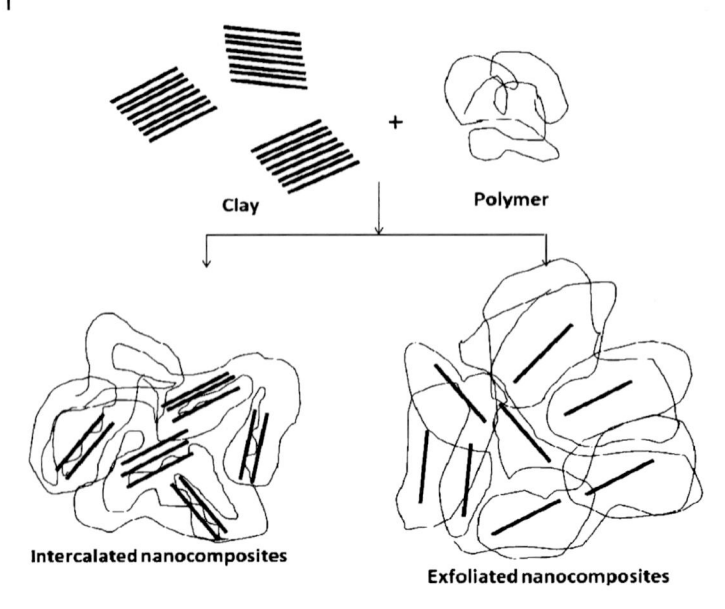

Figure 6.3 Clay–polymer composites showing intercalated and exfoliated nanostructures.

6.3.4
Processing and Characterization

The nature of the components and processing conditions plays a vital role in the formation of conventional composites or nanocomposites (Figure 6.3), when the clay layers are filled into a polymer matrix. Conventional composites are obtained only if the polymer is unable to intercalate into the galleries of the clay minerals.

The properties of such composites are similar to those of polymer composites reinforced with microparticles. There are two extreme nanostructures, resulting from the mixing of clay minerals and a polymer, under favorable conditions. One is intercalated nanocomposites, in which the polymer chains are intruded into the clay galleries, thereby increasing the spacing between adjacent layers by a few nanometers resulting in a well-ordered multilayer morphology stacking polymer layers and clay platelets alternately. The other one is the exfoliated or delaminated nanocomposites, in which the clay platelets are completely and uniformly dispersed in a continuous polymer matrix. However, it should be noted that, in most of the cases, the cluster (so-called partially exfoliated) nanocomposites is common in polymer nanocomposites.

Montmorillonite (MMT) is a multilayer silicate mineral that naturally possesses inorganic cations within its galleries to balance the charge of the oxide layers in a hydrophilic environment. The ion exchange of these interlayer cations with the organic ammonium ions affords a hydrophobic environment within the galleries of the organically modified MMT clay (OMMT) [5]. The organophilic galleries of

OMMT enhance the compatibility of the clay with the polymers, improve the dispersion of the silicate layers into the matrix [6], and assist the penetration of the monomers or polymers into the galleries [7]. In addition, the organic ammonium cations can provide functional groups that react or interact with the monomer or polymer units, to improve the interfacial strength between the reinforcement and the polymer matrix [8]. The degree of dispersion of the clay nanolayers and the resulting morphology of the nanocomposites depend on a number of factors, including the mixing method (melt or solvent), temperature, time, the choice of the solvent, its concentration, the steric size of the monomer or polymer, the choice of intercalating agent, and the yield of the ion exchange process.

6.4
Mechanical and Thermal Properties of Clay–UP Nanocopomposites

The data on thermal and mechanical properties of various clay–UPR nanocomposites are presented in Table 6.1. The MMT clay-reinforced unsaturated polyester resin has been widely studied in the literature, as given in Table 6.1. MMT was derived from bentonite, purified, activated by sodium ions, and mixed with reacting UP. The delamination of the clay layers leading to the formation of nanocomposites is usually indicated by XRD [12, 14] and TEM [10] studies. It was observed that in MMT content of only 1.5 v/v, the fracture energy of the nanocomposites, was doubled, $138 J/m^2$ as compared with that of $70 J/m^2$ for the neat UP. A higher loading of the MMT in unsaturated polyester resin of up to five parts per hundred resin (phr), drastically altered flexural modulus by about 120%, compared with that of neat resin. On the other hand, interestingly, the dispersion of clay slurry with water in UP resin is reported. The cross-linking reaction was influenced by the clay in such a way that the gelation temperature and the thermal stability were increased [18]. Using transmission and scanning electron microscopy, the resin intercalation into the MMT silicate layers was observed, and the fracture morphology indicated the effect of the clay slurry on the nanocomposites morphology. It is apparent that the addition of water to MMT clay allows better intercalation of the polymer chains in the interlamellar space. Because the clay is first suspended in water, it improves the dispersion and distribution of the particles in the resin matrix, and longer gelation periods lead to more uniform and mechanically stronger structures. As observed, the UP-MMT/W showed a shear-thinning viscosity, the presence of yield stress, and moduli behavior, indicative of a gel-like structure arranged between the clay and the polymer. The X-ray photoelectron spectroscopy determinations evidenced interactions arising between the carboxyl groups of the resin and the hydroxyl groups at the clay edges.

High-impact load-bearing UP-resin nanocomposites containing nanoclay in 0, 1.5, and 3 (wt%) were also prepared by the melt mixing method [13]. Results from the X-ray diffraction, transmission electron microscopy, and scanning electron microscope analyses, and viscosity changes in the liquid state resin, confirmed an exfoliation and intercalation of the nanoclay in the UP resin system used. The

6 Unsaturated Polyester Resin Clay Hybrid Nanocomposites

Table 6.1 Mechanical and thermal properties of UP resin with different clay composites.

UP/clay	% of clay	UTM	EF	FM	UTS	YM	FT	SM	H	CLD	IS	EM	T_g	Yc	Reference
EMS	0	0.25	0.32		1.76										[9]
	3	1.06	0.25		3.30										
MMT	0				59	2.87	70								[10]
	1.5				57	3.23	138								
	2.5				59	3.53	186								
	3.5				58	3.40	202								
	5				41	3.79	209								
Cloisite 30B	0	10.46	12.45		12.92										
	1	11.01	12.25		13.63										
	3	11.20	11.95		14.13										
	5	11.59	11.40		14.69										
	10	12.70	10.90		15.74										
	20	14.16	11.30		18.71										
	30	17.04	12.25		20.94										
Cloisite 10A	0	10.46	12.45		12.92										[11]
	1	10.82	12.00		13.49										
	3	11.30	12.15		13.75										
	5	11.50	13.05		14.24										
	10	11.66	11.99		15.67										
	20	11.82	12.30		14.97										
	30	11.97	12.30		13.01										
MMT	0							2693		3.52					[12]
	2							3792		4.76					
	4							4423		5.72					
	5							3651		4.60					
	6							3469		4.29					

6.4 Mechanical and Thermal Properties of Clay–UP Nanocopomposites

Clay	Loading (wt%)												Ref.
EMS	0	5.7	0.54										[9]
	10	5.4	0.68										
	20	3.5	0.90										
Cloisite 30B	0	32.4	2.82	29				58		36			[13]
	1.5	180.7	2.36	28				60		40			
	3	163.8	1.545	23				59		43			
MMT	0			66.97	130.201	22.210	1616			3.935			[14]
	2			61.59	153.241	21.221	2361			5.9			
	5			46.61	176.523	20.673	3034			5.932			
MMT-D	2				173.578	21.901	2541						
	8				165.231	21.565	2388						
	10				160.540	23.212	2104						
MMT-30B	0				195.214	17.568	3446				77.5	0	
	3				183.247	19.214	2610				83.6		
	5				175.214	21.020	2215				87.5	6	
	8										85.9		
	10										84.2		
Cloisite 11B	0	657	21.53										[15]
	2	662	24.09										
(K-10)	0								7.54				[16]
	5								7.192				
Nanoclay	0							301		5393			[17]
	1							387		6192			
	3							372		6074			
	5							343		6646			
MMT-resin	0		0.8306										[18]
	1		0.8847										
	5		1.1892										
MMT-slurry	0		0.8306										
	1		1.2968										
	5		1.8210										

(Continued)

Table 6.1 (Continued)

UP/clay	% of clay	UTM	EF	FM	UTS	YM	FT	SM	H	CLD	IS	EM	Tg	Yc	Reference
Organoclay	0	410		1.75	32										[19].
	1	650		2.8	34										
	3	500		2.75	36										
	5	440		1.75	29										
Nano clay	0				20.71						4.61		430		[20]
	1				24.31						4.92		432		
	2				26.29						5.15		433		
	3				29.71						5.49		433		
	5				25.28						5.04		429		
MMT clay	0	1023		780	32						23		33		[21]
	2.5	1182		853	37						27		130		
	5	1461		928	42						29		127		
VEO	0	1269		947	38						28		125		[21]
	2.5	1225		883	39						29		123		
	5	1507		940	44						31		121		
1.28MC organo-clay	0												6.3		[22]
	7.3												3.7		
	24												12.9		
MMA clay	0													1.0	[23]
	5													2.7	
	7.5													3.6	
	10													5.1	

Ammonium polyphosphate	0		1.0 [23]
	20		23.1
	30		28.0
MMT	0	54	[23]
	2	65	
	4	78.21	
	5	54.20	
	6	50	

Where,
UTM = ultimate tensile modulus (MPa)
EF = Elongation at failure (%)
FM = Flexural modulus (GPa)
UTS = Ultimate tensile strength (MPa)
YM = Young's modulus (MPa)
FT = Fracture toughness (MPa)
SM = Storage modulus (MPa)
H = Hardness (barcol)
CLT = Cross-link density × 10^5 (mol/m^3)
IS = Impact strength (J/m)
EM = Elastic modulus (MPa)
T_g = Glass transition temperature (°C)
Y_c = Char yield at 800 °C (%)
K-10 = Clay montmorillonite
EMS = Epoxidized methyl soyate
Organoclay = Clay modified with poly(ethylene oxide)
MMT = Montmorillonite
VEO = Organophilic clay-filled vinyl ester modified unsaturated polyester.

tensile modulus showed an increase with an increase in the nanoclay content. However, the tensile strength and elongation at break exhibited lower values with an increase in the nanoclay content. The izod impact test results indicated the better performance of the nanocomposites specimens, containing 1.5 wt% of nanoclay, showing the highest value. High-velocity impact tests were carried out using gas gun, in the velocity range of 20–100 m/s, and hardened steel hemispherical tip projectile with a diameter of 8.7 mm and weight of 11.54 g. The results of the high-velocity impact test indicated better performance by the specimens containing 1.5 wt% of nanoclay. The damage assessment of the impact area for all the specimens showed sapling-type brittle failure with punch out, and severe fragmentation patterns. The experimental results show that there is a strong correlation between the mechanical properties and the interlayer d-spacing of the clay particles in the nanocomposites system. It has been reported that an incorporation of 1%, 3%, and 5% by weight nanoclay into the polyester resin results in an improvement in the hardness by 29%, 24%, and 14%, respectively. The elastic modulus increased from 5393 to 6646 MPa (23% increase) for unreinforced polyester with the introduction of 5% by weight nanoclay [17].

Cost-competitiveness and high-strength nanocomposites were obtained using clay and unsaturated polyester resin derived from plastic waste; this is an example of the waste recycling technique. For example, poly(ethylene terephthalate) waste bottles were glycolized [12] and aminolized [14] and used as precursors for the synthesis of unsaturated polyester resin. The glycolysis product (hydroxyl-terminated oligomers) was converted into UP resin with varying acid moieties. These resins were miscible with styrene, and can be cured with peroxide initiators to produce thermosetting unsaturated polyester. Montmorillonite clay-filled nanocomposites was also prepared using the UP resin from PET waste. With an increase in the unsaturated acid content (for a fixed content of clay), the value of the storage modulus varied from 2737 to 4423 MPa. The glass-transition temperature of these nanocomposites ranged from 54 to 78 °C, and the cross-link density varied from 3.70×10^5 to 5.72×10^5 mol/m^3. The nanocomposites showed higher modulus values (2737–4423 MPa) than those of neat polymer (2693 MPa). The thermogravimetric analysis indicates that all the nanocomposite specimens were stable up to 200 °C, and showed a two-stage degradation. It was confirmed by the TGA that, unlike pristine polymer, all samples were stable up to 200 °C and showed two-stage degradation. The first stage was observed between 250 and 410 °C, and the second stage was observed between 410 and 560 °C. Unsaturated polyester resins based on aminolized PET waste and nano-MMT may be used to greatly enhance the performance of polymer composites at a relatively low cost [14]. The MMT (montmorillonite)–UP nanocomposites were developed by intercalating the UP resin into the silicate layers of MMT. The structures of the UP–MMT nanocomposites were investigated by X-ray diffraction and transmission electron microscopy (TEM). The test results indicate that the mechanical properties and thermal stability of the UP–MMT nanocomposites are better than those of neat UP. The glass transition and main chain decomposition temperatures of the UP–MMT nanocomposites exceed those of neat UP. The values of tensile strength, toughness, and Young's modulus of the UP–MMT nanocomposites exceeded those of neat UP. The decrease in the water permeability of the UP–MMT nanocomposites and improved performance of UP are very important for polymer concrete. The mechanical and thermal properties of the UP–MMT–D did not show a significant change because the degree of exfoliation is less than that of UP–MMT–30B nanocomposites. The compressive strength, elastic modulus, and splitting tensile strength of polymer concrete using exfoliated UP–MMT–30B nanocomposites exceeded the corresponding properties of polymer concrete using neat UP. The use of exfoliated UP–MMT nanocomposites enhances the polymer concrete strength. It is important to note that the exfoliated MMT–UP nanocomposite greatly affects the performance of the polymer concrete. The strength and elastic modulus of the polymer concrete was found to be positively correlated with the tensile strength and tensile modulus of the PC–MMT–UP nanocomposites. Polymer concrete exhibits excellent mechanical and thermal properties.

6.5
Flame Retardance

With the extensive use of polymers, especially in domestic applications, there is a need to reduce their potential for ignition or burn in order to make them safer in applications. Conventionally, chemical additives are used as flame retardants to retard ignition and control burning. Layered silicate nanoclays are used as potential flame retardants in unsaturated polyester nanocomposites. The incorporation of nanoclays (5% w/w) reduces the peak heat release rate (PHRR) by 23–27% and total heat release (THR) values by 4–11%, and the fire growth rate index (FIGRA) is reduced by 23–30% [23]. While the incorporation of condensed-phase flame retardants such as ammonium polyphosphate, melamine phosphate, and alumina trihydrate reduce the PHRR and THR values of the polyester resin, inclusion of small amounts of nanoclay (5% w/w) in combination with these char-promoting flame retardants causes total reductions of the PHRR of polyester resin in the range of 60–70%. Ammonium polyphosphate, in particular, and in combination with polyester–nanoclay hybrids, show the best results when compared with those of other flame retardants.

The nanoclay reinforcement of biocomposites by liquid composite molding (LCM) processes is widely used to manufacture composite parts for the automotive industry. An appropriate selection of the materials, and proper optimization of the manufacturing parameters, is important to produce parts with improved mechanical properties. A soy-based unsaturated polyester resin was used as the synthetic matrix, and glass and flax fiber fabrics were used as reinforcement [15]. To improve the mechanical and flammability properties of reinforced composites, the introduction of nanoclay particles in the unsaturated polyester resin was tried. Four different mixing techniques were used to improve the dispersion of nanoclay particles in the bioresin in order to obtain intercalated or exfoliated structures. An experimental study was carried out to assess the adequate parameter combinations between the vacuum pressure, filling time, and resin viscosity. Mechanical properties, such as the flexural modulus and ultimate strength, were evaluated and compared for conventional glass fiber composites and flax fiber biocomposites. Finally, the smoke density analysis was performed to predict the effects and advantages of using an environment-friendly resin combined with nanoclay particles.

The kinetics of swelling and the sorption behavior of unsaturated polyester nanocomposites (based on glycolyzed PET, maleic anhydride, styrene, and montmorillonite), developed by two different mixing methods, simultaneous and sequential, was studied in acetic acid at different temperatures. The dependence of the diffusion coefficient on the mixing methods and temperature has also been studied for unsaturated polyester nanocomposites [16]. The diffusion coefficients of the simultaneous mixed samples decrease with an increase in the mixing time in the samples. It was found that the diffusion coefficients increase with an increase in the mixing time and also with increase in temperature for the unsaturated polyester nanocomposites samples. The sorption coefficient increases with an increase in the mixing time for all the samples developed by *in situ* mixing method.

In recent years, polymer clay nanocomposites have been attracting considerable interest in polymer science, because of their advantages. There are many scientists who researched this kind of material and demonstrated that when polymer matrix was incorporated with a little weight of the clay, properties were enhanced considerably. Because clay is a hydrophilic substance it is difficult to use it as a filler in a polymer matrix having hydrophobic nature; so, clay needs to be modified to become compatible with polymer [19]. Poly(ethylene oxide) was used as a modifier for clay to replace some traditional ionic surfactants such as primary, secondary, tertiary, and quaternary alkyl ammonium or alkylphosphonium cations having some disadvantages, such as disintegration at high temperature and catalyzing polymer degradation. Then, organoclay was used to prepare nanocomposites based on unsaturated polyester. The morphology and properties of nanocomposites were measured by X-ray diffraction and tensile strength, and thermal stability was measured by transmission electron microscopy. The results showed that clay galleries changed to an intercalated state in the nanocomposites. The properties of nanocomposites improved to a significant extent when the loading of the organoclay was used at 1 phr.

The hybrid nanocomposites of clay with toughened UP resin were found to be promising composites material in recent days. There are some research articles published on clay blended with UP-epoxy, UP-vinyl ester, UP-alkyd styrene, and UP-epoxydized soybean oil [20, 21, 24–27]. The inter-cross-linked networks of UP-toughened epoxy blends were developed and characterized. Blended nanocomposites were fabricated by high shear mechanical mixing, followed by the ultrasonication process, to get homogeneous mixing using *in situ* polymerization. The mechanical properties were studied as per ASTM standards. Data obtained from the mechanical studies indicated that the epoxy/UP/clay combination improved the mechanical properties to an appreciable extent. The ILSS, TS, and IS of the 3 wt% clay-filled blend nanocomposites system was increased by 52%, 30%, and 16%, respectively, compared with that of the neat blend (0 wt% clay) resin system. The homogeneous morphologies of the UP-toughened epoxy and epoxy/UP/clay nanocomposites systems were ascertained using scanning electron microscope (SEM) and transmission electron microscope (TEM) studies. The objective of this study is to identify suitable nanocomposites that offer a low-cost, high strength material, which can be used for making lightweight components for automobile parts, transportation systems, and consumer products.

Organophilic montmorillonite clay–unsaturated polyester (UP) and vinyl ester oligomer (VEO) nanocomposites were prepared and the formation of nanocomposites was evaluated by X-ray diffraction (XRD), dynamic mechanical analysis (DMA), and scanning electron microscopy (SEM). The VEO was prepared by reacting commercially available epoxy resin LY556 and acrylic acid and was used as a toughening agent for unsaturated polyester resin. Organoclay-filled hybrid VEO–UP matrices, developed in the form of castings, were characterized for their thermal and mechanical properties. The dynamic mechanical measurements indicated the presence of higher cross-link density for the clay-filled systems than that of unfilled systems. The XRD analysis implies the intercalation of the polymer molecules between the clay layers, which, in turn, restricts the mobility of polymer in the vicinity of clay layers. A shift in T_g toward lower temperature was observed for the nanocomposites. Significant improvement in the mechanical properties was also observed in the case of nanocomposites when compared with that of neat resin matrix. The nanocomposites developed based on organophilic clay-filled VEO–UP systems exhibited a significant improvement in the mechanical properties and a negligible tendency to water absorption in comparison with the unmodified unsaturated polyester resin system. The hybrid matrix systems developed in the present study can be used to fabricate advanced composites components for engineering and aerospace applications for better performance and enhanced longevity.

Unsaturated polyester systems give rise to numerous possible approaches in the development of nanocomposites. A simultaneous mixing method was used to prepare UP-resin/organoclay nanocomposites. The effects of various mixing processes, using several organically modified clay types, were investigated. The incorporation of these organoclays resulted in an intercalated structure, the extent of which depended mainly on the type of the clay organic treatment. Organoclays that exhibited the highest intercalation levels were further studied using a sequential mixing method. The UP-alkyd (without styrene) was mixed with different types of organoclays. The processing parameters, such as mixing modes, applied shearing levels, clay contents, and mixing temperatures were investigated. Prolonged high shear levels promoted the intercalation and exfoliation of the silicate layers, resulting in a better dispersion of the clay particles. The high shear level effects were achieved by vigorous mechanical mixing and were intensified by using large amounts of clay and optimized matrix viscosity. Rheological studies of the nanocomposites were found complementary, and in correlation with morphological and thermal characterization. This methodological approach provides a basis for understanding the structuring processes involving the formation of the UP/clay nanocomposites and establishing the materials–processing–structure interrelations [22].

6.6
Bio-Derived Unsaturated Polyester–Clay Nanocomposites

Hybrid bio-based composites that exploit the synergy between the natural fibers in a nano-reinforced bio-based polymer can lead to improved properties, while maintaining the environmental friendliness. The drawbacks of the addition of bio-resins to the base polymer, the reduction of stiffness and strength, have been shown to be recoverable through nanoclay incorporation. Solvent-based processing of nanoclay-reinforced bio-based resin systems have been developed recently, but it has some limitations such as phase separation, thermal degradation of the resin system, and ratio of bio-resin and nanoclay content to base resin. The incorporation of higher concentrations of bio-resins and nanoclay, along with improvements in processing, will enable maximizing the enhanced thermo-mechanical properties of nanocomposites. The enhanced thermo-mechanical properties of bio-based materials are likely to increase their appeal for use in transportation and housing structural applications. Bio-based unsaturated polyester resins clay nanocomposites were obtained by partial substitution of unsaturated polyester with epoxidized soybean oil [9] and clay. The reinforcement of the bio-based resin with nanoclays permits the relation of stiffness without sacrificing the toughness, with enhancing barrier and thermal properties. The characterization of the different hybrid composites verified indicates synergistic behavior, in which systems with 10% soybean oil and 1.5 wt% nanoclay retained the original stiffness, strain to failure, and hygro-thermal properties of the neat resin, while improving the toughness. This indicates that the properties of the bio-nanocomposites can be tailored and influenced by the amount and distribution of their constituents.

Biodegradable nanocomposites are also available with unsaturated polyester and clay [11], and it was obtained using biodegradable aliphatic polyester (APES)/organclay through the melt intercalation method. The APES/Cloisite 30B hybrids possess higher tensile properties and lower biodegradability than those of APES/Cloisite 10A hybrid; increasing the content of organoclay in the APES hybrids resulting in high melt viscosity was reported. Bio-based clay/polymer nanocomposites, using blends of styrene-based unsaturated polyester and epoxidized methyl soyate, were manufactured using solvent-based processing techniques. It was understood from the literature that the solvent type, bio-resin addition sequence, and sonication energy govern the processing efficiency and composites quality. It was found that the addition of bio-resin after solvent removal minimizes the problems associated with the phase separation between the base polymer and the bio-resin additive. Acetone was found to be one of the good solvents for bio-resin–clay nanocomposites, and it provides nanocomposites with good nanoclay dispersion and exfoliation and high tensile modulus.

References

1 Komarneni, S., Vaia, R.A., Lu, G.Q., Matsushita, J.I., and Parker, J.C. (2002) *Nanophase and Nanocomposite Materials IV*, vol. 703, Materials Research Society, Pittsburgh.

2 Pinnavaia, T.J. and Beall, G.W. (2000) *Polymer–Clay Nanocomposites*, John Wiley & Sons Ltd, Chichester, UK, p. 173.

3 Krishnamoorti, R. and Vaia, R.A. (2001) *Polymer Nanocomposites: Synthesis, Characterization, and Modeling*, ACS symposium series: 804, American Chemical Society, Washington, DC.

4 Max Lu, G.Q., Zeng, Q.H., Yu, A.B., and Paul, D.R. (2005) Clay-based polymer nanocomposites: research and commercial development. *J. Nanosci. Nanotech.*, **5**, 1574–1592.

5 Akelah, A. (1995) Nanocomposites of grafted polymers onto layered silicates, in *Polymers and Other Advanced Materials: Emerging Technologies and Business Opportunities* (eds P.N. Prasad, J.E. Mark, and J.F. Tung), Plenum Press, New York, pp. 625–644.

6 Kawasumi, M., Hasegawa, N., Kato, M., Kojima, Y., Usuki, A., and Okado, A. (1997) Preparation and mechanical properties of polypropylene clay- hybrids. *Macromolecules*, **30**, 6333–6338.

7 Lebaron, P.C., Wang, Z., and Pinnavaia, T.J. (1999) Polymer-layered silicate nanocomposites: an overview. *Appl. Clay. Sci.*, **15**, 11–29.

8 Giannelis, E.P. (1996) Polymer layered silicate nanocomposites. *Adv. Mater.*, **8**, 29–35.

9 Haqa, M., Burgueño, R., Mohanty, A.K., and Misra, M. (2008) Hybrid bio-based composites from blends of unsaturated polyester and soybean oil reinforced with nanoclay and natural fibers. *Composit. Sci. Technol.*, **68**, 3344–3351.

10 Kornmann, X., Berglund, L.A., and Sterte, J. (1998) Nanocomposites based on montmorillonite and unsaturated polyester. *Polym. Eng. Sci.*, **38** (8), 1351–1358.

11 Lee, S., Park, H.-M., Lim, H., Kang, T., Li, X., Cho, W.-J., and Ha, C.-S. (2002) Microstructure, tensile properties, and biodegradability of aliphatic polyester/clay nanocomposites. *Polymer*, **43**, 2495–2500.

12 Katoch, S. and Kundu, P.P. (2011) Thermal and mechanical behavior of unsaturated polyester [derived from poly(ethylene terephthalate) waste] and montmorillonite-filled nanocomposites synthesized by *in situ* polymerization. *J. Appl. Polym. Sci.*, **122**, 2731–2740.

13 Esfahani, J.M., Sabet, A.R., and Asfandeh, M. (2012) Assessment of nanocomposites based on unsaturated polyester resin/nanoclay under impact loading. *Polym. Adv. Technol.*, **23**, 817–824.

14 Elsaeed, S.M. and Farag, R.K. (2012) Mechanical, thermal and barrier properties of unsaturated polyester nanoomposite based on pet-waste for polymer concrete. *J. Eng. Technol. Innov.*, **1** (1), 16–24.

15 Bensadoun, F., Kchit, N., Billotte, C., Bickerton, S., Trochu, F.O., and Ruiz, E. (2011) A study of nanoclay reinforcement of biocomposites made by liquid composite molding. *Int. J. Polym. Sci.*, **2011**, 1–10, article ID: 964193, 10 pages.

16 Katoch, S., Sharma, V., and Kundu, P.P., (2010) Swelling kinetics of unsaturated polyester-layered silicate nanocomposite depending on the fabrication method, Diffusion-fundamentals.org 13(2010) 1, pp. 1–31.

17 Dhakal, H.N., Zhang, Z.Y., and Richardson, M.O.W. (2006) Nanoindentation behaviour of layered silicate reinforced unsaturated polyester nanocomposites. *Polym. Test.*, **25**, 846–852.

18 Rivera-Gonzaga, A., Sanchez-Solis, A., Sanchez-Olivares, G., Calderas, F., and Manero, O. (2012) Unsaturated polyester–clay slurry nanocomposites. *J. Polym. Eng.*, **32**, 1–5.

19 Thanh, T., Nguyen, N., Ngan, T.K., Nhan, H.T.C., Huy, H.T., and Grillet, A.C. (2012) Study structure and properties of nanocomposite material based on unsaturated polyester with clay modified by poly(ethylene oxide). *J. Nanomater.*, **2012**, 1–5, article ID: 841813, 5 pages.

20 Chakradhar, K.V.P., Venkata Subbaiah, K., Ashok Kumar, M., and Ramachandra Reddy, G. (2011) Epoxy/polyester blend nanocomposites: effect of nanoclay on mechanical, thermal and morphological properties. *Malays. Polym. J.*, **6** (2), 109–118.

21 Sharmila, R.J., Premkumar, S., and Alagar, M. (2010) Preparation and characterization of organoclay-filled, vinyl ester-modified unsaturated polyester nanocomposites. *High Perform. Polymer.*, **22**, 16–27.

22 Mironi-Harpaz, I., Narkis, M., and Siegmann, A. (2005) Nanocomposite systems based on unsaturated polyester and organo-clay. *Polym. Eng. Sci.*, **45** (2), 174–186.

23 Nazare, S., Kandola, B.K., and Horrocks, A.R. (2006) Flame-retardant unsaturated polyester resin a. Incorporating nanoclays. *Polym. Adv. Technol.*, **17**, 294–303.

24 Chozhan, C.K., Alagar, M., Sharmila, R.J., and Gnanasundaram, P. (2007) Thermo mechanical behaviour of unsaturated polyester toughened epoxy–clay hybrid nanocomposites. *J. Polym. Res.*, **14** (4), 319–328.

25 Dinakaran, K. and Alagar, M. (2002) Preparation and characterization of bismaleimide(N,N'-bismaleimido–4,4′-diphenyl methane)–vinyl ester oligomer modified unsaturated polyester interpenetrating matrices for advanced composites. *J. Appl. Polym. Sci.*, **86**, 2502–2508.

26 Josephine Sharmila, R., Alagar, M., and Suresh Kumar, R. (2006) Synthesis and characterization of bismaleimide (1,3-bismaleimido benzene)–vinyl ester oligomer modified unsaturated polyester interpenetrating matrices. *Prog. Rubber Plast. Recycl. Technol.*, **22** (3), 147–159.

27 Josephine Sharmila, R., Premkumar, S., and Alagar, M. (2007) Toughened polyester matrices for advanced composites. *J. Appl. Polym. Sci.*, **103** (1), 167–177.

7
Hyperbranched Polymers as Clay Surface Modifications for Nanocomposites

Teresa Corrales, Fernando Catalina, Iñigo Larraza, Gema Marcelo, and Concepción Abrusci

7.1
Introduction

Hyperbranched polymers (HBPs) are dendritic polymers characterized by a tree-like architecture that confers upon them very different properties compared with their linear counterparts. HBPs are built from multifunctional monomers AB_n, where the function A can couple with the function B as proposed by Flory [1]. Majority of the early hyperbranched polymer studies were focused on the synthesis of dendrimers via iterative synthetic approaches. However, more efficient routes, based on a noniterative polymerization procedure, to the generation of branched polymers have been developed, driving to the formation of irregular hyperbranched polymers. Compared with dendrimers, the synthesis of HBPs is easier and more cost-effective. Although some scientists have pointed out that the poorly defined structure and the relatively high polydispersity of HBPs might restrict their applications, novel synthetic methods have recently been developed, in which substantially better-defined HBPs have been obtained with narrow molecular distributions [2]. These procedures allow the synthesis of a wide variety of hyperbranched architectures, including polyurethanes, polyesters, polycarbonates, and polyamides. In comparison to dendrimers, HBPs are polydisperse systems in terms of molecular weight and incomplete branch points dispersed throughout the structure [3]. The structural control in hyperbranched polymers through the adjustment of their degree of branching or by the modification of their end-groups allows the fine-tuning of their physical properties and optimizing them for a great variety of applications, such as multifunctional initiators and for rheology control [4], nanofillers for polymer nanocomposites [5], sensors, homogeneous catalysts, or medical applications [6]. HBPs are characterized by their high chemical reactivity, enhanced solubility, and also higher compatibility with others polymers, when compared to linear analogs. The use of hyperbranched polymers blended with linear polymers has resulted in new polymeric materials, for example, improved thermal stability and mechanical properties was observed for polystyrene when blended with hyperbranched polyphenylenes [7].

Thermoset Nanocomposites, First Edition. Edited by Vikas Mittal.
© 2013 Wiley-VCH Verlag GmbH & Co. KGaA. Published 2013 by Wiley-VCH Verlag GmbH & Co. KGaA.

Aliphatic hyperbranched polyesters has been studied by applications such as liquid or solid resins, resin additives, or UV-cured thin films, and the results showed the advantages of hyperbranched polyesters as thermoset coating resins [8]. Thermoset nanocomposites are often prepared by swelling organically modified clays with a monomer precursor; intercalation and/or exfoliation can be achieved by combining functionalized monomers or prepolymers with layered silicate [9]. In this sense, dendritic hyperbranched polyester polyol is of particular interest because exfoliated clay nanocomposites may be obtained without the aggregation associated with linear water-soluble polyurethane precursors [10]. Aromatic–aliphatic hyperbranched polyesters with vinylic end groups of different length have been reported as modifiers of epoxy/anhydride thermosets [11], in order to improve the toughness of the materials. It has been found that HBPs with a large number of chain ends enhanced flexibilities and increased toughness to thermosets without affecting other properties, such as glass transition temperature, modulus, or hardness. The enhanced toughness has been related to local inhomogeneities in the cross-linked network caused by the incorporation of HBPs, and the compatibilization between the modifier and the polymer matrix through physical or chemical interactions was seen to play a crucial role [12]. Also, the addition of HBPs significantly increased the curing rate with respect to the neat formulation, and depended on the length of the aliphatic chain and the amount of modifier, probably due to the different viscosities of the mixtures.

Other applications of HBPs have been explored, such as hyperbranched polyesters synthesized via the ring-opening polymerization of a ε-caprolactone derivative for nanoporosity templating agents in organosilicates [13]; hyperbranched PEG doped with lithium perchlorate for ion-conducting materials [14]; or hyperbranched polyurethane intercalated into organically modified montmorillonite and applied in Li-ion batteries, which contributes to obtain nanocomposites polymer electrolytes with improved electrochemical and mechanical properties [15]. Recently, the use of hyperbranched polymers for applications at surfaces and interfaces has expanded; HBPs are prepared from the flat surfaces of porous particles for many applications [16, 17].

Although many works have been reported on the hyperbranched polymer grafted onto the surfaces of silica nanoparticles [18], there are very few examples of the hyperbranched polymer grafted clay in the literature [19]. HBP aliphatic polyester grafted attapulgite has been prepared via a melt polycondensation, the process has proved to be very effective and has provides a higher percentage of surface grafting of hyperbranched polymers, which could be used as selective absorbents for certain compounds or reactive fillers for polymers [20]. Systems comprising mixtures of polymers and other materials, such organic as inorganic additives, nanoparticles, nanofibers, and mineral sheets, are increasingly touted for their superior properties. Three main methods are used to incorporate nanofillers into polymers, including *in situ* polymerization of monomer/clay intercalates, solution-intercalation, and melt-compounding. Previous works have shown that when organoclays are incorporated into the nanocomposite systems, their properties often depend on the nature of modifier used, although no significant changes in the polymerization reaction are induced.

Compatibility between the matrix and the filler in the nanocomposite preparation is a key factor, which will affect the nanostructure (intercalated/exfoliated) and consequently the properties of the nanocomposite. Since most polymers are organophilic, organomodified layered silicates are used to obtain better affinity between the filler and matrix. Cation exchange reaction constitutes one of the more common modification methods for the introduction of an ammonium or a phosphonium salt, bearing a suitable organic functionality, inside the interlayer space. The organic salt substitutes the metal ions present inside the clay mineral galleries, enhances the interlayer between the platelets, and for instance, facilites the intercalation of polymer within galleries, and also is responsible of changes in the surface properties of the clay. For this purpose, the attention is on designing new strategies to enhance interactions occurring at the interface polymer/nanoclay. For example, free radical polymerization is initiated from quaternized molecules that are exchanged into the gallery region of a layered silicate and favored intragallery initiation instead of intergallery [21–23]. Surfmers are used to facilitate exfoliation during *in situ* polymerization by taking part in the growing step of the polymer chain by copolymerization [24]. An emerging application is the incorporation of positively charged polymers, which can be intercalated in the interlayer space of montmorillonite (MMT) by an ion-exchange mechanism. Polications have been used for MMT modification and used as fillers for preparation of polymer/MMT nanocomposites [25].

Nowadays, new approaches are being developed for clay modification to improve interactions between polymer and organoclays to further prepare nanocomposites. Knowledge of the factors controlling the interface polymer/nanoclay would contribute to the design of systems for highly valued-added applications. This chapter presents the recent advances on applications of hyperbranched as clay surface modifications, with particular reference to the preparation of antimicrobial surface and adsorbents for Cr(VI) water treatment.

7.2
Hyperbranched Polymers for Antimicrobial Surface

Recently, montmorillonite clay has been modified by hyperbranched polymers. The influence of the organoclay on the photopolymerization reaction of acrylate systems as well as on the thermal properties of nanocomposites has been investigated. The potential properties of these new materials for the preparation of antimicrobial surface have been evaluated. Photopolymerization constitutes a very promising alternative method to prepare nanocomposites [26–29], which is expanding due to the unique advantages (ultrafast reactions, solvent-free systems, temporal and spatial control), because it takes place at an ambient temperature avoiding any thermal degradation of the alkylammonium salt, which is a serious problem in the melt polymerization or by *in situ* thermal polymerization.

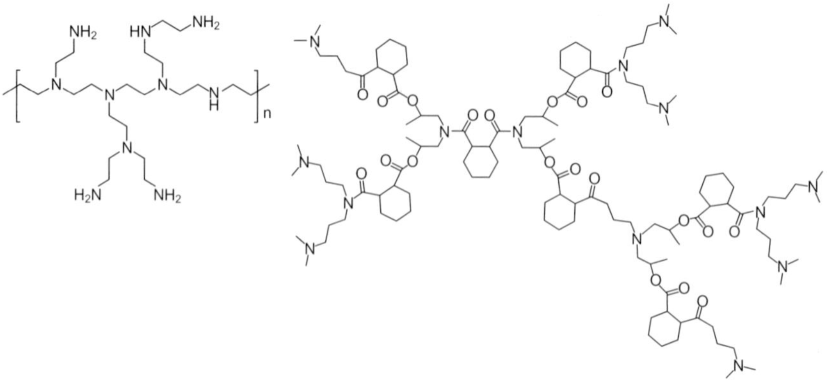

Figure 7.1 Structure of hyperbranched PEI800, PEI25000, and HA1690.

First, montmorillonite clay MMTk10 has been modified by hyperbranched polymers based on polyethylenimine, PEI800 and PEI25000 (M_w 800 and branched M_w 25,000, respectively) and the aliphatic polyestheramide Hybrane® HA1690. For that purpose, quaternary amination of hyperbranched polymers by reaction with MeI was previously required, and subsequently incorporated in the clay by cationic interchange. Structure of hyperbranched polymers are shown in Figure 7.1.

The X-ray diffraction patterns for MMTk10 showed a primary silicate (001) reflection at 8.9°, which corresponds to a *d*-spacing of 1 nm. For the organomodified clays, MMT-PEI800Q and MMT-PEI25000Q, independently of their molecular weight, an increase in *d*-spacing to 1.45 nm was observed when compared with the unmodified clay, which indicates the successful intercalation of the polymeric ammonium salts. Although the chain length did not affect the basal spacing, the influence of the structure was more significant, and for montmorillonite modified with the aliphatic polyestheramide Hybrane, MMT-HA1690Q, the basal spacing was 1.74 nm. In general, the organomodified clays showed a significantly broadened reflection, which reflected an intercalated structure less spatially ordered with respect to MMTk10. That fact has been related to the structure of the organomodifiers, and the presence of several charged groups that could interact electrostatically with different but neighboring platelets surfaces [30].

The thermogravimetric analysis (TGA) of pure clay in nitrogen showed a high thermal stability, and a total weight loss of 9% was observed when heated, due to the water being physically adsorbed in the montmorillonite clay (Figure 7.2). For the intercalated clays, the weight loss is significantly decreased at temperatures below 200 °C, indicating the lower content of adsorbed water due to the reduced hydrophilicity. When heated up to 800 °C, there is a pronounced weight loss above 200 °C, which indicates the presence of the intercalated organomodifiers on clay platelets, since the decomposition and combustion occurs at the same temperature range, and the overall degradation behavior is quite similar to that of pure hyperbranched polymers. The content of the organic components in the modified clays was 9.7%, 17.2%, and 32.3% for MMT-PEI800Q, MMT-HA1690Q, and MMT-

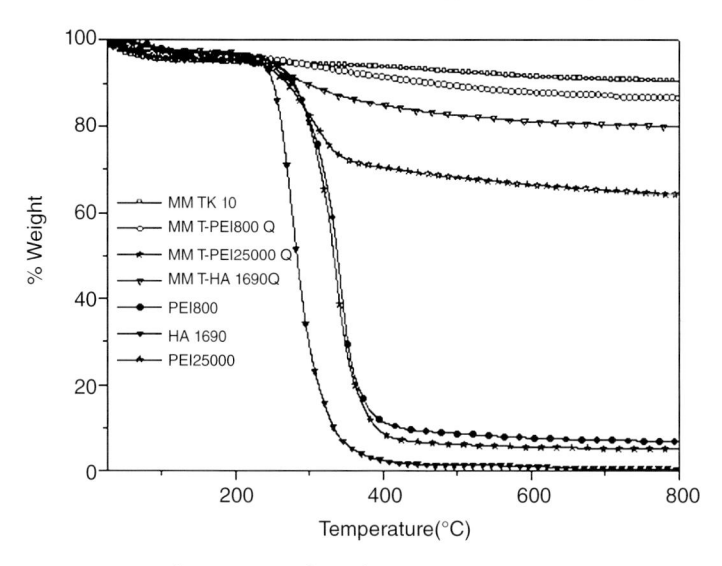

Figure 7.2 TGA thermograms of MMTk10, MMT-PEI800Q, MMT-PEI25000Q, and MMTHA1690Q, and the hyperbranched polymers PEI800, PEI25000, HA1690.

PEI25000Q, respectively. It has been determined as the weight loss from 200 to 800 °C, where the influence of water content is eliminated and the weight loss corresponding to the intercalated organomodifier takes place.

The pure and organomodified clays were added to a mixture of monomers, HEMA and PEGDMA (4 wt% with respect to HEMA), in different proportions (1, 2, and 3 wt% with respect to HEMA), and the kinetics of photoinitiated polymerization were monitored by differential scanning calorimetry.

The addition of organoclay was shown to have no detrimental on the curing reaction. Kinetic profiles for photopolymerization are conventional, showed an autoacceleration period and then slowed down rate because T_g of system ($T_g = 103\,°C$) is well above the polymerization temperature ($T = 25\,°C$) and vitrification of the material takes place (Figure 7.3). In general, the clay incorporation caused the increase of photopolymerization rate at the peak maximum (R_p^{max}), associated with the low diffusivity of the monomer into the polymer matrix, and this fact is more noticeable for MMT-PEI25000Q, which showed a higher degree of modification, and better dispersion in the matrix would be expected (Table 7.1). As the clay content increased, the polymerization rate and final degree of double bound conversion (α_F) decreased, which may be due to a reduced penetration of UV radiation into the sample as a consequence of the presence of large particle aggregates comprised of dozens of platelets.

The glass transition temperature decreased with the addition of 1% silicate platelets, from 104 °C for the pure polymer, to 97–99 °C for the nanocomposite materials. For higher clay content, a T_g value comparable to that of the polymer matrix was obtained; this effect is ascribed to the confinement of the intercalated

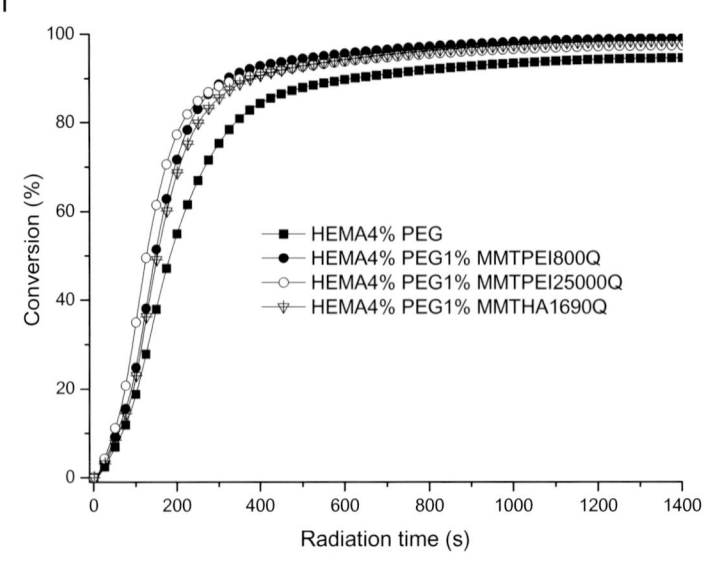

Figure 7.3 The double bond conversion versus irradiation time of the neat HEMA/PEGDMA, and with 1% mass of MMT-PEI800Q, MMT-PEI25000Q, and MMTHA1690Q. $I = 0.08\,\text{mW/cm}^2$.

Table 7.1 R_p^{max} and α_F of HEMA/PEGDMA in the presence of 1–3% of MMT-PEI800Q, MMT-PEI25000Q, and MMT-HA1690Q; and glass transition temperature (T_g) determined by differential scanning calorimetry.

HBP surfactant	wt% clay	α_F (%)	R_p^{max} (s^{-1})	T_g (°C)
*	0	94	0.41	104
PEI800Q	1	97	0.54	98
	2	93	0.48	98
	3	93	0.42	103
PEI25000Q	1	99	0.59	97
	2	98	0.58	103
	3	96	0.59	104
HA1690Q	1	98	0.55	99
	2	87	0.43	95
	3	88	0.35	100

polymer chains within the clay galleries that restrict polymer chain motions as a result of the strong interactions between the clay layers and the polymer chains.

The morphology of intercalated/exfoliated structure in the nanocomposites was revealed by the X-ray diffraction patterns (Figure 7.4). The absence of the (001) peak would suggest that the d_{001} spacing between the layered silicates is intercalated to a spacing greater than the measurable range or clay layers are disorderly dispersed in the HEMA/PEGDMA matrix, and also the lack of sensitivity at the lower amount of clay loading or random orientation of clay tactoids may be con-

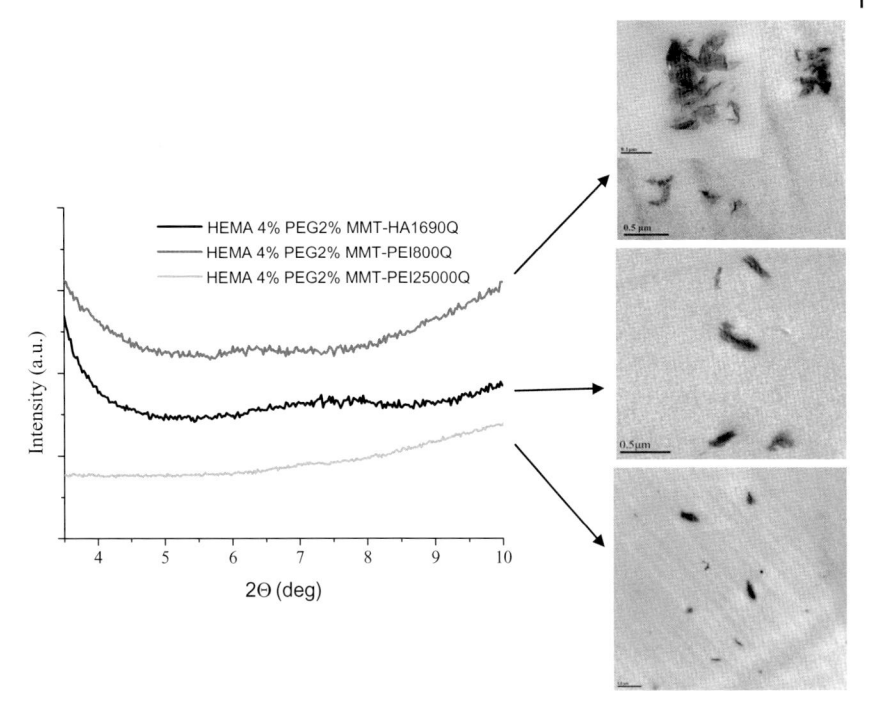

Figure 7.4 XRD patterns of the nanocomposites containing 2 mass % of MMT-PEI800Q, MMT-PEI25000Q, and MMTHA1690Q; TEM images of the nanocomposites.

sidered. These results were confirmed by TEM analysis, which revealed that the clay was dispersed uniformly and randomly during the curing process in the polymer matrix, and exfoliated into thin tactoids that contained few clay layers. The clay treatment leads to a widening of the clay interlamellar spacing, and together with the organophilic character of the treated clay, allows an easy penetration of the UV-curable monomers into the lamellar structure inducing intercalation and eventually exfoliation.

The thermal behavior of the nanocomposites was studied by thermogravimetric analysis and was similar to HEMA/PEGDMA (Figure 7.5). The TGA curve indicates that there are two stages of decomposition. The first one is due to the loss of adsorbed water, and the addition of clay significantly decreased this weight loss, indicating a lower content of adsorbed water due to the reduced hydrophilicity. With regard to the second stage, at temperature above 300 °C, the nanocomposites have a higher decomposition temperature than pure polymer. This enhancement in the thermal stability is due to the presence of clay nanolayers with high thermal stability and the great barrier properties of the nanolayers dispersed in the nanocomposites to maximize heat insulation. Comparing the different systems, a better stability is achieved by the samples containing MMT-PEI25000Q, which would be related to the enhanced compatibility with the polymer matrix and better dispersion.

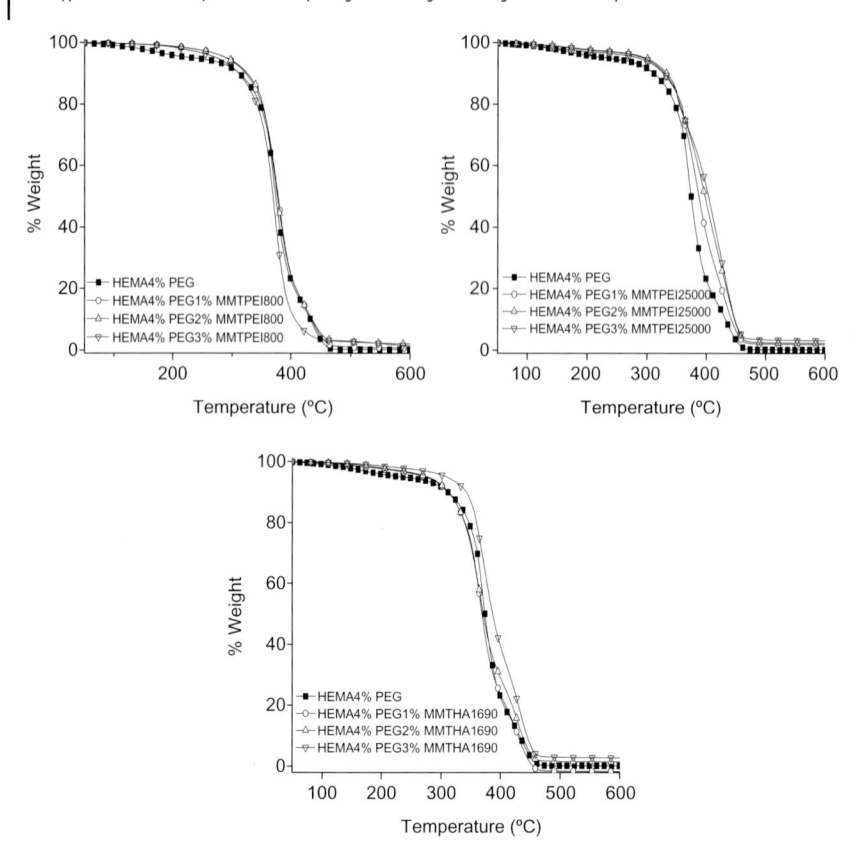

Figure 7.5 TGA thermograms of neat HEMA/PEGDMA and nanocomposites containing 1, 2, 3 mass % of MMT-PEI800Q, MMT-PEI25000Q, and MMTHA1690Q.

By analyzing the chemiluminescence emission of HEMA/PEGDMA sample and nanocomposites, influence of the nature of the modifier and the content of clay in the thermal stability is established. The chemiluminescence emission (CL) in polymers is due to the light emission that accompanies the thermal decomposition of the thermooxidative degradation products (hydroperoxides), which are formed during processing or service life of the material under ambient conditions, and can be related to the hydroperoxide (POOH) content of a polymer [31]. All the clays exhibited chemiluminescence emission above 100 °C under oxygen, and the intensity was markedly enhanced and the onset decreased in the presence of the hyperbranched polymers confirming the presence of the modifiers, where hydroperoxide can been generated due to the aliphatic carbons present in their structures. However, only in bulk powder the chemiluminescence may be detectable, and these emissions are negligible when the organomodified clays are included in a polymer formulation at the concentrations used in the nanocomposites.

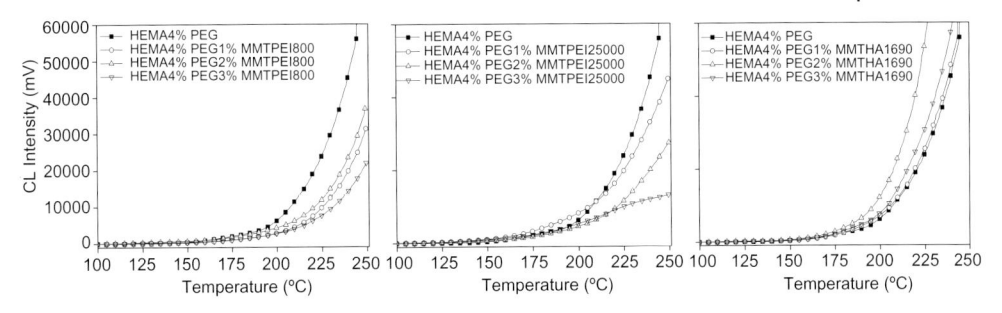

Figure 7.6 CL temperature-ramping curves obtained for neat HEMA/PEGDMA and nanocomposites containing 1, 2, 3 mass % of MMT-PEI800Q, MMT-PEI25000Q, and MMTHA1690Q.

The emission of CL for the neat polymer was detected from temperatures above 100 °C (Figure 7.6). This temperature corresponds to the glass transition temperature of the material, where the efficiency of the quantum yield of chemiluminescence emission was enhanced by the increase in mobility of the hydroperoxides, which favor the disproportion reaction responsible for the emission. A high intensity of chemiluminescence was observed under oxygen, since in such conditions, the samples are highly oxidized in a diffusion-controlled reaction simultaneously to the emission.

The nanocomposites that contained MMTPEI800 and MMTPEI25000 presented an improved thermal stability regarding the unloaded material. In general, the CL intensity decrease and the onset delay to higher temperature with the content of clay. It would indicate that the presence of reinforcement hinders the diffusion of oxidative species and makes the materials more stable under thermal oxidative conditions. For nanocomposites with MMTHA1690, a slight increase in the emission was detected for higher clay content, which may be associated with the presence of large particle aggregates.

The properties of these new materials for the preparation of antimicrobial surface have been evaluated. Polymers have been used as antimicrobial coatings in many areas, including additives for antifouling paints, biomedical devices, and food processing [32, 33]. On the other hand, quaternary ammonium compounds are widely used as cationic disinfectants or biocidal agents to prevent the growth of microorganism [34]. The activity of the quaternary ammonium compounds has been proposed to involve a general perturbation of lipid bilayer membranes, which constitute the bacterial cytoplasmic membrane of Gram-negative bacteria [16, 35]. The excess protonated amino groups of hyperbranched polymers that do not interact electrostatically act as effective agents against colonization of bacteria and may play an important role on safe materials for fighting biofilm. Adhesion of microorganism to a surface is one step of the biofilm formation, which depends on factors such as surface hydrofobicity/hydrophilicity, roughness, charge and functional groups, but the molecular and physical interactions that are involved have not been completely understood. In addition, because biofilm formation requires the initial stable attachment of a viable bacteria population on a surface, another promising approach to limiting microbial colonization will be provided by regulation of mechanical stiffness of substrata material through incorporation of nanoclay.

The bacteria attachment onto the photogenerated materials with a content of 1% of the organomodified clays has been analyzed by using *Bacillus subtilis* and *Pseudomonas aeuruginosa*. The SEM micrograph showed that the nanocomposites with MMTPEI800Q, MMT-PEI25000Q, and MMT-HA1690Q markedly inhibited the adhesion of the two bacteria, as compared to the HEMA/PEGDMA surface, where the colonization by both microorganisms was evidenced in SEM images, reflecting no antibacterial activity (Figure 7.7).

Because of the surface stiffness, hydrophobic termination and high roughness properties of surfaces can regulate adhesion of bacteria [36], microhardness (MH) of nanocomposites was analyzed as a function of clay content (Table 7.2).

The differences of stiffness observed in the nanocomposites suggested that is not the main property that determines the extent of bacterial attachment. The nanocomposites including MMT-HA1690 exhibited similar values to pure polymer, and in presence of MMT-PEI25000Qa slight increase is observed as the clay weight fraction is raised. Otherwise, the contribution of the MMT-PEI800Q is noticeable and MH values increase from 155 to 280 MPa when 3 wt% of clay is

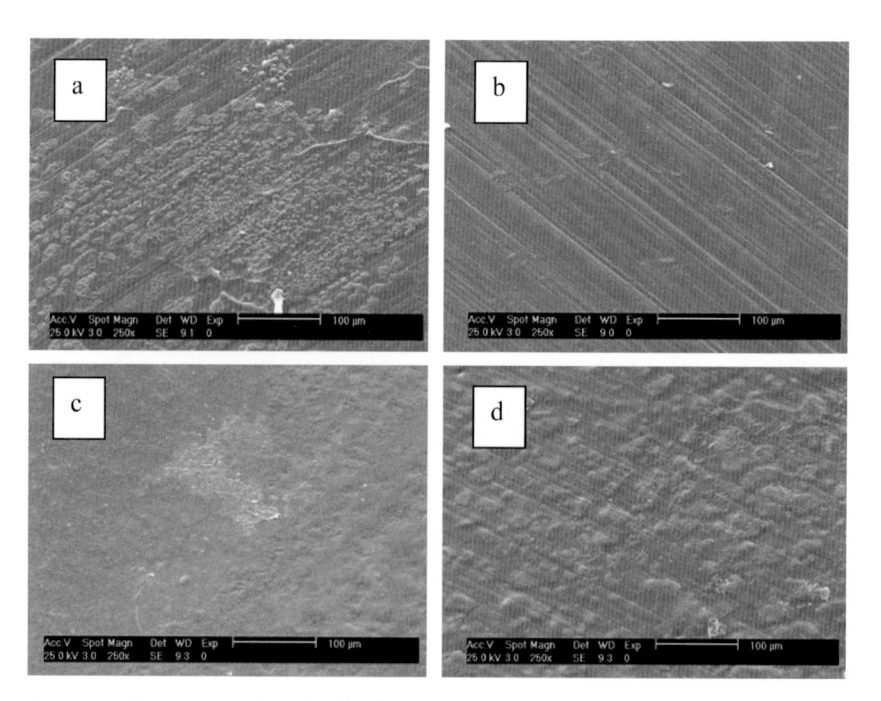

Figure 7.7 SEM micrographs of biofilm formation with *B. subtilis* on (a) neat HEMA/PEGDMA samples and nanocomposites with 1% (b) MMT-PEI800Q, (c) MMT-PEI25000Q, and (d) MMTHA1690Q.

Table 7.2 Microhardness of neat HEMA/PEGDMA and nanocomposites containing 1, 2, 3 mass % of MMT-PEI800Q, MMT-PEI25000Q, and MMTHA1690Q.

HBP surfactant	wt% clay	MH (MPa)
None	0	156.19
PEI800Q	1	172.59
	2	191.08
	3	286.1
PEI25000Q	1	164.22
	2	179.73
	3	181.25
HA1690Q	1	126.07
	2	167.5
	3	150.72

added, which may be ascribed to the favorable interfacial characteristics reached during the preparation of materials. Consequently, it would indicate that the nature of organomodifier in the clay plays a more important role in *B. subtilis* and *P. aeuruginosa* adhesion processes than stiffness, and the high efficiency against microorganism may be related to the presence of hyperbranched polymers with a high density of quaternary ammonium groups attached to the clay. The results obtained suggest a new application of these materials in biomedical area.

7.3
Hyperbranched Polymers on Adsorbents for Cr(VI) Water Treatment

Chromium(VI) is one of the heavy metals usually generated from various industrial activities such as textile dyeing, leather tanning process, electroplating, wood, and water treatment. Removal of chromium(VI) in drinking water is important because this metal poses serious health risks [37]. Different technologies have been employed for the removal of Cr(VI) from aqueous solutions, including electrochemical precipitation, phytoextraction, reverse osmosis, ultrafiltration, and evaporative recovery. It has been proved that the most effective and economical technique to remove chromium from water is adsorption and several types of adsorbents such as bacteria, fungal biomass, algal biomass, or chemicals components from agricultural products. By using traditional separation techniques, the adsorbents are difficult to be separated from the solution. In the past few years, magnetic sorbents [38] have emerged as a new generation of materials for environmental decontamination, which can be separated by application of an external magnetic field [39]. In this sense, iron and iron oxide nanostructures have been proved to be highly efficient materials [40]; however, these materials present the inconvenience of easy oxidation/dissolution, difficulty in recycling such small-sized NPs, and the coaggregation of the nanoparticles, which decreases the effective surface area of nanoparticles and thus reduces their reaction activities. To minimize the coaggregation of the nanoparticles and improve their manipulation, magnetite nanoparticles are supported on polymers or inorganic matter [41–43], such as MMT due to its excellent surface properties [44]. Hyperbranched polymers based on polyethylenimine (PEI) possess primary and secondary amine groups in a molecule, which exhibit good sorption ability of heavy metals [45]. The combination of polymers containing these functional groups and magnetic nanoparticles affords magnetic adsorbencey [46].

Recently, hybrid materials consisting of MMT sheets as support of magnetite nanoparticles (Fe_3O_4 NPs) (40 nm) coated with hyperbranched PEI (PEI800 and PEI25000) were prepared by using a cationic exchange strategy [47].

A better dispersion of the magnetite nanoparticles, as well as a reduced coaggregation of nanoparticles, was achieved when the hybrid material was prepared with the polyethylenimine of higher molecular weight, PEI25000. TEM images of Fe_3O_4-PEI800-MMT and Fe_3O_4-PEI25000-MMT (Figure 7.8) showed a morphology with individual exfoliated platelets and intercalation tactoids composed by a few sheets. The hyperbranched polymer on magnetite surface penetrated into the clay platelets producing intercalation and facilitated the exfoliation of the layered silicates.

The X-ray diffraction patterns for the Fe_3O_4-PEI800-MMT and the Fe_3O_4-PEI25000-MMT materials showed characteristic signals corresponding to the layered silicates and Fe_3O_4 NPs. For the hybrid materials, an increase in d-spacing to 1.42 and 1.45 nm for Fe_3O_4-PEI800-MMT and Fe_3O_4-PEI25000-MMT, respectively, was observed as a result of the hyperbranched polymer intercalated within the layered structures. Therefore, the available surface area of MMT to support the magnetite nanoparticles was increased.

The thermogravimetric analysis for the hybrid materials, Fe_3O_4-PEI800-MMT and Fe_3O_4-PEI25000-MMT, showed a slight weight loss at temperatures below 200 °C, indicating the lower content of adsorbed water due to the reduced

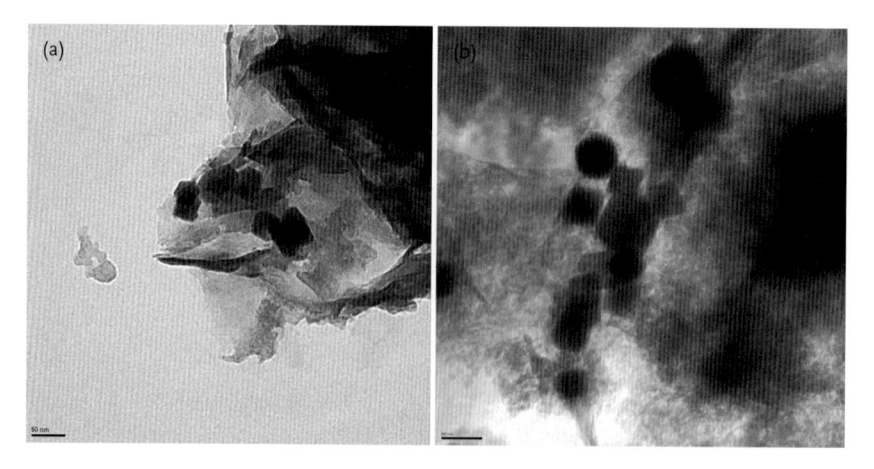

Figure 7.8 TEM images of (a) Fe_3O_4-PEI800-MMT and (b) Fe_3O_4-PEI25000-MMT hybrid.

hydrophilicity, and a pronounced weight loss above 800 °C, which indicates the presence of the Fe_3O_4/PEI particles on clay platelets. The content of the organic components in the hybrid materials was estimated on 4.9% and 3.6% for Fe_3O_4-PEI800-MMT and Fe_3O_4-PEI25000-MMT, respectively.

The magnetic behavior of the synthesized Fe_3O_4 NPs along with the hybrid material Fe_3O_4-PEI800-MMT at 300 K was analyzed (Figure 7.9). Both systems showed hysteresis loops in the $M–H$ curves, indicating the ferrimagnetic nature of the materials. Saturation magnetization values obtained from the M-H curves for the Fe_3O_4 NPs was 87.4, very close to the bulk values. Incorporation of Fe_3O_4 NPs in the Fe_3O_4-PEI800-MMT hybrid material led to a decrease of the saturation magnetization value, 4.2 emu/g, attributed to the lower content of the magnetic component in the hybrid material.

Coercivity values were found to be 15.0 and 58.1 Oe for the Fe_3O_4 NPs and the Fe_3O_4-PEI800-MMT hybrid material, respectively. The increase in coercivity value in the hybrid material is a consequence of the change in the magnetic dipole coupling interaction after the dispersion of the Fe_3O_4 NPs among the MMT sheets. It would confirm that Fe_3O_4 NPs had been successfully dispersed among MMT layers, and the coaggregation problem in magnetite nanoparticles was minimized. The magnetization property enables the hybrid material to be easily separated from aqueous solution under an external magnetic field in less than 30 s.

The effect of the initial Cr(VI) concentration on the Cr removal efficiency of both Fe_3O_4-PEI800-MMT and Fe_3O_4-PEI25000-MMT was analyzed (Figure 7.10). The removal efficiency of Cr(VI) decreased with the increase of the initial Cr(VI) concentration, which was attributed to the insufficient total available adsorption sites of adsorbent with fixed dosage when high Cr(VI) concentrations were used. Fe_3O_4-PEI800-MMT presented slightly high values, which is in agreement with the higher amount of PEI determined by TGA, with respect to Fe_3O_4-PEI800-MMT hybrid material.

The effect of initial solution pH on Cr(VI) percentage removal by the Fe_3O_4-PEI800-MMT materials was studied (Figure 7.11). The pH value has very little

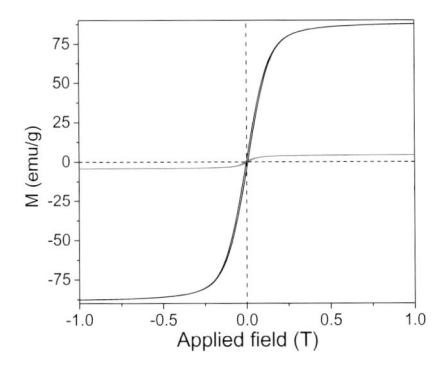

Figure 7.9 Magnetic behavior at 300 K: (—) $M–H$ curves for the Fe_3O_4 NPs and (—) $M–H$ curves for Fe_3O_4-PEI800-MMT hybrid material.

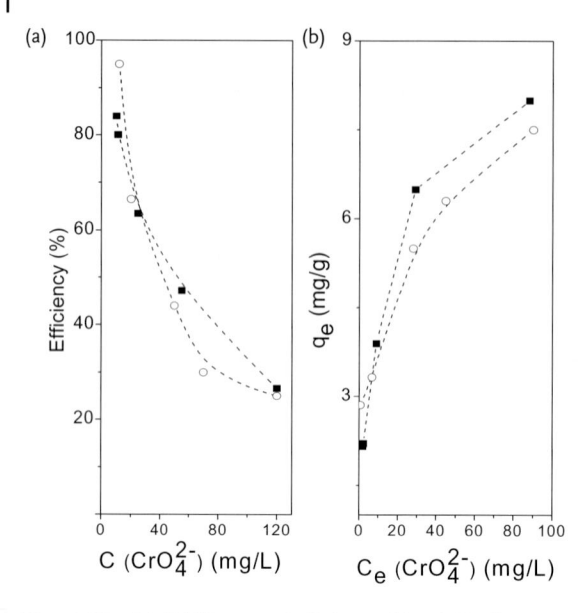

Figure 7.10 (a) Cr(VI) removal efficiency; (b) Cr(VI) adsorption isotherms, as a function of initial chromate concentration for both Fe_3O_4-PEI800-MMT (■) and Fe_3O_4-PEI25000-MMT (○).

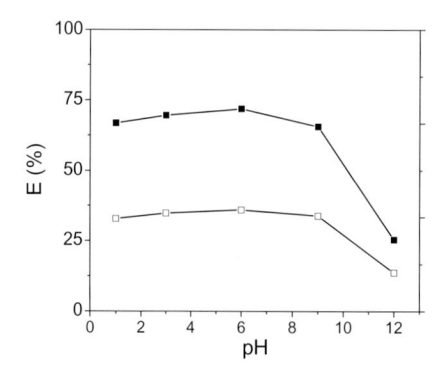

Figure 7.11 Cr(VI) removal efficiency as a function of pH for two initial chromate concentrations 10 mg/L (■) and 25 mg/L (□).

effect on the adsorption of Cr(VI) in a wide range of pH, from 1.0 to 9, unlike other magnetic adsorbents described in literature. A dramatic drop occurred at pH 12. It can be explained considering that pH is a parameter that affects crucially the sorption of metal ions, not only by influencing the surface properties of the sorbent, but also by affecting the metal speciation in solution [48, 49]. Regarding PEI, it is well known that PEI is highly effective for the neutralization of excess anionic colloidal charge, especially under acidic and neutral pH. The amine groups ($-NH$, $-N<$, $-NH_2$) in PEI would be protonated at pH below 10.4 and would adsorb anionic hexavalent chromium (CrO_4^{2-}, $HCrO_4^-$, and $Cr_2O_7^{2-}$) via electrostatic attraction, and at pH above 10.4, the electrostatic attractions with the CrO_4^{2-} anions would be reduced, decreasing the removal efficiency considerably. Also, it should be considered that the surface of magnetite nanoparticles in aqueous solution is positively charged and the anion adsorption is favored at a pH value below the pH_{ZPC} (the zero point of charge, 8.3) of magnetite. Therefore, the magnetite nanoparticles might also contribute to keep stable the adsorption capacity of the hybrid materials from pH 1 to pH 9.

The role of each component in Cr(VI) adsorption was evaluated. The removal efficiency of Cr(VI) by MMT material determined at a pH = 3 was 40%, much lower than that for Fe_3O_4-PEI800-MMT hybrid material at the same experimental conditions (80%). This lower efficiency value can be explained taking into account that the negatively charged MMT layers do not favor the adsorption of the anionic Cr(VI) species. In case of magnetite nanoparticles, the adsorption capacity was reported to be 28.16 mg/g in a pH range from 1 to 3, it would due to the coaggregation of magnetite that reduced the stability. On the other hand, PEI has been established as a good adsorbent for Cr(VI) at any range of pH (0–10.4) below its pK_a, but presents the inconvenience of the difficulty to be separated from the reaction medium.

The adsorption isotherms of Cr(VI) removal by both Fe_3O_4-PEI800-MMT and Fe_3O_4-PEI25000-MMT hybrid materials at 25 °C are shown in Figure 7.10B. The equilibrium data for Cr(VI) adsorption using both Fe_3O_4-PEI800-MMT and Fe_3O_4-PEI25000-MMT hybrid material were applied to two adsorption isotherm models, Langmuir [50], which consider that sorption takes place at specific homogeneous sites within the adsorbent, indicating a monolayer adsorption process, and Freundlich [51] based on multilayer adsorption on heterogeneous surface. The relative parameters of both models were determined at 25 °C and showed that both hybrid materials, Fe_3O_4-PEI800-MMT and Fe_3O_4-PEI25000-MMT, follow Langmuir isotherm more closely. The adsorption capacity of Fe_3O_4-PEI800-MMT and Fe_3O_4-PEI25000-MMT were calculated to be 8.77 and 7.69 mg g^{-1}, respectively.

These results suggested that the hybrid materials can be consider as potential magnetic sorbents for the Cr(VI) water treatment. The combination of the three components resulted in a new magnetic material with a number of advantages: high removal efficiency in a pH range from 0 to 11, chemical stability at any pH (due to the protection of Fe_3O_4 NPs given by both the PEI coating and MMT sheets), solved coaggregation and easy magnetic separation.

Acknowledgment

The authors would like to thank the MICINN (Spain) for financial support (MAT2009-09671). C.A. also thank the Ramon y Cajal Program.

References

1. Flory, P.J. (1952) *J. Am. Chem. Soc.*, **74**, 2718.
2. Stiriba, S.E., Slagt, M.Q., Kautz, H., Kleingebbink, R.J., Thomman, R., and Frey, H. (2004) *Chemistry (Easton)*, **10**, 1267.
3. Inoue, K. (2000) *Prog. Polym. Sci.*, **25**, 453.
4. Kim, Y.H. and Webster, O.W. (1990) *J. Am. Chem. Soc.*, **112**, 4592.
5. Xu, J., Wu, H., Mills, O.P., and Heiden, P.A. (1999) *J. Appl. Polym. Sci.*, **72**, 1065.
6. Romagnoli, B. and Hayes, W. (2002) *J. Mater. Chem.*, **12**, 767.
7. Kim, Y.H. and Webster, O.W. (1992) *Macromolecules*, **25**, 5561.
8. Johansson, M., Glauser, T., Janson, A., Hult, A., Malmstrom, E., and Claesson, H. (2003) *Prog. Org. Coat.*, **48**, 194.
9. Deka, H. and Karak, N. (2011) *Polym. Adv. Technol.*, **22**, 973.
10. Rodler, M., Plummer, C.J.G., Garamszegi, L., Leterrier, Y., Grunbauer, H.J.M., and Manson, J.E. (2004) *Polymer*, **45**, 949.
11. Tomuta, A., Ferrando, F., Serra, A., and Ramis, X. (2012) *React. Funct. Polym.*, **72**, 556.
12. Xie, X.L., Tjong, S.C., and Li, R.K.Y. (2000) *J. Appl. Polym. Sci.*, **77**, 1975.
13. Nguyen, C., Hawker, C.J., Miller, R.D., Huang, E., Hedrick, J.L., and Gauderon, R. (2000) *Macromolecules*, **33**, 4281.
14. Hawker, C.J., Chu, F.K., Pomery, P.J., and Hill, D.J.T. (1996) *Macromolecules*, **29**, 3831.
15. Madhuryya, D., Kumar, A., Hareckrishna, D., and Karak, N. (2012) *Ionics*, **18**, 181.
16. Yates, C.R. and Hayes, W. (2004) *Eur. Polym. J.*, **40**, 1257.
17. Crooks, R.M. (2001) *ChemPhysChem*, **2**, 644.
18. Liu, P., Liu, W.M., and Xue, Q.J. (2004) *J. Mater. Sci.*, **39**, 3825.
19. Larraza, I., Peinado, C., Abrusci, C., Catalina, F., and Corrales, T. (2011) *J. Photochem. Photobiol. A Chem.*, **224**, 46–54.
20. Liu, P. (2007) *Appl. Clay Sci.*, **35**, 11.
21. Stadtmueller, L.M., Ratinac, K.R., and Ringer, S.P. (2005) *Polymer*, **45**, 9574.
22. Di, J. and Sogah, D.Y. (2006) *Macromolecules*, **39**, 1020.
23. Qin, X., Wu, Y., Wang, K., Tan, H., and Nie, J. (2009) *Appl. Clay Sci.*, **45**, 133.
24. Sangermano, M., Lak, N., Malucelli, G., Samakand, A., and Sanderson, R.D. (2008) *Prog. Org. Coat.*, **61**, 89.
25. Huskic, M., Zagar, E., Zigon, M., Brnardic, I., Macan, J., and Ivankovic, M. (2009) *Appl. Clay Sci.*, **43**, 420.
26. Owusu-Adom, K. and Allan Guymon, C. (2009) *Macomolecules*, **42**, 180.
27. Decker, C., Keller, L., Zahouily, K., and Benfarhi, S. (2005) *Polymer*, **46**, 6640.
28. Kim, S. and Guymon, C.A. (2011) *J. Polym. Sci. Part A Polym. Chem.*, **49**, 465.
29. Lv, S., Zhou, W., Li, S., and Shi, W. (2008) *Eur. Polym. J.*, **44**, 1613.
30. Fan, X., Xia, C., and Advincula, R.C. (2003) *Langmuir*, **19**, 43481.
31. Billingham, N.C., Then, E.T.H., and Gijman, P.H. (1991) *Polym. Deg. Stab.*, **42**, 263.
32. Meng, N., Zhou, N.-L., Zhang, S.-Q., and Shen, J. (2009) *Appl. Clay Sci.*, **42**, 667.
33. Galya, T., Sedlarik, V., Kuritka, I., Novotny, R., Sedlarikova, J., and Saha, P. (2008) *J. Appl. Polym. Sci.*, **110**, 3178.
34. Hong, S.-I. and Rhim, J.-W. (2008) *J. Nanosci. Nanotech.*, **8**, 5818.
35. Gilbert, P. and Moore, L.E. (2005) *J. Appl. Microbiol.*, **99**, 703.
36. Lichter, J.A., Todd, M., Delgadillo, M., Nishikawa, T., Rubner, M., and Van Vliet, K.J. (2008) *Biomacromolecules*, **9**, 1571.
37. Costa, M. (2003) *Reg. Toxicol. Pharmacol.*, **188**, 1.

38 (a) Oliveira, L.C.A., Petkowiczb, D.I., Smaniottob, A., and Pergherb, S.B.C. (2004) *Water Res.*, **38**, 3699; (b) Booker, N.A., Keir, D., Priestley, A.J., Sudarmana, D.L., and Woods, M.A. (1991) *Water Sci. Technol.*, **23**, 1703; (c) Orbell, J.D., Godhino, L., Bigger, S., Nguyen, T.M., and Ngeh, L.N. (1997) *J. Chem. Educ.*, **74**, 1446.

39 Hua, X., Wang, J., Liua, Y., Li, X., Zeng, G., Bao, Z., Zeng, X., Chen, A., and Long, F. (2011) *J. Hazard. Mater.*, **185**, 306.

40 (a) Zhang, W.X. (2003) *J. Nanopart. Res.*, **5**, 323; (b) Hu, J., Zhong, L., Song, W., and Wan, L. (2008) *Adv. Mater.*, **20**, 2977.

41 (a) Novakova, A.A., Lanchinskaya, V.Y., Volkov, A.V., Gendler, T.S., Kiseleva, T.Y., Moskvina, M.A., and Zezin, S.B. (2003) *J. Magn. Magn. Mater.*, **258**, 354–357; (b) Kang, Y.S., Risbud, S., Rabolt, J., and Stroeve, P. (1996) *Langmuir*, **12**, 4345.

42 (a) Zhang, M.J., Zhang, Q., Itoh, T., and Abe, M. (1994) *IEEE Trans. Magn.*, **30**, 4692; (b) Bruce, I.J., Taylor, J., Todd, M., Davies, M.J., Borioni, E., Sangregorio, C., and Sen, T. (2004) *J. Magn. Magn. Mater.*, **284**, 145.

43 Arruebo, M., Fernandez Pacheco, R., Irusta, S., Arbiol, J., Ibarra, M.R., and Santamaria, J. (2006) *Nanotechnology*, **17**, 4057.

44 (a) Park, J., Kim, I., Cheong, I., Won Kim, J., Bae, D., Cho, J.W., and Yeum, J.H. (2010) *Colloid Polym. Sci.*, **288**, 115; (b) Podsiadlo, P., Rouillard, J.M., Zhang, Z., Lee, J., Lee, J.W., Gulari, E., and Kotov, N.A. (2005) *Langmuir*, **21**, 11915.

45 (a) Ghoul, M., Bacquet, M., and Morcellet, M. (2003) *Water Res.*, **37**, 729; (b) Deng, S.B. and Ting, Y.P. (2005) *Water Res.*, **39**, 2167.

46 (a) Huang, G., Zhang, H., Shi, J.X., and Langrish, A.G. (2009) *Ind. Eng. Chem. Res.*, **48**, 2646; (b) Hu, X., Wang, J., Liu, Y., Li, X., Zeng, G., Bao, Z., Zeng, X., Chen, A., and Long, F. (2011) *J. Hazard. Mater.*, **185**, 306; (c) Bhaumika, M., Maity, A., Srinivasu, V.V., and Onyango, M.S. (2011) *J. Hazard. Mater.*, **190**, 381.

47 Larraza, I., López-Gónzalez, M., Corrales, T., and Marcelo, G. (2012) *J. Colloid Interface Sci.*, **385**, 24.

48 (a) Nomanbhay, S.M. and Palanisamy, K. (2005) *Electron. J. Biotechnol.*, **8**, 43; (b) Park, S.J. and Jang, Y.S. (2002) *J. Colloid Interface Sci.*, **249**, 458.

49 Hu, J., Lo, I.M.C., and Chen, G. (2004) *Water Sci. Technol.*, **50**, 139.

50 Langmuir, I. (1916) *J. Am. Chem. Soc.*, **40**, 1361.

51 Freundlich, H.M.F. (1906) *J. Phys. Chem.*, **57**, 385.

8
New Methods for the Preparation of Metal and Clay Thermoset Nanocomposites

Kübra Doğan Demir, Manolya Kukut, Mehmet Atilla Tasdelen, and Yusuf Yagci

8.1
Introduction

Thermoset polymers are covalently bonded, three-dimensional cross-linked materials that cannot be melted by heating or dissolved completely in any solvents. The polymerization process for thermosets generally takes place in two stages: (i) monomers are partially polymerized into linear chains and (ii) cross-linking is completed through the addition of chemicals, heat, and/or electromagnetic radiation. Because of their desirable properties, thermoset polymers form the basis of many important advanced materials, such as computer chip packaging (insulation), protective coatings, and adhesives and aerospace composites, based on their great strength, high-temperature stability, good processability, and good chemical resistance. However, thermoset materials may have several limitations in terms of their final properties, notably their brittleness and lack of durability. Therefore, thermoset nanocomposites have been developed to circumvent some of these property shortcomings, primarily by redesigning the nanostructure and underlying morphology of the system. Thermoset nanocomposites can be fabricated from a variety of nanoparticles, including one-dimensional nanoparticle (e.g., clay platelets), two-dimensional nanofibers (e.g., nanotubes and whiskers), and three-dimensional spherical and polyhedral nanoparticles (e.g., colloidal silica, metal, and POSS). The incorporation of nanoparticles in thermoset resins leads to significant improvements in several physical properties as illustrated in Figure 8.1.

This chapter will present about the most recent researches on the preparation of metal and clay thermoset nanocomposites, prepared by various nanoparticles via *in situ* polymerization technique. Special emphasize is devoted to the synthetic routes and structures of the thermoset nanocomposites rather than to their practical properties.

Thermoset Nanocomposites, First Edition. Edited by Vikas Mittal.
© 2013 Wiley-VCH Verlag GmbH & Co. KGaA. Published 2013 by Wiley-VCH Verlag GmbH & Co. KGaA.

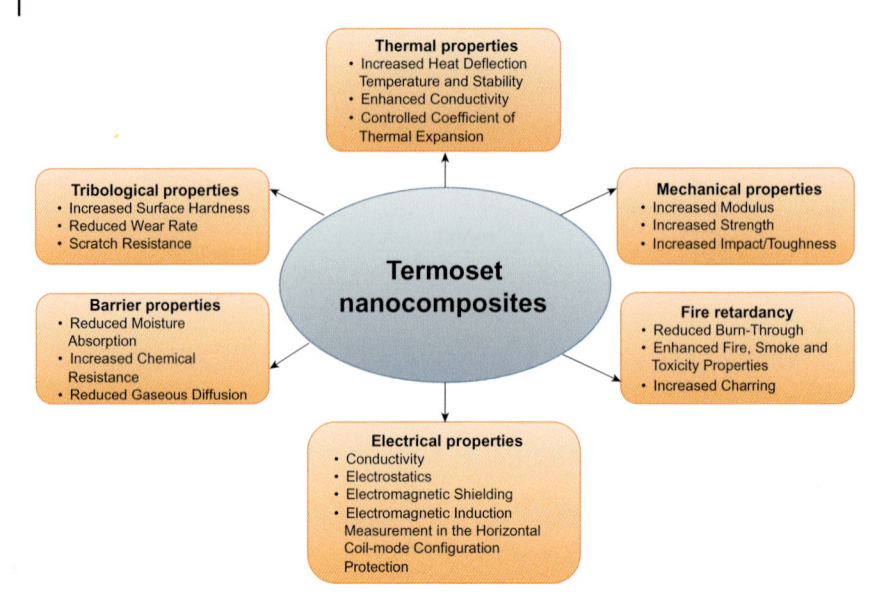

Figure 8.1 Effects of nanoparticles on various properties of thermosets.

8.2
Thermoset Nanocomposites Based on Nanoclays

Polymer/clay nanocomposites have exhibited immensely enhanced properties and higher performance as compared to both their conventional polymer composites and pure polymers. Currently, polymer/clay nanocomposites can be prepared by three ways such as solution mixing, melt blending, and *in situ* polymerization [1]. Solution mixing method consists to solubilize polymer in an organic solvent, then the clay is dispersed in the obtained solution, and subsequently either the solvent is evaporated or the polymer is precipitated. The large quantities of volatile solvent necessary for this approach make it less attractive as an industrial process. Melt blending is a solvent-free method that enables mixing of the layered silicate with the polymer matrix in the molten state. However, very careful attention has to be paid to finely tune the processing conditions to increase the compatibility of clay layer surfaces with the polymer matrix [2]. In the *in situ* polymerization technique, the monomer, together with the initiator and/or catalyst, is intercalated within the silicate layers and the polymerization is initiated by external stimulation such as thermal, photochemical, or chemical activation [3–11]. The chain growth in the clay galleries triggers the clay exfoliation and hence the nanocomposite formation.

Although *in situ* polymerization has proved successful in the preparation of various polymer-layered silicate nanocomposites, it has some drawbacks: (i) it is a time-consuming preparation route (the polymerization reaction may take more than 24 h); (ii) exfoliation is not always thermodynamically stable; and the platelets may re-aggregate during subsequent processing steps; and (iii) the process is available only to the resin manufacturer who is able to dedicate a production line for this purpose. Despite the limitations described above, *in situ* polymerization is the only viable nanocomposite preparation route to support the use of thermosetting polymers in commercial applications as melt intercalation methods cannot support the balance between intra- and extragallery cross-linking reactions. Many classes of thermoset polymer matrices used in composites are epoxy, melamine, and phenolic resins, polyurethane, and unsaturated polyester.

Epoxy resins are the most commonly used engineering thermosets because of their outstanding mechanical and thermal properties, such as high strength, high modulus, low creep, and reasonable elevated-temperature performance. However, their intrinsic brittleness (i.e., no large-scale plastic deformation before cracking) has imposed severe limitations on their structural applications. Most successful strategies concerning the toughening of epoxy resins have involved the incorporation of organically modified montmorillonite (MMT) clay into the resin matrix, which results in nanocomposite formations (Scheme 8.1). Because of the broad diversity of molecular structures available in organic–inorganic hybrid materials, these nanocomposites frequently exhibit unexpected hybrid properties synergistically derived from both components; that is, a desirable balance of stiffness, strength, and toughness, as well as improved thermal stability, most of which are unique and different from any other traditional composites or blends. The key to achieve an exfoliated epoxy/clay nanocomposite formation is first to load the clay gallery with hydrophobic onium ions, and then intercalation of epoxy compounds into the intergallery region of the clay. Interestingly, the acidic onium ions catalyze intragallery polymerization at a rate that is competitive with the cross-linking of the epoxy compounds with the amine-curing agents in the intergallery region. However, the relative rates of reagent intercalation, chain formation, and network cross-linking can be controlled by the nature of onium ions in order to form the gel state and, eventually, the fully cured epoxy-exfoliated clay nanocomposites [12–19]. The influence of various parameters such as type of epoxy resin, type of curing agents, and type of clays; the effect of curing conditions, intercalating, and

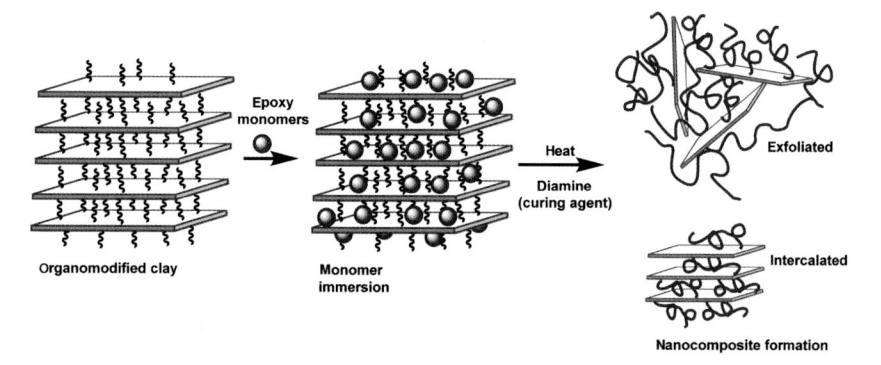

Scheme 8.1 *In situ* synthesis of epoxy/clay nanocomposites containing intercalated and exfoliated structures.

exfoliation mechanisms; and corresponding morphological, physical and mechanical, thermal, and rheological properties have been intensively investigated. Studies of both rubbery and glassy thermoset epoxy/clay nanocomposites formed using different types of curing agents such as aliphatic and aromatic diamines, anhydrides and catalytic curing agents have been done, and the clay nanolayers are more effective in improving mechanical and thermal properties when the polymer is in its rubbery state. Epoxy–clay nanocomposites also show high water-resistant ability, which arises from the hydrophobic character of clays and the increase in the diffusion path of water in the nanocomposites. The use of nanoclay as reinforcing filler also shows potential for use in more highly cross-linked resin systems with high glass transition temperatures, as commonly used in aerospace and other high performance applications.

A new class of green nanocomposites has been also prepared from renewable resource, epoxidized plant oil, and organo-modified nanoclays via *in situ* polymerization. Recently, Wool *et al.* [20] have prepared bio-based nanocomposites by dispersing an organoclay in acrylated epoxidized soybean oil and styrene, followed by *in situ* free radical polymerization. Various groups have also developed the thermoset nanocomposites by acid-catalyzed curing and anhydride curing of epoxidized linseed and soybean oils, respectively, in the presence of an organoclay. The effect of epoxides content on the mechanical properties of the bionanocomposites has been also investigated. While tensile modulus increases with clay content, only slight difference was observed on tensile and tear strength. Because the epoxidation causes cross-linking in polymer matrix by functional groups, elongation at break of the nanocomposites with lower epoxide content (40%) is higher than the nanocomposites with higher epoxide content (100%). The morphology observed by TEM images and X-ray diffraction of bioplastic nanocomposites also confirmed the successful dispersion and unambiguous intercalation in these composites. The resulting thermoset nanocomposites show improved mechanical properties [20–23], thermal stability [20–23], and barrier properties [21], when compared to the pure polymer alone. These nanocomposites are anticipated to become a new class of coating materials derived from inexpensive renewable resources, which will contribute to global sustainability.

In the cationic ring-opening polymerization of epoxides, the alcohol acts as a chain transfer agent [24]. The presence of hydroxyl groups in the clay fillers also creates such an effect by increasing polymerization rate and temperature. Moreover, the interlamellar distance of clays and flexibility of resulting nanocomposites have been increased by the addition of the diols in the curing conditions. The diol favors the interactions between organoclay and matrix leading to increase exfoliated structures in the nanocomposites. The 4-(N-maleimido) phenylglycidylether is a hybrid monomer composed of both epoxy and maleimide reactive groups that forms high-performance thermosets [25, 26]. Like other polymeric materials, the incorporation of nanoclays into these epoxy resins dramatically enhances their barrier properties as well as their thermal and mechanical resistance [27].

Phenolic resins are exceptional materials for a wide range of industrial applications such as adhesives, coatings, laminates, and composites due to their excellent ablative properties, structural integrity, thermal stability, and solvent resistance. However, only a few studies have been performed with phenolic resins in the nanocomposite field because of their three-dimensional molecular structure even before cure, which may avoid the exfoliation of the clay. Moreover, the formation of water as a by-product of cross-linking is also another problem of this type of resins. Currently, only limited studies on phenolic resin/clay nanocomposites via the *in situ* polymerization of phenol and formaldehyde in the presence of organo-modified clays have been published [28–30]. In fact, it was shown that it is more difficult to obtain exfoliated structures employing resole-type phenolic resins than using novalac-type phenolic resins (Scheme 8.2). This can be explained by their more cross-linked structures even before curing [31].

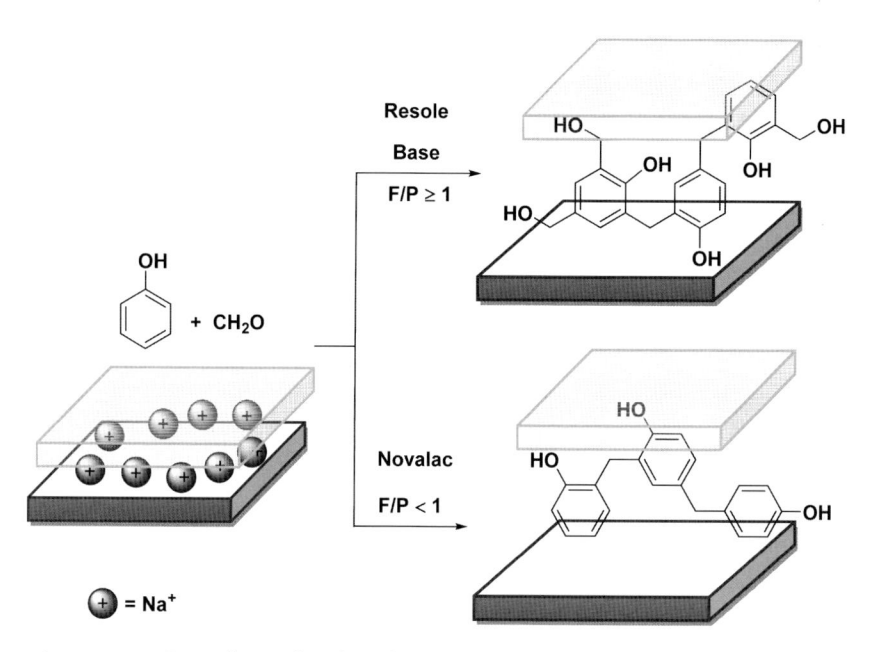

Scheme 8.2 *In situ* synthesis of resole and novalac types phenol–formaldehyde/clay nanocomposites.

Polybenzoxazines are newly developed addition polymerized phenolic systems in recent years as attractive alternatives to traditional phenolic resins for a variety of high-performance applications. Polybenzoxazine/clay nanocomposite using a polybenzoxazine precursor and organically modified montmorillonites via melt blending was first reported by Agag and Takeichi [32]. However, the well-dispersed homogeneous-layered silicates in the polybenzoxazine matrix could not be prepared by melt blending or solution mixing techniques due to partial incompatibility of organo-modified clays. But as in the case of the other polymers, an *in situ* polymerization method is the best way to prepare the polybenzoxazine/MMT nanocomposites. In a recent study sporting polybenzoxazine/clay nanocomposites,

MMT was modified with quaternized benzoxazine monomer with fluid benzoxazine monomers [9]. *In situ* simultaneous polymerization of intercalated and fluid benzoxazine monomers leads to partially exfoliated and intercalated nanocomposites with improved thermal resistance (Scheme 8.3).

Scheme 8.3 Intercalation of quaternized benzoxazine monomer/MMT and *in situ* synthesis of polybenzoxazine/MMT nanocomposites. Reproduced from Ref. [9].

Silicate layers catalytically affect the ring-opening behavior of the benzoxazines and lower the polymerization temperature by increasing clay content. In the thermogravimetric analysis, although the first degradation temperature of the nanocomposites was improved by the inclusion of the benzoxazine modified MMT, the clay content did not affect this onset. But, at elevated temperatures, thermal stabilities of nanocomposites were increased by clay content. Moreover, the char yield at 900 °C is increased up to 20% with 10% benzoxazine modified clay loading. These nanocomposite materials offer highly improved comprehensive properties in both stiffness and toughness, which is difficult to attain from individual components, especially when silicate clay layers are in the exfoliated structures.

A number of studies have been reported for the preparation of thermoset polyurethane/clay nanocomposites by the *in situ* polymerization method. Clays provide good compatibility with polyols, which are used for the synthesis of polyurethane when they modified with long chain onium anions. There are two different routes: either organoclays are intercalated by polyol segments followed by reaction with diisocyanate or by diisocyanate functional prepolymer followed by chain extension with diols [33–36]. However, polyurethane nanocomposites are formulated commonly by premixing the inorganic filler with the polyol and then curing resulting mixture with diisocyanate [37]. Organo-modified MMT are easily solvated by several polyols, which are commonly used in polyurethane synthesis as cross-linkers and chain extenders. The extent of clay gallery expansion depends on the chain length of the modifier but not on the molecular weight or the functionality of the polyols and the charge density of the clay. The degree of hydroxyl modifying of the clay also affects the clay dispersion, and modifier bearing more hydroxyl groups inspires higher degree of intercalated structures. The strength, modulus, and strain at break all are increased by more than 100% at a loading of 10% organoclay. The improvement on strength and modulus is related to the reinforcement of the dispersed silicate layers whereas that of the elasticity attributed to plasticizing effect of gallery onium ions. The effect of compatibility between nanoclay and polymer matrix on the mechanical properties of the resulting nanocomposites was studied by analyzing Cloisite Na+ (Na-MMT) or Cloisite 30B (diol-functionalized MMT)-modified polyether- or polyester-based polyurethane nanocomposites [38]. The modifiers have functional groups, which can react with the polymer matrix and strengthening the interfacial interactions between the inorganic and organic phases as presented in Scheme 8.4. The coherence of these phases is described using the proximity of their solubility parameters and Cloisite 30B modifier possesses the closer solubility value to virgin polyurethane. This justifies higher improvements in elastic modulus of the polyester-based nanocomposites filled by 1.5 wt% Cloisite 30B. But above this critical clay content, other compositions gave a reduction on modulus due to aggregation of clay layers. The higher viscosity of the polyester prepolymers comparing to polyethers depending on their molecular weights also causes an increment on modulus and exfoliation degree.

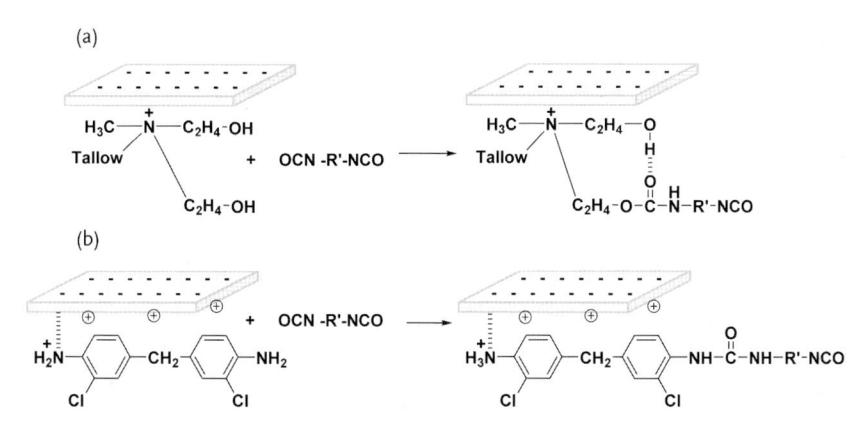

Scheme 8.4 Interactions between polyurethane chains; (a) diol-functionalized MMT and (b) Na-MMT. Reproduced from Ref. [38].

Hydrogen bonding as well as chemical linkages between clay layers and polyurethane matrix at the interface also causes an improvement in the tensile strength while the value for elongation at break decreases as a result of the restricted movement of the clay layers (Scheme 8.5) [39]. These interface interactions are enhanced by the formation of exfoliated structures from higher clay contents, and thermal stability is improved due to the restricted motion of the polymer chain. After clay loading, the molecular composition of the thermoset polyurethane matrix with large numbers of the end functionality also provide an improved adhesive strength due to strong interactions between hydroxyl group of the substrate and polar groups of the system. The MMT layers increases the cross-linking density, thereby reducing the free volume in the cured matrix consequently, the water absorption property decreases.

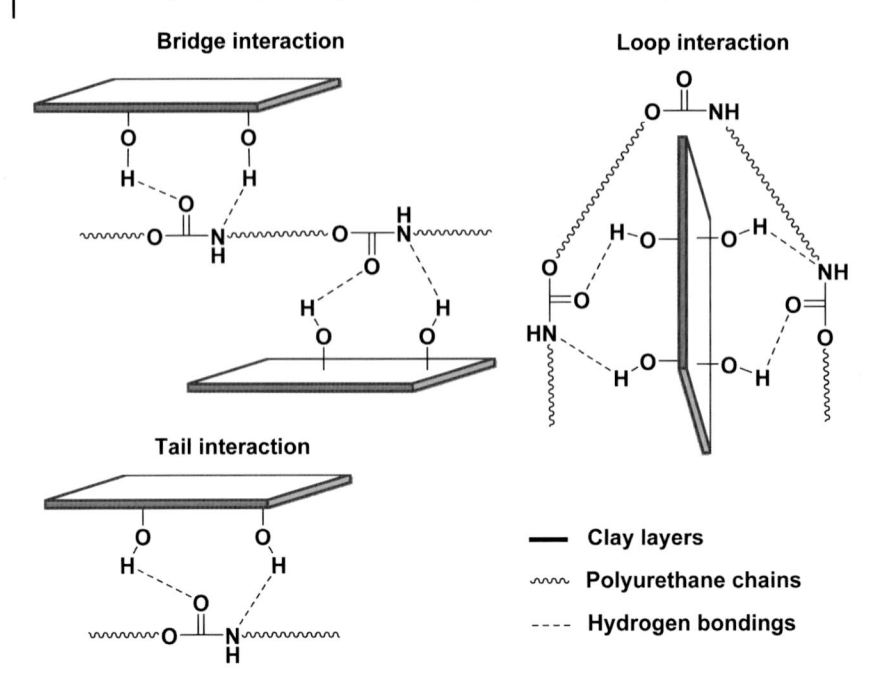

Scheme 8.5 Various interactions between nanoclay and matrix. Reproduced from Ref. [2].

Unsaturated polyester resins are the most commonly used thermoset resins where polar unsaturated prepolymers are dissolved in styrene monomer (usually around 30 wt%) and then cured by free-radical polymerization. The concept of nanoscale reinforcement is a novel opportunity for the synthesis of exfoliated nanocomposites based on clay and unsaturated polyester. In a typical example, Bharadwaj et al. [40] described the preparation of cross-linked polyester/clay nanocomposites by dispersing organically modified MMT in prepromoted polyester resin and subsequently cross-linking the system using methyl ethyl ketone peroxide as catalyst at several clay concentrations. Although, the results showed the formation of a nanocomposite structure of mixed type, containing intercalated and exfoliated regions, the tensile modulus and the loss and storage moduli are found to exhibit a gradually decreasing trend with increasing clay concentration of 1–10 wt%. These trends are explained on the basis of a progressive decrease in the degree of cross-linking due to the presence of organoclay.

In another study, Suh et al. [41, 42] studied the formation mechanism of unsaturated polyester/clay nanocomposites by using two different mixing methods (simultaneous and sequential). In the simultaneous mixing process, the unsaturated polyester chains, styrene monomers, and organo-modified MMTs were simultaneously mixed and cured. In the sequential mixing process, preintercalates of the unsaturated polyester/MMT nanocomposites are first prepared and then cured with styrene monomer, initiator, and catalyst at 80 °C for 3 h and postcured at 120 °C for 4 h. Hence, the styrene monomers are more easily dispersed inside and outside the silicate layers as mixing time increases. Therefore, in the sequential mixing method, the cross-linking reaction takes place homogeneously inside and outside the silicate layers, and cross-linking density reaches to cured pure unsaturated polyester (Scheme 8.6).

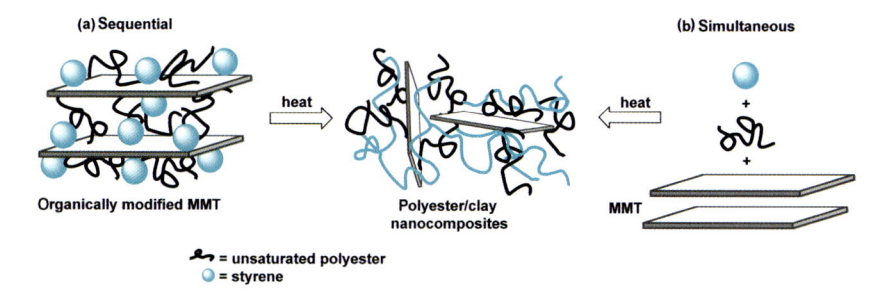

Scheme 8.6 *In situ* synthesis of polyester/clay nanocomposites by using two different mixing methods; (a) sequential and (b) simultaneous.

Hydrogels are three-dimensional hydrophilic, insoluble, and cross-linked networks, which absorb large amount of water. The incorporation of nanoclay into hydrogel structures produces new materials possessing different functional groups, which are employed in novel applications. Swelling properties of hydrogel in water and dye solutions are significantly affected especially by the type of nanoclay due to its hydrophilic structure [43]. Generally, an appropriate addition of clay nanoparticles can improve the mechanical and thermal properties of polymeric networks and contribute to the higher swelling ratio of a hydrogel in aqueous solution, which is beneficial for the adsorption. However, high clay loadings formed additional cross-linking junctions decreasing the swelling ratio as well as adsorption capacity of hydrogels. The hydrogel-based nanocomposites can be useful as a novel high-capacity adsorbent material for removal of cationic dyes especially in the textile industry (Figure 8.2).

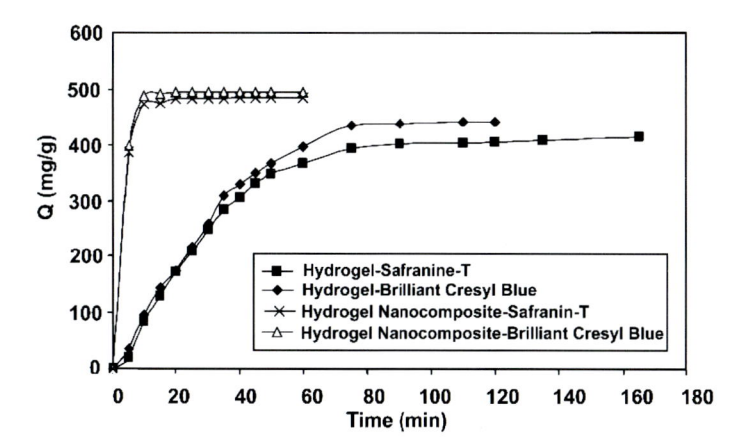

Figure 8.2 Plots of adsorbed Safranine-T and Brilliant Cresyl Blue dyes amounts by the hydrogel and hydrogel nanocomposite samples. Reproduced from Ref. [43].

Incorporation of clay layers into polysulfone matrix has been usually applied by solution exfoliation process up to the study of Dizman et al. who prepare the cross-linked polysulfone/clay nanocomposites using the *in situ* method [10]. Polysulfone dimethacrylate macromonomer and intercalated methacrylate monomer were polymerized at the same time by photoinduced cross-linking reactions leading to delaminated clay tactoids (Scheme 8.7). Attachment of monomeric sites into clay layers and subsequent photoinduced cross-linking of immersed monomers with polysulfone dimethacrylate macromonomers facilitate propagation and exfoliation processes concomitantly, leading to the formation of homogeneous clay/polymer nanocomposites.

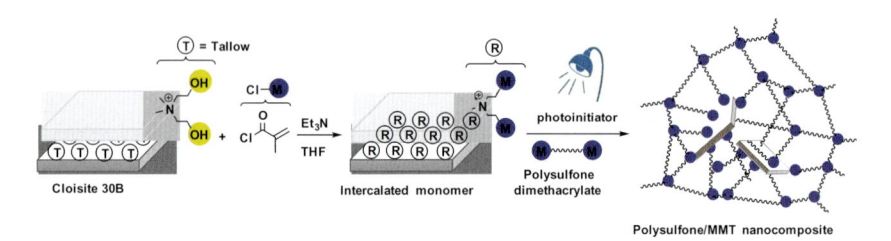

Scheme 8.7 Preparation of polysulfone/clay nanocomposites by *in situ* photoinitiated cross-linking polymerization. Reproduced from Ref. [10].

These nanomodified polysulfone matrixes have mixed morphology of the well-dispersed exfoliated and intercalated structures in 1, 3, and 5% clay contents (Figure 8.3). Increasing organoclay concentration supplied an enhancement in thermal resistance, which can be attributed to the barrier properties of clay mineral. Despite all, nanocomposites show a higher T_g value compared with the virgin PSU, with the increase of the clay content T_g slightly decreased.

The thermoset nanocomposites can be also obtained by ring-opening metathesis polymerization of dicyclopentadiene followed by cross-linking in a one-pot reaction using surface-initiation concept [44] as presented in Scheme 8.8. The Grubbs catalyst was first immobilized on the surface of MMT clay modified with vinylbenzyl dimethyloctadecyl ammonium chloride, and dicyclopentadiene was polymerized from the clay surface while simultaneously cross-linking to form a thermoset nanocomposite in a one-pot reaction. For comparison, organically modified MMT was first dispersed in dicyclopentadiene and then the Grubbs catalyst was added to the monomer/clay mixture, and finally a bulk-initiated composite was formed via conventional bulk *in situ* polymerization. X-ray diffraction and TEM analysis indicated that the resulting nanocomposites exhibited exfoliated morphologies with heterogeneous clay platelet distribution. XRD and TEM analysis indicated that the intergallery-surface-initiated nanocomposites were exfoliated with a heterogeneous platelet distribution, whereas the conventional bulk-initiated polym-

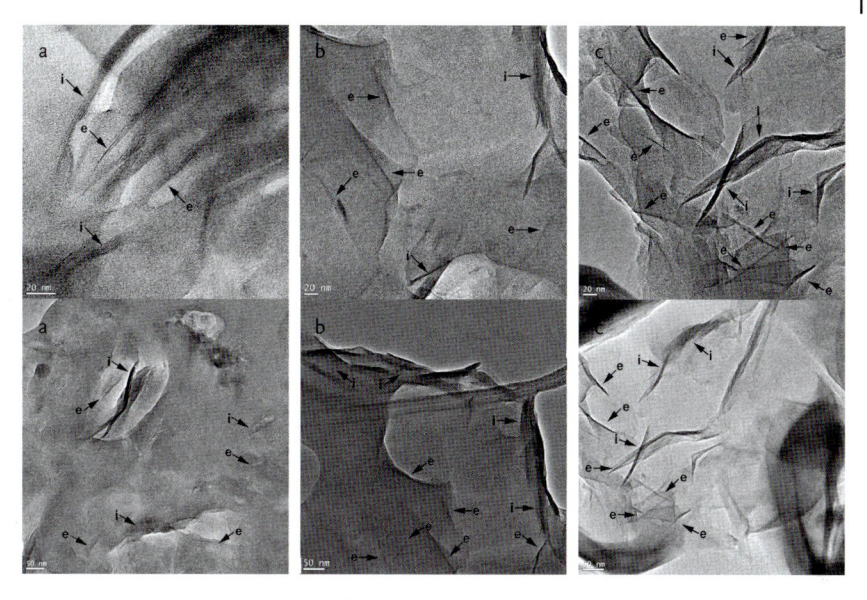

Figure 8.3 TEM micrographs of PSU/MMT nanocomposites with various clay contents: (A, %1), (B, %3), and (C, %5) in high (scale bar: 20 nm, upper images) and low magnification (scale bar: 50 nm). Reproduced from Ref. [10].

erization resulted in nanocomposites with intercalated morphologies. The differences between the morphologies demonstrated that growing polymer chains from the initiator sites on the intergallery surface of the clay platelets pushed the platelets apart during the polymerization, aiding in the exfoliation process.

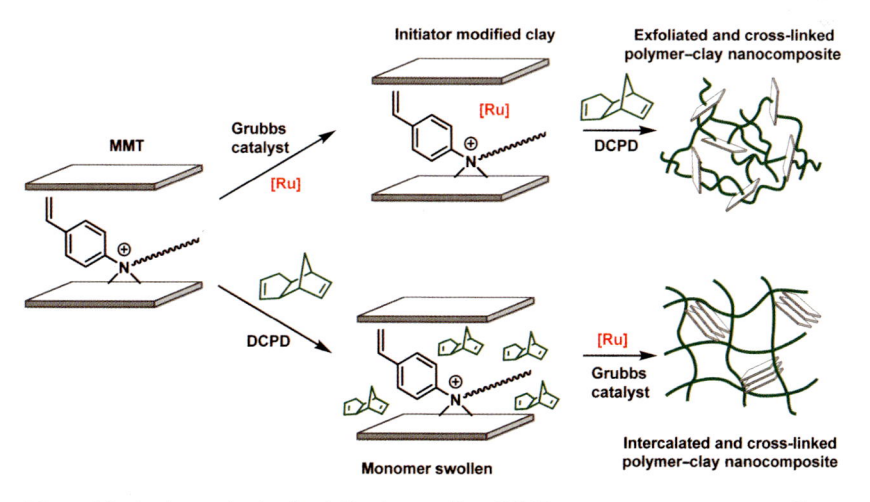

Scheme 8.8 *In situ* synthesis of polydicyclopentadiene/MMT nanocomposites via intergallery-surface-initiated (a) and bulk-initiated (b) methods. Reproduced from Ref. [44].

8.3
Thermoset Nanocomposites Based on Metal Nanoparticles

There is a growing interest on polymer/metal nanocomposites according to their multifunctionality, ease in processing, potential of a large-scale fabrication, and being lighter than metals [45, 46]. In these nanocomposites, metal nanoparticles are dispersed in the polymer matrix to combine the mechanical, electrical, and chemical properties of metals and polymers. The novel nanomaterials can be used in various areas such as catalysis, sensors, electronics, optics, medicine, and biotechnology. The most studied noble metals are gold, silver, copper, and platinum due to their stable dispersions and applicability [47]. Moreover, some of them are good choromophores where gold, silver, and copper nanoparticles show plasmonic band in the visible region [48]. There are two different types of metal nanoparticles namely, monometallic, formed from single metal elements and bimetallic, composed of alloyed or core–shell structured of two different metals [49].

Polymer/metal nanocomposites can be obtained by two different approaches, namely, *ex situ* and *in situ* techniques [50]. In the *ex situ* approach, polymerization of monomers and formation of metal nanoparticles were separately produced, and then they were mechanically mixed to form nanocomposites. In this approach, a wide size distribution of metal nanoparticles and poor dispersion in the polymer matrix were usually observed. In the *in situ* methods, metal particles are generated inside a polymer matrix by decomposition (e.g., thermolysis, photolysis, radiolysis, etc.) or by chemical reduction of a metallic precursor dissolved into the polymer. A commonly employed *in situ* method is the dispersion process, in which the solutions of the metal precursor and the protective polymer are combined, and the reduction is subsequently performed in solution [51]. The *in situ* preparation has proven to be more effective and lower cost to the performance improvements on polymers, in comparison with that of mechanical mixing described above. Similar cationic mode utilizing epoxy-based thermoset have also been reported in the literature [52]. The overall process for the *in situ* synthesis of polymer–metal nanocomposite is represented in Scheme 8.9.

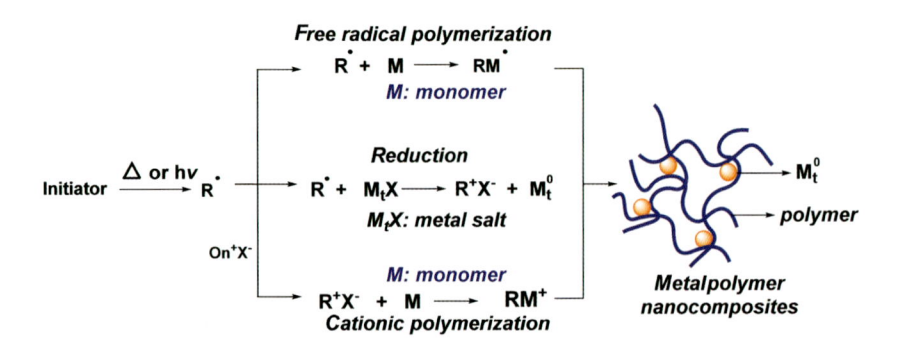

Scheme 8.9 *In situ* synthesis of polymer/metal nanocomposites by free radical and cationic polymerizations.

The main challenge in both techniques is to obtain a good dispersion of metal nanoparticles due to their easy agglomeration arising from their high surface free energy [53]. To access good dispersion, stability of nanoparticles and prevention of practical aggregation must be considered in design of nanocomposites [54]. In this part of the chapter, novel synthetic methodologies utilizing *in situ* polymerization methods for the preparation of the thermoset nanocomposites containing metal nanoparticles are discussed.

8.3.1
Polymer–Silver Nanocomposites

Silver nanocomposites have received much attention due to their wide application fields such as catalysts, wound dressings, optical information storage, and electrochromic devices [55–57]. Silver nanoparticles are usually prepared in the presence of a stabilizing agent in order to avoid their aggregation, which could lead to the loss of the unique activities associated with the nanoscale; typically, polyelectrolyte solutions can be used to control the formation and the long-term stability of nanoparticles [58].

Various strategies have been employed to prepare nanocomposites containing silver nanoparticles with controlled size. One successful approach that has been employed is the reductive processes in homogeneous matrices. However, this requires separated processes leading to low homogeneity and more effort [59–61]. Mainly, two different methods have been employed for the preparation of polymer networks containing silver nanoparticles, which involve two distinct steps; conventional polymerization for the formation of supportive and stabilizing matrix, and the reduction of silver salts for the preparation of silver nanoparticles [59, 61]. In the first method, polymer matrix is firstly synthesized, and then formerly prepared silver nanoparticles are incorporated into the polymeric material. This is called breath-in and breath-out technique [59]. In the second approach, previously formed cross-linked polymer is swollen with a solution of silver salts and reducing agent, and then reduction takes place within the polymer [61]. As stated previously, these techniques involve at least two distinct processes emerging some complications and difficulties. Although nanocomposites containing metal nanoparticles, including thermoset polymers with silver nanoparticles, have elegant features, the homogeneous dispersion of metal nanoparticles is not easy using a simple process because of their high surface-free energy, which may cause agglomeration. Therefore, the preparation of such nanocomposites with desired properties (i.e., silver nanoparticles with convenient size) is an important issue.

For example, silver nanoparticles were prepared via chemical reduction of silver nitrate ($AgNO_3$) as precursor in the presence of chitosan as a stabilizing agent and sodium borohydride ($NaBH_4$) as a reducing agent [62]. Subsequently, chitosan/silver nanocomposites are formed by cross-linking reaction with glutaraldehyde at 37 °C. The size of the microstructure of nanoparticles depends on the concentration of precursor ($AgNO_3$) and the ratio of oxidant and reducing agent ($AgNO_3$: $NaBH_4$). The color of the obtained nanocomposites was changed by $AgNO_3$

concentration from bright yellow to light red, and the absorption of intensity was increased according to the formation of more silver nanoparticles. Moreover, the chitosan/silver nanocomposites display a blue shift of the absorption bands. It is well known that metal nanoparticles display surface plasmon resonance bands in the UV–Vis region due to the excitation of electrons [63]. The antibacterial efficiency of the silver nanoparticles and the biodegradability of the chitosan matrix, these nanocomposites have practical medical applications, for example, wound dressing and tissue engineering scaffolds [64, 65].

Recently, Yagci et al. have introduced a novel approach for the preparation of metal–polymer nanocomposites, in which nanoparticle formation and UV cross-linking process were accomplished in one pot by simply irradiating appropriate formulations, obtaining the homogeneous distribution of the nanoparticles within the polymer network without any macroscopic agglomeration [7, 66–70]. The simultaneous polymerization and reduction process were formed in a single redox process. The overall strategy is represented in Scheme 8.10.

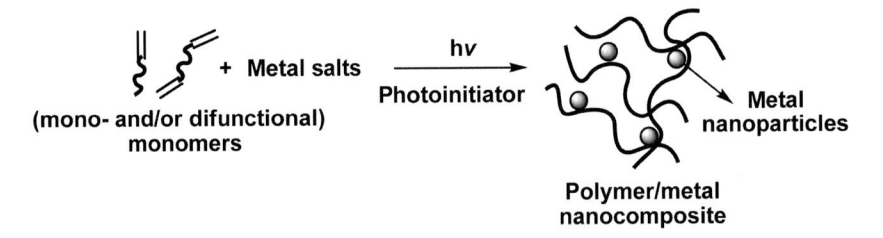

Scheme 8.10 General mechanism for the light-induced synthesis of polymer/metal nanocomposites.

The nature of the photochemically generated radicals and the redox properties of the salt are quite crucial for the success of the process. Appropriate combinations of electron donor radicals with certain metal salts induce metal formation and chain propagation concomitantly. For example, photoinduced cleavage of 2,2-dimethoxy-2-phenyl acetophenone (DMPA) is a good source of electron donor radicals, namely alkoxy benzyl radicals [70]. Thus, irradiation of DMPA in the system in the presence of $AgPF_6$ leads to its reduction with rapid generation of both metallic silver and initiating cations without any undesirable side reactions (Scheme 8.11).

TEM analysis of the cured films confirmed the formation of silver particles in nanometer range size. TEM micrographs show the well dispersion of silver nanoparticles without any macroscopic agglomerations (Figure 8.4).

The process is not limited to the cationically polymerizing epoxy resins. Acrylates and methacrylates that polymerize by a free radical mechanism can also form nanocomposites by the described *in situ* method. In this case, while the silver nanoparticles are formed in a similar manner, the free radical chain growth occurs from the benzoyl radicals derived from DMPA. This method is applicable to any

Scheme 8.11 Light-induced synthesis of silver–epoxy nanocomposites. Reproduced from Ref. [70].

other metal salts, which undergo similar redox reactions with photochemically generated electron donor radicals. Table 8.1 summarizes epoxy- and acrylate-based cross-linked networks containing silver nanoparticles prepared by light-induced processes.

For the potential use of such nanocomposites as coating materials, Yagci et al. prepared oil-based nanocomposites containing silver nanoparticles. The composites showed good adhesion, flexibility, water, acid, and alkali resistances. For further extension of their use as antimicrobial coatings, antibacterial properties were also studied. The composite film sample had good antibacterial effect against Gram-positive (*Staphylococcus aureus*), Gram-negative (*Pseudomonas aeruginosa*), and spore-forming (*Bacillus subtilis*) bacteria. All kinds of bacteria were killed and an inhibition zones were formed due to the antibacterial effect of the silver in the surrounding of the film samples [67, 68]. On the other hand, without silver no inhibition zone was detected.

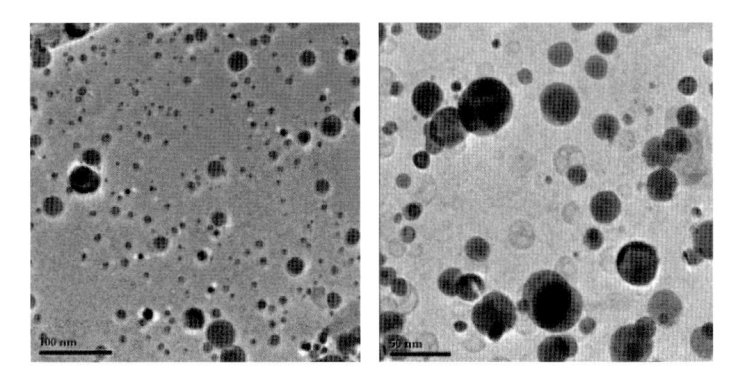

Figure 8.4 TEM micrographs of epoxy/silver nanocomposites containing 5 wt% Ag^0.

Table 8.1 Photoinduced synthesis of metal/cross-linked polymer nanocomposites by simultaneous electron transfer and polymerization processes.

Polymer matrix	Photoinitiator	Polymerization mechanism	Reference
Acrylate	DMPA-(AgNO$_3$)	Radical	[7]
Acrylate	2,7-diaminofluorene/amine-(AgNO$_3$)	Radical	[71]
Acrylate	Eosin dye/amine-(AgNO$_3$)	Radical	[72, 73]
Acrylamide	Irgacure 2959-(AgNO$_3$)	Radical	[67]
Epoxy	DMPA-(AgSbF$_6$)	Cationic	[70]
Silicon epoxy	CQ/amine-(AgSbF$_6$)	Cationic	[69]

The cross-linked matrix was also obtained by gamma irradiation instead of UV light while *in situ* reduction of Ag$^+$ ion takes place. 2-propanol was used as a scavenger to protect poly(N-vinyl-2-pyrrolidone) macromolecules against gamma irradiation-induced macro radical formation [74]. Fast reduction of Ag$^+$ ions was observed during cross-linking process that resulted in small spherical-shaped nanoparticles. The size of silver nanoparticles was determined by TEM analysis and found between 10 and 60 nm. Furthermore, these nanocomposites showed high elasticity caused by the high gel percentage, which was found as 94%.

Another one pot synthesis of silver/polymer nanocomposites is thiol-ene chemistry, described by Colucci *et al.* [75]. The mechanism of UV-induced thiol-ene polymerization consists two main steps, that is, the addition of a thiyl radical to double bond followed by a chain-transfer reaction [76, 77]. The thiolene mixture consists of trimethylolpropane tris(mercaptopropanoate), allylpentaerythritol, silver hexafluoroantimonate, and benzophenone as photoiniator. Dynamic-mechanical analysis of cured films analysis indicated the slight increase of T_g probably due to the interaction between the thiol-ene matrix and the metal nanoparticles. Antimicrobial tests of UV cured thiol-ene hybrid films were carried out by the modified Kirby–Bauer technique. According to the test results silver nanocomposites embedded in the thiol-ene films show a clear antimicrobial activity.

In a recent study, rubber fibers containing silver nanoparticles with high morphological stability were prepared by combination of electrospinning and *in situ* chemical cross-linking [78]. For this purpose, silver nanoparticles were first generated through reducing the Ag$^+$ ions in rubber (butadiene, silicon, or isobutylene–isoprene rubber)/N,N-dimethyformamide/tetrahydrofuran solutions. The resulting mixture was directly used for *in situ* chemical cross-linking combined with electrospinning to obtain rubber composite fibers with uniform diameters from hundreds of nanometers to several micrometers. The *in situ* chemical cross-linking was conducted during and shortly after electrospinning; camphorquinone was used as the photoinitiator for butadiene and isobutylene-isoprene rubber, whereas a curing agent containing the catalyst of platinum was used for silicon rubber. The results from UV and optical spectroscopies, and TEM revealed that

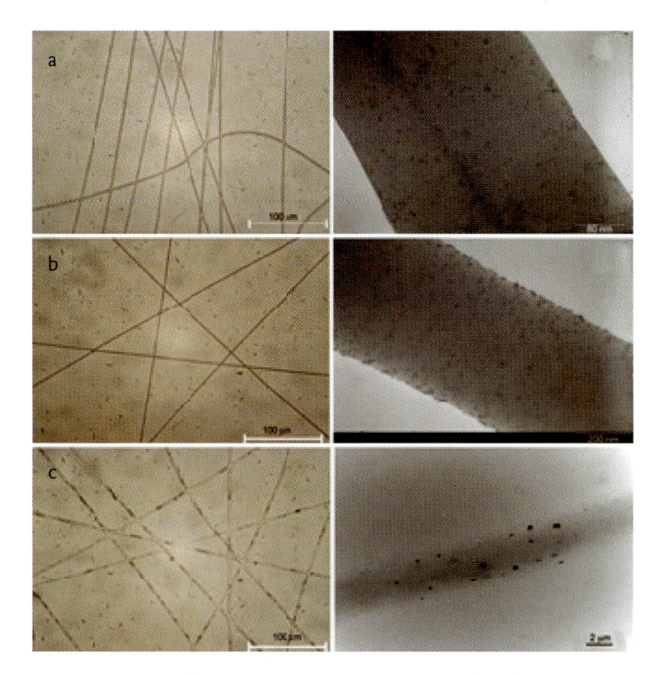

Figure 8.5 Optical microscope and TEM images of rubber/silver nanocomposite fibers including butadiene (a), isobutylene–isoprene (b), and silicon (c) rubbers. Reproduced from Ref. [78].

silver nanoparticles with the average size of 10–20 nm were dispersed uniformly in the rubber fibers (Figure 8.5).

To determine the antimicrobial activity, the obtained composites were tested against *Escherichia coli* according to the nonwoven fabric attachment method, and it was shown that a small amount of the fibers introduce very strong antimicrobial activity.

8.3.2
Polymer–Gold Nanocomposites

Polymer/gold nanocomposites are considered to have great potential in emerging technologies such as drug delivery, cancer therapy, catalysis, chemical and biosensing, and microelectronics devices due to their chemical stability, nontoxicity, optical properties, and biocompatibility [79]. Polymer/gold nanocomposites have been prepared under a variety of conditions, including citrate reduction [80], thiol stabilization [81], and several thermal and photochemical techniques using thiols [82], amines [83], micelles [84], dendrimers [85], and polymers [86] as protective agents to aid in their stabilization in aqueous or organic media.

For example, a synthetic method to obtain thermoset/gold nanocomposites involves preparing nanoparticles separately and then combining the components physically or chemically. In this case, gold metal precursor is dissolved in the polymer solution and the ions are subsequently reduced to nanoparticles chemically, sono-, or photochemically (by using X or γ-rays, UV, or microwaves) and finally polymer solution is cross-linked by externally stimulus. In another example, well-dispersed gold nanoparticle/poly(N-isopropylacrylamide) hydrogel nanocomposite is prepared by the copolymerization of vinyl-functionalized gold nanoparticles with N-isopropylacrylamide. In this way, the electrical conductivity of the nanocomposite is adjusted by controlling the interparticle distance under temperature stimuli [87].

A second approach consists of the utilization of polymer hydrogels as reactors for the synthesis and stabilization of gold nanoparticles [88]. Colloidal gold nanoparticles have been embedded in cross-linked polymers by suitable ligands that have affinity to both Au^{3+} ions and colloidal gold nanoparticles into the polymer matrix. The functionalized polymer matrix then modulates the formation of colloidal gold nanoparticles after the addition of sodium borohydride as reductant [89, 90].

In a recent study, thermoresponsive hydrogels containing gold nanoparticles were prepared by 1,3-dipolar cycloaddition click reaction and *in situ* reversible addition-fragmentation chain-transfer polymerization (Scheme 8.12). First, gold nanoparticles decorated with azide groups were prepared through ligand exchange. Then, click reactions between azido-gold nanoparticles and dialkynetrithiocarbonate yielded cross-linked gold nanoparticle aggregates. The size and cross-linking

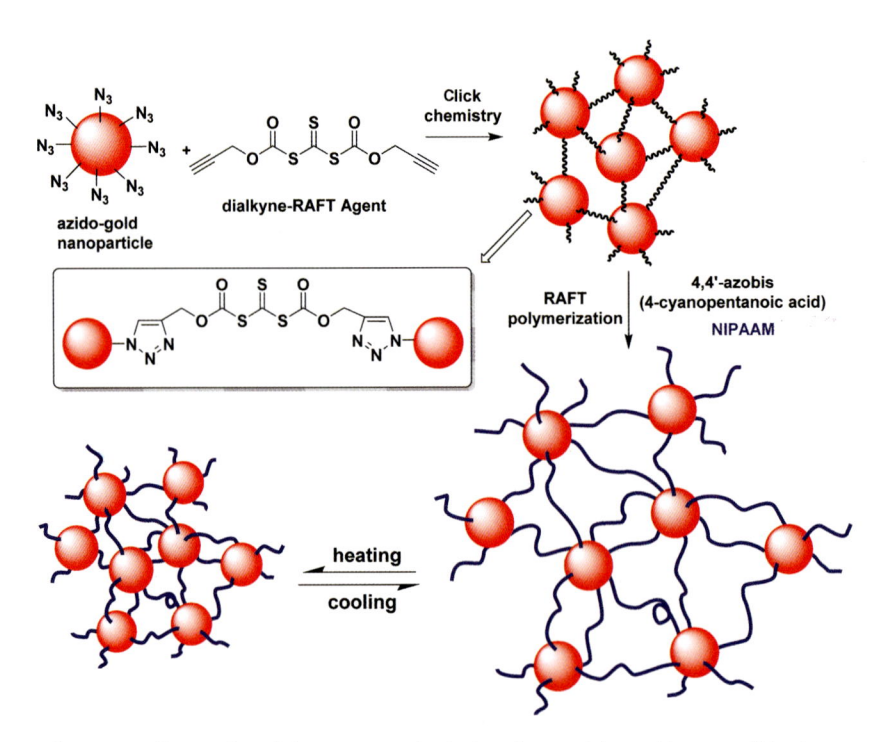

Scheme 8.12 Preparation of thermoresponsive hydrogels containing gold nanoparticles by combination of 1,3-dipolar cycloaddition click reaction and RAFT polymerization.

density of these aggregates were adjusted by changing the molar ratio of acetylene groups to azide groups. The hydrogels containing gold nanoparticles were prepared by *in situ* RAFT polymerization of N-isopropylacrylamide using trithiocarbonate in the cross-linked gold nanoparticle aggregates as chain-transfer agents [91].

Although, the preparation of nanoparticles by photochemical means is direct excitation of metal ions and complexes by hard UV photons below 300 nm [92–94] has many potential applications, it has several serious drawbacks that limit their use. For example, some transparency to high-energy radiation and stability against extensive photochemical degradation is often difficult to achieve [95]. To circumvent existing problems associated with the absorption characteristics, metal complexes that absorb light at wavelengths above 300 nm were used. However, there exists only limited number of metallic species absorbing at long wavelengths and undergoing efficient photoreduction. Several other procedures involving photoreactions of aromatic carbonyl compounds to initiate the formation of nanoparticles have been developed [96]. Photolysis of aromatic ketones leads to the formation of electron donor radicals capable of reducing gold complexes to nanoparticles. Both cleavage (Type I) and H-abstraction type (Type II) photoinitiators were successfully used for the photochemical generation of reducing radicals [97, 98].

Quite recently, Yagci and coworkers proposed [99–101] an elegant *in situ* synthesis of gold nanoparticles achieved by simultaneous photoinduced electron transfer and free-radical polymerization of an acrylic formulation. These results established a novel approach for the preparation of gold nanoparticles within acrylic matrix and demonstrated that the nanoparticles are homogenously distributed in the network without macroscopic agglomeration. The approach was also successfully applied to epoxy-based systems [102] as well production of biosensor [100]. *In situ* synthesis of poly(ethylene glycol) hydrogels containing gold nanoparticles and glucose oxidase enzyme was performed by taking advantage of two reactions, namely photoinduced electron transfer and polymerization processes applied in electrochemical glucose biosensing as the model system (Scheme 8.13).

The bionanocomposite matrix by simple one-step fabrication offered a good contact between the active site of the enzyme and gold nanoparticles inside the network that caused the promotion in the electron transfer properties that was evidenced by cyclic voltammetry as well as higher amperometric biosensing responses in comparing with response signals obtained from the matrix without gold nanoparticles. The particle size was found to be between 2 and 5 nm in the absence of biomolecule, whereas bigger sizes in the range of 4–20 nm were attained in the presence of the biomolecule due to the nanoparticle agglomeration. The resulting nanocomposites were applied for glucose analysis for some beverages and obtained data were compared with HPLC as the reference method to test the possible matrix effect due to the nature of the samples. The results showed that the use of the nanocomposite as biosensor provided very similar results with the HPLC data, which indicated the analysis without any sample matrix effect. The combination of small-sized gold nanoparticles with the redox enzyme inside the polymeric material yielded a nanobiocomposite structure with unique and versatile properties and can be a useful candidate as biocatalytic nanoscale-device or

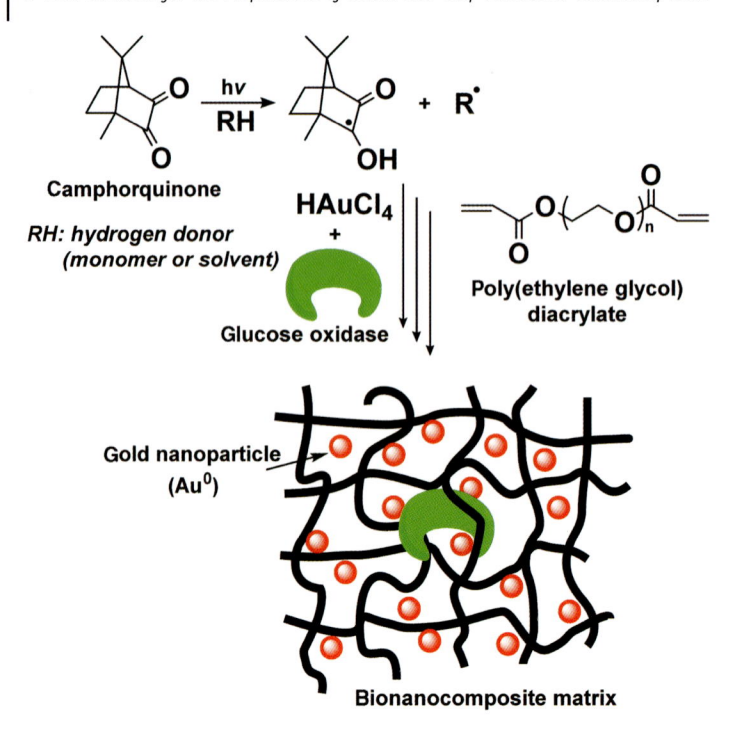

Scheme 8.13 Synthesis of poly(ethylene glycol)/gold nanocomposite containing glucose oxidase by simultaneous polymerization and metal ion reduction processes induced by light.

nanorectors based on glucose oxidation reaction in various biotechnological applications such as biofuel cell design as well as biological analysis.

The *in situ* preparation of polysulfone-based thermoset nanocomposites containing metal nanoparticles (silver and gold) was also achieved by simultaneous UV-induced radical polymerization and nanoparticle formation (Figure 8.6). In the presence of a suitable radical photoinitiator, the UV-generated radicals started the polyaddition reaction of the acrylic functionalized PSU oligomer, and at the same time, they were capable of reducing Ag^+ to Ag^0 and Au^{3+} to Au^0. Thus, silver or gold NPs were formed *in situ* during the polymer network formation. The morphology of the cured systems, investigated by FESEM analysis on the surface fracture of the films, demonstrated that the nanoparticles of the noble metals are homogeneously distributed in the network without macroscopic agglomeration [103].

8.4
Concluding Remarks

To conclude, it is clear that over the past two decades much attention has been devoted to the preparation of metal/clay thermoset nanocomposites. The develop-

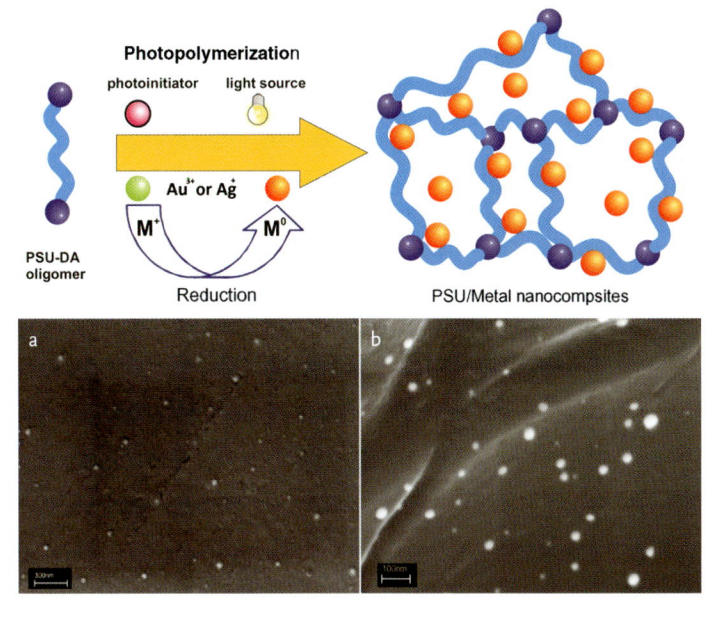

Figure 8.6 Photoinduced synthesis of polysulfone/metal/nanocomposites and TEM images of the obtained nanocomposites containing 5 wt% of (a) $AgSbF_6$ and (b) $HAuCl_4$.

ment of *in situ* methods has clearly facilitated the major advances in their synthesis in a one-pot manner combining polymerization processes leading to the network formation with intercalation/exfoliation or nanoparticle formation, respectively. This way, the existing final properties of thermosets can further be improved to expand their use as advanced materials. In this chapter, an overview has been provided for the synthesis of such materials.

References

1 Ray, S.S. and Okamoto, M. (2003) *Prog. Polym. Sci.*, **28**, 1539–1641.
2 Pavlidou, S. and Papaspyrides, C.D. (2008) *Prog. Polym. Sci.*, **33**, 1119–1198.
3 Tasdelen, M.A., Kreutzer, J., and Yagci, Y. (2010) *Macromol. Chem. Phys.*, **211**, 279–285.
4 Akat, H., Tasdelen, M.A., Du Prez, F., and Yagci, Y. (2008) *Eur. Polym. J.*, **44**, 1949–1954.
5 Oral, A., Tasdelen, M.A., Demirel, A.L., and Yagci, Y. (2009) *J. Polym. Sci. Part A Polym. Chem.*, **47**, 5328–5335.
6 Yenice, Z., Tasdelen, M.A., Oral, A., Guler, C., and Yagci, Y. (2009) *J. Polym. Sci. Part A Polym. Chem.*, **47**, 2190–2197.
7 Eksik, O., Tasdelen, M.A., Erciyes, A.T., and Yagci, Y. (2010) *Compos. Interfaces*, **17**, 357–369.
8 Altinkok, C., Uyar, T., Tasdelen, M.A., and Yagci, Y. (2011) *J. Polym. Sci. Part A Polym. Chem.*, **49**, 3658–3663.
9 Demir, K.D., Tasdelen, M.A., Uyar, T., Kawaguchi, A.W., Sudo, A., Endo, T., and Yagci, Y. (2011) *J. Polym. Sci. Part A Polym. Chem.*, **49**, 4213–4220.

10. Dizman, C., Ates, S., Uyar, T., Tasdelen, M.A., Torun, L., and Yagci, Y. (2011) *Macromol. Mater. Eng.*, **296**, 1101–1106.
11. Nese, A., Sen, S., Tasdelen, M.A., Nugay, N., and Yagci, Y. (2006) *Macromol. Chem. Phys.*, **207**, 820–826.
12. Massam, J. and Pinnavaia, T.J. (1998) *Mater. Res. Soc. Symp. Proc.*, **520**, 223–232.
13. Messersmith, P.B. and Giannelis, E.P. (1994) Novel Techniques in Synthesis and Processing of Advanced Materials, 497–506.
14. Pinnavaia, T.J., Lan, T., Kaviratna, P.D., and Wang, M.S. (1994) Better Ceramics through Chemistry VI, **346**, 81–88.
15. Pinnavaia, T.J., Lan, T., Wang, Z., Shi, H.Z., and Kaviratna, P.D. (1996) *ACS Sym. Ser.*, **622**, 250–261.
16. Wang, Z. and Pinnavaia, T.J. (1998) *Chem. Mater.*, **10**, 1820–1826.
17. Wang, M.S. and Pinnavaia, T.J. (1994) *Chem. Mater.*, **6**, 468–474.
18. Lan, T., Kaviratna, P.D., and Pinnavaia, T.J. (1995) *Chem. Mater.*, **7**, 2144–2150.
19. Lan, T. and Pinnavaia, T.J. (1994) *Chem. Mater.*, **6**, 2216–2219.
20. Lu, J., Hong, C.K., and Wool, R.P. (2004) *J. Polym. Sci. Part B Polym. Phys.*, **42**, 1441–1450.
21. Uyama, H., Kuwabara, M., Tsujimoto, T., Nakano, M., Usuki, A., and Kobayashi, S. (2004) *Macromol. Biosci.*, **4**, 354–360.
22. Uyama, H., Kuwabara, M., Tsujimoto, T., Nakano, M., Usuki, A., and Kobayashi, S. (2003) *Chem. Mater.*, **15**, 2492–2494.
23. Miyagawa, H., Misra, M., Drzal, L.T., and Mohanty, A.K. (2005) *Polymer*, **46**, 445–453.
24. Bongiovanni, R., Malucelli, G., Sangermano, M., and Priola, A. (2002) *Macromol. Symp.*, **187**, 481–492.
25. Liu, Y.-L., Chen, Y.-J., and Wei, W.-L. (2003) *Polymer*, **44**, 6465–6473.
26. Liu, Y.-L., Wei, W.-L., Chen, Y.-J., Wu, C.-S., and Tsai, M.-H. (2004) *Polym. Degrad. Stab.*, **86**, 135–145.
27. Liu, Y.-L., Wang, Y.-H., and Chen, H.-S. (2005) *Macromol. Chem. Phys.*, **206**, 600–606.
28. Wu, Z., Zhou, C., and Qi, R. (2002) *Polym. Compos.*, **23**, 634–646.
29. Wang, H., Zhao, T., Yan, Y., and Yu, Y. (2004) *J. Appl. Polym. Sci.*, **92**, 791–797.
30. Jiang, W., Chen, S.-H., and Chen, Y. (2006) *J. Appl. Polym. Sci.*, **102**, 5336–5343.
31. Byun, H.Y., Choi, M.H., and Chung, I.J. (2001) *Chem. Mater.*, **13**, 4221–4226.
32. Agag, T. and Takeichi, T. (2000) *Polymer*, **41**, 7083–7090.
33. Okada, A., Kawasumi, M., Usuki, A., Kojima, Y., Kurauchi, T., and Kamigaito, O. (1990) *Polym. Based Mol. Compos.*, **171**, 45–50.
34. Fukushima, Y., Okada, A., Kawasumi, M., Kurauchi, T., and Kamigaito, O. (1988) *Clay Miner.*, **23**, 27–34.
35. Ni, P., Li, J., Suo, J., and Li, S. (2004) *J. Appl. Polym. Sci.*, **94**, 534–541.
36. Ni, P., Wang, Q., Li, J., Suo, J., and Li, S. (2006) *J. Appl. Polym. Sci.*, **99**, 6–13.
37. Woods, G. (1990) *The ICI Polyurethanes Book*, Published jointly by ICI Polyurethanes and Wiley.
38. Joulazadeh, M. and Navarchian, A.H. (2011) *Polym. Advan. Technol.*, **22**, 2022–2031.
39. Deka, H. and Karak, N. (2009) *Nanoscale Res. Lett.*, **4**, 758–765.
40. Bharadwaj, R.K., Mehrabi, A.R., Hamilton, C., Trujillo, C., Murga, M., Fan, R., Chavira, A., and Thompson, A.K. (2002) *Polymer*, **43**, 3699–3705.
41. Suh, D.J. and Park, O.O. (2002) *J. Appl. Polym. Sci.*, **83**, 2143–2147.
42. Suh, D.J., Lim, Y.T., and Park, O.O. (2000) *Polymer*, **41**, 8557–8563.
43. Kasgoz, H. and Durmus, A. (2008) *Polym. Advan. Technol.*, **19**, 838–845.
44. Simons, R., Guntari, S.N., Goh, T.K., Qiao, G.G., and Bateman, S.A. (2012) *J. Polym. Sci. Part A Polym. Chem.*, **50**, 89–97.
45. Caseri, W. (2000) *Macromol. Rapid Commun.*, **21**, 705–722.
46. Avasthi, D.K., Mishra, Y.K., Kabiraj, D., Lalla, N.P., and Pivin, J.C. (2007) *Nanotechnology*, **18**, No: 125604.
47. Sharma, V.K., Yngard, R.A., and Lin, Y. (2009) *Adv. Colloid Interface Sci.*, **145**, 83–96.
48. Kelly, K.L., Coronado, E., Zhao, L.L., and Schatz, G.C. (2003) *J. Phys. Chem. B*, **107**, 668–677.

49 Toshima, N. and Yonezawa, T. (1998) *New J. Chem.*, **22**, 1179–1201.

50 Carotenuto, G., Nicolais, L., Martorana, B., and Perlo, P. (2005) in *Metal–Polymer Nanocomposites* (eds L. Nicolais and G. Carotenuto), John Wiley & Sons, Inc., New York, pp. 155–181.

51 Pomogailo, A.D. and Kestelman, V.N. (2005) in *Metallopolymer Nanocomposites* (eds A.D. Pomogailo and V.N. Kestelman), Springer, Berlin Heidelberg, pp. 135–236.

52 Yagci, Y.J. (2012) *J. Coat. Technol. Res.*, **9**, 125–134.

53 Lee, J.Y., Liao, Y.G., Nagahata, R., and Horiuchi, S. (2006) *Polymer*, **47**, 7970–7979.

54 Lee, W.F. and Tsao, K.T. (2010) *J. Mater. Sci.*, **45**, 89–97.

55 Namboothiry, M.A.G., Zimmerman, T., Coldren, F.M., Liu, J.W., Kim, K., and Carroll, D.L. (2007) *Synth. Met.*, **157**, 580–584.

56 Faupel, F., Zaporojtchenko, V., Strunskus, T., and Elbahri, M. (2010) *Adv. Eng. Mater.*, **12**, 1177–1190.

57 Thomas, V., Namdeo, M., Mohan, Y.M., Bajpai, S.K., and Bajpai, M. (2008) *J. Macromol. Sci. Part A Pure Appl. Chem.*, **45**, 107–119.

58 Travan, A., Marsich, E., Donati, I., Benincasa, M., Giazzon, M., Felisari, L., and Paoletti, S. (2011) *Acta Biomater.*, **7**, 337–346.

59 Thomas, V., Yallapu, M.M., Sreedhar, B., and Bajpai, S.K. (2009) *J. Appl. Polym. Sci.*, **111**, 934–944.

60 Lu, Y., Spyra, P., Mei, Y., Ballauff, M., and Pich, A. (2007) *Macromol. Chem. Phys.*, **208**, 254–261.

61 Murali Mohan, Y., Lee, K., Premkumar, T., and Geckeler, K.E. (2007) *Polymer*, **48**, 158–164.

62 An, J., Luo, Q.Z., Yuan, X.Y., Wang, D.S., and Li, X.Y. (2011) *J. Appl. Polym. Sci.*, **120**, 3180–3189.

63 Li, T., Park, H.G., and Choi, S.H. (2007) *Mater. Chem. Phys.*, **105**, 325–330.

64 Rao, K., Reddy, P.R., Lee, Y.I., and Kim, C. (2012) *Carbohydr. Polym.*, **87**, 920–925.

65 Liu, B.S. and Huang, T.B. (2008) *Macromol. Biosci.*, **8**, 932–941.

66 Yagci, Y., Sahin, O., Ozturk, T., Marchi, S., Grassini, S., and Sangermano, M. (2011) *React. Funct. Polym.*, **71**, 857–862.

67 Uygun, M., Kahveci, M.U., Odaci, D., Timur, S., and Yagci, Y. (2009) *Macromol. Chem. Phys.*, **210**, 1867–1875. doi: 10.1002/macp.200900296

68 Eksik, O., Erciyes, A.T., and Yagci, Y. (2008) *J. Macromol. Sci. Part A Pure Appl. Chem.*, **45**, 698–704.

69 Yagci, Y., Sangermano, M., and Rizza, G. (2008) *Polymer*, **49**, 5195–5198.

70 Sangermano, M., Yagci, Y., and Rizza, G. (2007) *Macromolecules*, **40**, 8827–8829.

71 Balan, L., Jin, M., Malval, J.-P., Chaumeil, H.L.N., Defoin, A., and Vidal, L.C. (2008) *Macromolecules*, **41**, 9359–9365.

72 Balan, L., Malval, J.-P., Schneider, R., Le Nouen, D., and Lougnot, D.-J. (2010) *Polymer*, **51**, 1363–1369.

73 Balan, L., Turck, C., Soppera, O., Vidal, L.C., and Lougnot, D.J. (2009) *Chem. Mater.*, **21**, 5711–5718.

74 Jovanovic, Z., Krkljes, A., Stojkovska, J., Tomic, S., Obradovic, B., Miskovic-Stankovic, V., and Kacarevic-Popovic, Z. (2011) *Radiat. Phys. Chem.*, **80**, 1208–1215.

75 Colucci, G., Celasco, E., Mollea, C., Bosco, F., Conzatti, L., and Sangermano, M. (2011) *Macromol. Mater. Eng.*, **296**, 921–928.

76 Cramer, N.B. and Bowman, C.N. (2001) *J. Polym. Sci. Part A Polym. Chem.*, **39**, 3311–3319.

77 Hoyle, C.E., Lee, T.Y., and Roper, T. (2004) *J. Polym. Sci. Part A Polym. Chem.*, **42**, 5301–5338.

78 Hu, Q., Wu, H., Zhang, L., Fong, H., and Tian, M. (2012) *Express Polym. Lett.*, **6**, 258–265.

79 Uehara, N. (2010) *Anal. Sci.*, **26**, 1219–1228.

80 Brust, M., Walker, M., Bethell, D., Schiffrin, D.J., and Whyman, R. (1994) *Chem. Commun.*, 801–802.

81 Jana, N.R., Gearheart, L., and Murphy, C.J. (2001) *J. Phys. Chem. B*, **105**, 4065–4067.

82 Thomas, K.G. and Kamat, P.V. (2003) *Acc. Chem. Res.*, **36**, 888–898.

83 Gandubert, V.J. and Lennox, R.B. (2005) *Langmuir*, **21**, 6532–6539.

84 Mandal, M., Ghosh, S.K., Kundu, S., Esumi, K., and Pal, T. (2002) *Langmuir*, **18**, 7792–7797.

85 Esumi, K., Matsumoto, T., Seto, Y., and Yoshimura, T. (2005) *J. Colloid Interface Sci.*, **284**, 199–203.

86 Sakamoto, M., Tachikawa, T., Fujitsuka, M., and Majima, T. (2006) *Chem. Phys. Lett.*, **420**, 90–94.

87 Zhao, X.L., Ding, X.B., Deng, Z.H., Zheng, Z.H., Peng, Y.X., and Long, X.P. (2005) *Macromol. Rapid Commun.*, **26**, 1784–1787.

88 Tarnavchyk, I., Voronov, A., Kohut, A., Nosova, N., Varvarenko, S., Samaryk, V., and Voronov, S. (2009) *Macromol. Rapid Commun.*, **30**, 1564–1569.

89 Wang, C., Flynn, N.T., and Langer, R. (2004) *Adv. Mater.*, **16**, 1074.

90 Jiang, X.W., Xiong, D.A., An, Y.L., Zheng, P.W., Zhang, W.Q., and Shi, L.Q. (2007) *J. Polym. Sci. Part A Polym. Chem.*, **45**, 2812–2819.

91 Lian, X.M., Jin, J., Tian, J., and Zhao, H.Y. (2010) *ACS Appl. Mater. Interfaces*, **2**, 2261–2268.

92 Han, M.Y. and Quek, C.H. (2000) *Langmuir*, **16**, 362–367.

93 Dong, S.A., and Zhou, S.P. (2007) *Mater. Sci. Eng. B*, **140**, 153–159.

94 Eustis, S., Hsu, H.Y., and El-Sayed, M.A. (2005) *J. Phys. Chem. B*, **109**, 4811–4815.

95 Korchev, A.S., Shulyak, T.S., Slaten, B.L., Gale, W.F., and Mills, G. (2005) *J. Phys. Chem. B*, **109**, 7733–7745.

96 Nishioka, K., Niidome, Y., and Yamada, S. (2007) *Langmuir*, **23**, 10353–10356.

97 McGilvray, K.L., Decan, M.R., Wang, D.S., and Scaiano, J.C. (2006) *J. Am. Chem. Soc.*, **128**, 15980–15981.

98 Marin, M.L., McGilvray, K.L., and Scaiano, J.C. (2008) *J. Am. Chem. Soc.*, **130**, 16572–16584.

99 Amici, J., Sangermano, M., Celasco, E., and Yagci, Y. (2011) *Eur. Polym. J.*, **47**, 1250–1255.

100 Odaci, D., Kahveci, M.U., Sahkulubey, E.L., Ozdemir, C., Uyar, T., Timur, S., and Yagci, Y. (2010) *Bioelectrochemistry*, **79**, 211–217.

101 Yagci, Y., Sangermano, M., and Rizza, G. (2008) *Chem. Commun.*, 2771–2773.

102 Yagci, Y., Sangermano, M., and Rizza, G. (2008) *Macromolecules*, **41**, 7268–7270.

103 Sangermano, M., Roppolo, I., Camara, V.H.A., Dizman, C., Ates, S., Torun, L., and Yagci, Y. (2011) *Macromol. Mater. Eng.*, **296**, 820–825.

9
Bio-Based Epoxy Resin/Clay Nanocomposites
Mitsuhiro Shibata

9.1
Introduction

Epoxy resin is one of the most popular thermosetting resins, which is used in combination with epoxy hardener in various fields: coatings, adhesives, electrical uses such as printed circuit board and encapsulate for semiconductor, composite materials for sports/leisure equipment and aircraft, and so on, because the cross-linked polymers of epoxy resins and hardeners have superior adhesiveness toward various substrates, excellent chemical resistance and electrical insulating properties, and moderate heat resistance and mechanical properties [1, 2].

Figure 9.1 shows the petroleum-based epoxy resins which are widely used in various applications. Diglycidyl ether of bisphenol A (DGEBA) is the most common epoxy resin prepared from bisphenol A and epichlorohydrin (ECH). Polyglycidyl ether of o-cresol novolac (EOCN) and N,N,N′,N′-tetraglycidyl-4,4′-diaminodiphenylmethane (TGDDM) are also produced by the reactions of ECH with o-cresol novolac (OCN) and 4,4′-diaminodiphenylmethane (DDM), respectively. EOCN is mainly used for insulating sealants for semiconductor. TGDDM is used for carbon fiber-reinforced plastics. Tetrahydrophthalic acid diglycidyl ester (THPA-DGE) is prepared from THPA and ECH, which is used for weatherproof electric insulation materials by combining with acid anhydride hardeners. An alicyclic epoxy resin, 3,4-epoxycyclohexylmethyl-3′,4′-epoxycyclohexane carboxylate (ECHM-ECHC) is prepared from the corresponding biscyclohexenyl compound and peracetic acid, which is used for weatherproof electric insulation materials and UV-curing coatings.

Polyamines (diethylenetriamine [DETA], triethylenetetramine [TETA], etc.), polyphenols (phenol novolac [PN], dicyclopentadiene phenol novolac, etc.), and acid anhydrides (hexahydrophthalic anhydride, methyltetrahydrophthalic anhydride [MTHPA], etc.) are used as hardeners of epoxy resins (Figure 9.2). 2-Ethyl-4-methylimidazole and triphenylphosphine are typical curing catalyst that accelerates the curing reaction of epoxy resin and hardener. Dicyandiamide (DICY) is latent curing agent that is normally used in the form of fine powder dispersed in the epoxy resin. Although the epoxy resin containing DICY has a long pot life

Thermoset Nanocomposites, First Edition. Edited by Vikas Mittal.
© 2013 Wiley-VCH Verlag GmbH & Co. KGaA. Published 2013 by Wiley-VCH Verlag GmbH & Co. KGaA.

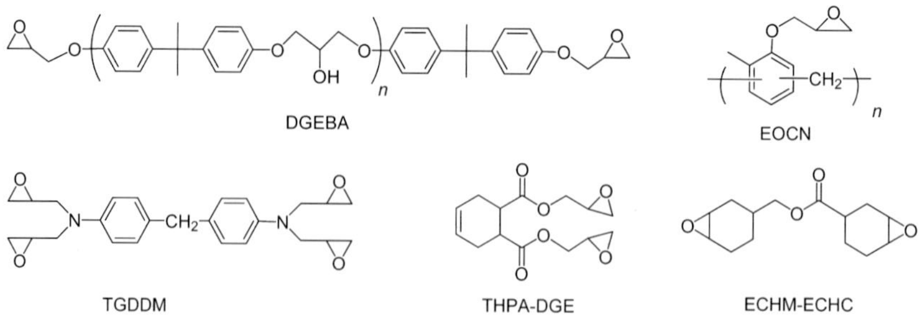

9 Bio-Based Epoxy Resin/Clay Nanocomposites

Figure 9.1 Conventional petroleum-based epoxy resins.

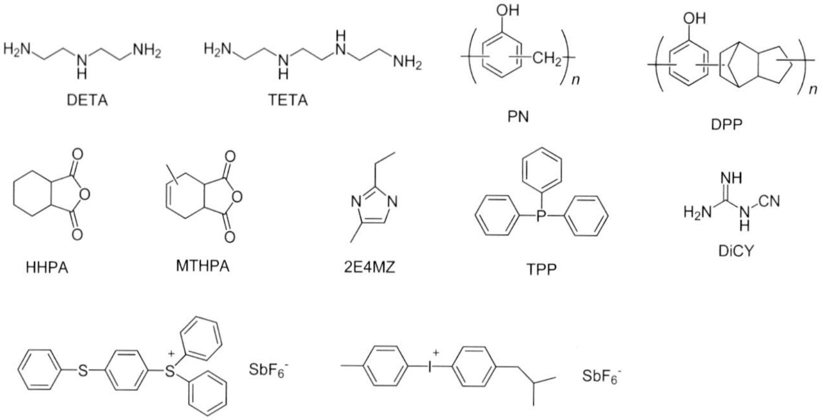

Figure 9.2 Conventional petroleum-based epoxy hardeners and catalysts.

at room temperature, the curing reaction occurs at 160–180 °C. DICY is commonly used together with basic curing catalyst such as imidazole and tertiary amine. Aromatic sulfonium and iodonium ions are mainly used as a UV-curing catalyst for cationic homopolymerization of epoxy resins.

In recent years, renewable resources-derived polymers (bio-based polymers) are attracting a great deal of attention because of the advantages of these polymers such as conservation of limited petroleum resources, possible biodegradability, and the control of carbon dioxide emissions that lead to global warming [3–7]. Most recently, much focus is being placed on bio-based thermosetting resins such as epoxy resins, phenol resins, unsaturated polyester resins and their composites, because their materials are hard to be recycled due to the infusible and insoluble properties [8–12].

In the next section, after the studies on the bio-based epoxy resins and the bio-based epoxy hardeners are reviewed, our previous results on the properties of the cured materials of bio-based epoxy resin/bio-based hardener are summarized. In order to improve the thermal and mechanical properties of the bio-based epoxy resins cured with bio-based hardener without spoiling the environmental friendliness, the reinforcement with naturally occurring layered silicates such as montmorillonite (MMT) and hectorite should be one of the best methods because small amount of residual clay puts back to nature after the biodegradation [13–16]. In the subsequent section, the preparation and properties of the bio-based epoxy resin/hardener/layered silicate nanocomposites are described based on the references that our group and other groups previously reported.

9.2
Bio-Based Epoxy Resins and Hardeners

9.2.1
Bio-Based Epoxy Resins

Bio-based epoxy resins such as glycerol polyglycidyl ether (GPE), polyglycerol polyglycidyl ether (PGPE), sorbitol polyglycidyl ether (SPE), epoxidized soybean oil (ESO), epoxidized linseed oil (ELO), diglycidyl ester of dimer aid (DGEDA), and limonene diepoxide (LMDE) are industrially available in large volumes at a reasonable cost (Figure 9.3). Glycerol and sorbitol are abundant and inexpensive bio-based aliphatic polyols, which can be derived from triglyceride vegetable oil and corn starch, respectively. Recently, glycerol is gathering much attention as a 10% by-product of biodiesel production *via* the transesterification of vegetable oils [17]. Glycerol-based GPE and PGPE are derived from glycerol and ECH, and have been used in textile and paper-processing agents and reactive diluents, etc.

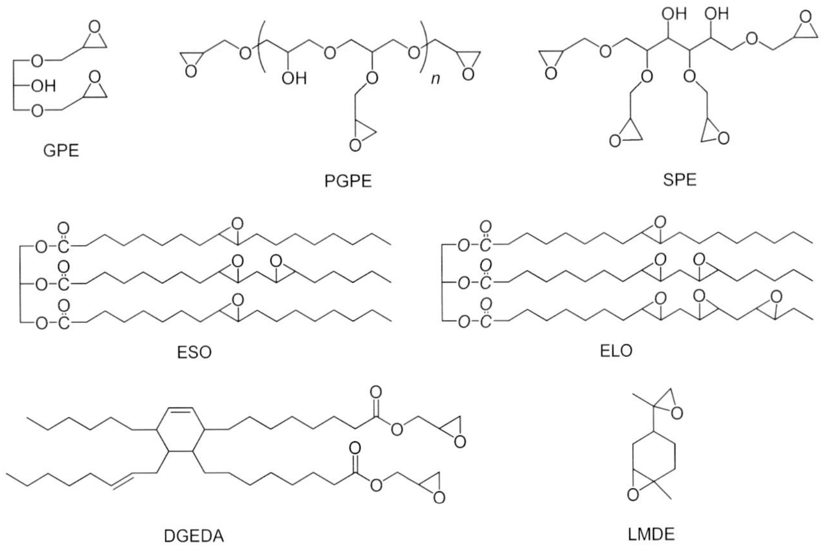

Figure 9.3 Examples of bio-based epoxy resins.

Although ECH is prepared from petroleum-based propylene and chlorohydrin (HOCl) at present, it can be derived from bio-based glycerol. Sorbitol-based SPE, which is prepared by the reaction of sorbitol and ECH, has been mainly used for tacifier, coatings, paper and fiber modifier, and so on. Epoxidized vegetable oils such as ESO and ELO are manufactured by the epoxidation of the double bonds of vegetable oils with hydrogen peroxide, either in acetic acid or in formic acid [18–20], and have mainly been used as plasticizer or stabilizer to modify the properties of plastic resins such as poly(vinyl chloride). Because ELO has a 30% more oxirane content than ESO does, the cured ELO has a higher cross-linking density, which results in a better performance. Dimer acid-based DGEDA is a flexible epoxy resin produced from dimer acid and ECH. Although dimer acid can be also obtained from animal fats or vegetable oils, most of the dimer acids appearing in the market are synthesized from the crude tall oil provided as a by-product of kraft pulp [21]. The commercially available dimer acid usually contains monomer (1–5%) and trimer or more (14–16%) in addition to dimer. Limonene-based LMDE (1-methyl-4-(2-methyl-2-oxiranyl)-7-oxabicyclo[4.1.0]heptane) is commercially produced by the reaction of limonene and peracetic acid, which is used as cationically curable resins and reactive diluents. (+)-d-Limonene is a popular monoterpene that is commercially obtained from citrus fruits.

Recently, diglycidyl ether of isosorbide (DGEDAS) has been synthesized *via* allyl etherification of isosorbide and subsequent epoxidation as shown in Figure 9.4 [22]. Isosorbide is commercially produced by the double dehydration of sorbitol. The glass transition temperature (T_g) measured by differential scanning calorimetry (DSC) for the DGEDAS cured with isophorone diamine is 108 °C, which is a little lower than that for the DGEBA cured with isophorone diamine.

However, their bio-based aliphatic epoxy resins had not been versatile materials because of inferior mechanical and thermal properties to the bisphenol A or novolac-based epoxy resins. Therefore, the selection of hardener and the addition of nanofiller are important for the use of the bio-based aliphatic epoxy resins in wide applications.

On the other hand, recently, lignin-based epoxy resins are attracting worldwide attention as bio-based aromatic epoxy resins [23–26]. The lignin contained in wood is a complex macromolecular compound with three-dimensional network, which is basically composed of methoxy-substituted propylphenol framework. When lignin is appropriately used as a raw material of epoxy resins and takes a full advantage of the aromatic structure, it is expected that the lignin-based epoxy resin

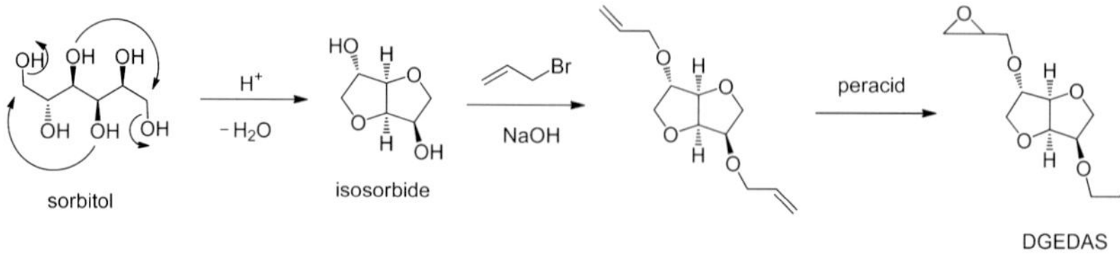

Figure 9.4 Synthetic scheme of diglycidyl ether of isosorbide.

can express the performance equivalent to those of DGEBA. There are various methods to synthesize lignin-based epoxy resins. The epoxy resins can be cured with general epoxy hardeners. The lignin-based epoxy resins are mainly synthesized from kraft lignin or lignin-like compounds derived from wood. However, the chemical structure of the lignin-based epoxy resins has not yet been elucidated in detail, because these compounds are complex mixtures containing fragmented lignin moieties.

The synthetic methods are roughly classified into the two categories by the raw material to synthesize the lignin-based epoxy resins as follows:

1) Methods preparing from kraft lignin

   Lignin-based epoxy resins are prepared from kraft lignin by mainly four methods as follows:

   a) Kraft lignin reacts with phenol/hydrochloric acid, acetone/sulfuric acid, or bisphenol A/boron trifluoride etherate to form low-molecular weight phenolic compounds by the cleavage of the cross-linked bond between the guaiacyl moieties of lignin, followed by epoxidizing the phenolic hydroxy group with ECH to provide the lignin-based epoxy resins [27–31].

   b) Ozone-oxidized kraft lignin with carboxylic acid groups is dissolved in aqueous sodium hydroxide. The obtained mixture with carboxylate groups is cured with water-soluble GPE [32, 33].

   c) Kraft lignin is dissolved in 1% sodium hydroxide aqueous solution at 60 °C, and then water-soluble diglycidyl ether of poly(ethylene glycol) and/or emulsified DGEBA are added. The resulting mixture is cured with TETA [34, 35]. In this case, the formed phenolate groups react with epoxy groups.

   d) Alcoholysis lignin or lignin sulfuric acid is dissolved in ethylene glycol and/or glycerin, and then the resulting hydroxy group is reacted with succinic acid to produce $HOOCCH_2CH_2CO$-grafted lignin. The polycarboxylic acid is cured with GPE [36, 37].

2) Methods preparing from wood

   As there are several methods preparing bio-based epoxy resins from wood, representative two methods are described below.

   a) After wood chip is treated with a high-pressure steam (3.6 MPa, 5 min) in a steam-explosion apparatus, the mixture is exploded by a rapid release of the pressure. The obtained steam-exploded mixture is separated into the water-insoluble component containing cellulose and lignin and the water-soluble component containing monosaccharide, oligosaccharide, polyphenol, and so on. Further, lignin-based polyphenol compounds are extracted with methanol from the water-insoluble component, epoxidized with ECH in the presence of sodium hydride or tetramethylammonium hydroxide to produce a lignin-based epoxy resin (Figure 9.5) [38–40]. It is reported that the use of tetramethylammonium hydroxide in the epoxidation reaction prevents the formation of polymerized product to produce the epoxidized lignin with M_w 2100 [40]. The lignin-based epoxy resin is cured with its precursor, the lignin-based polyphenol or phenol novolac, to produce the network polymer which has a higher T_g than the conventional cured materials of DGEBA, although water absorption of the former is higher than the latter.

b) After wood powder is impregnated into p-cresol, 72% concentrated sulfuric acid is added to swell and decompose the cellulose component in the wood. The separated organic layer is composed of the lignocresol (Figure 9.6), which is then reacted with ECH to form epoxidized lignocresol resin [41, 42].

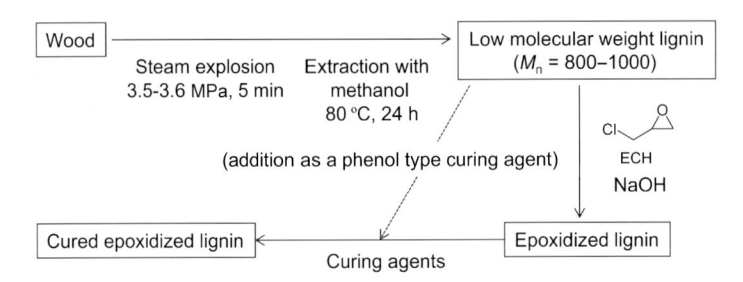

Figure 9.5 Preparative scheme of cured epoxidized lignin from wood by steam explosion method.

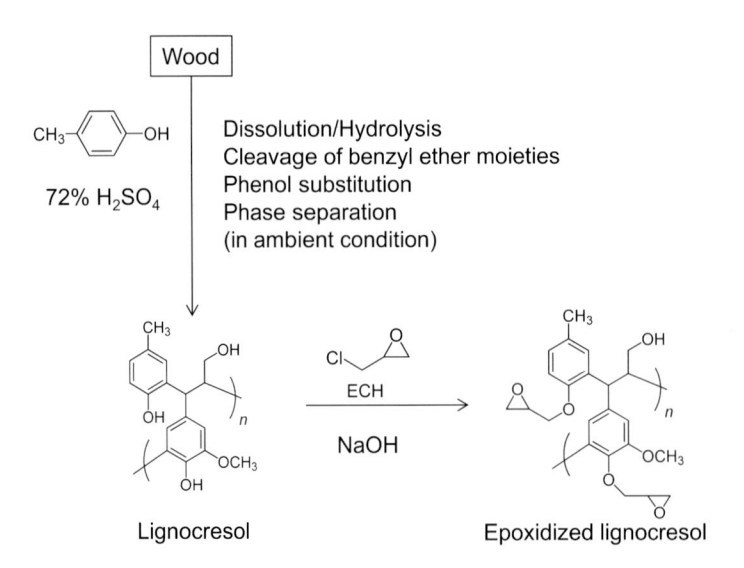

Figure 9.6 Synthetic scheme of epoxidized lignocresol.

9.2 Bio-Based Epoxy Resins and Hardeners

Recently, as a wood-based epoxy resin whose chemical structure is clear, diglycidyl ester of 2-pyrone-4,6-dicarboxylic acid (DGEPDC) has been reported by Hasegawa et al. [43]. The synthetic scheme is shown in Figure 9.7. Low-molecular weight lignin compounds containing vanillin, vanillate, syringaldehyde, and syringate are generated by dissolving wood with sodium hydroxide and nitrobenzene in an autoclave at 170 °C. The lignin-derived compounds are further decomposed biochemically to produce a uniform 2-pyrone-4,6-dicarboxylic acid (PDC) by the lignin-degrading bacteria, *Sphingomonas paucimobilis* SYK-6 strain. The PDC is esterified with glycidol at 0–5 °C in tetrahydrofuran to generate DGEPDC. The epoxy resin cured with maleic anhydride or phthalic anhydride has a higher tensile adhesive strength toward stainless steel or iron plate than the cured DGEBA does. Also, limonene-based aromatic epoxy resin (DGELBP) is a known compound. DGELBP is synthesized from ECH and limonene bisphenol (LBP) which is prepared by the reaction of limonene and phenol in the presence of boron trifluoride etherate as shown in Figure 9.8 [44]. The cured material of DGELBP with a conventional epoxy hardener has a better water-resistant property than that of DGEBA.

It is known that tensile strength and T_g of the layered silicate nanocomposites using the petroleum-based aromatic DGEBA often decrease with increasing clay content due to the hindrance of epoxy cross-linking reaction by the presence of clay, although tensile modulus increases with clay content [45, 46]. Therefore, the bio-based aromatic epoxy resins described here should be used as matrix resins for fiber-reinforced plastics in a similar manner to the conventional DGEBA. The methodology of layered silicate nanocomposite is rather effective for the improvement of the thermal and mechanical properties of the bio-based aliphatic epoxy resins such as GPE, PGPE, SPE, and ESO.

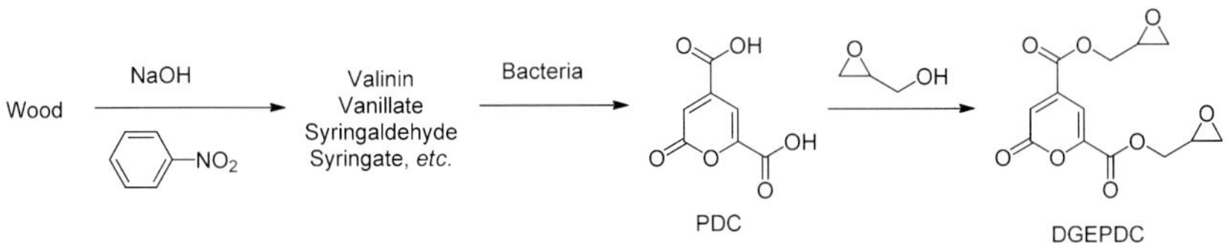

Figure 9.7 Synthetic scheme of diglycidyl ether of 2-pyrone-4,6-dicarboxylic acid.

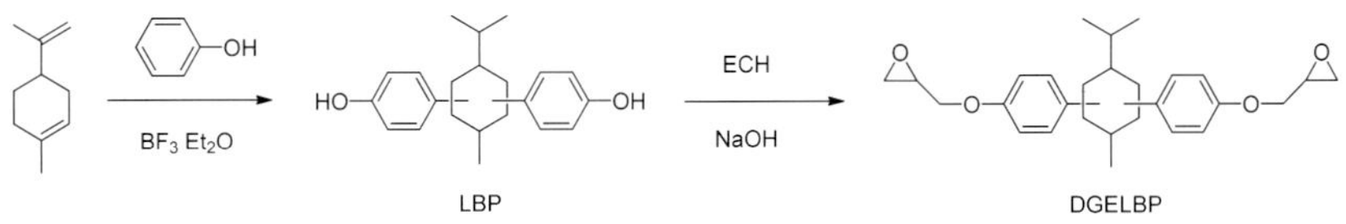

Figure 9.8 Synthetic scheme of diglycidyl ether of limonene bisphenol.

9.2.2
Bio-Based Hardeners

Most of epoxy hardeners such as polyamines, polyphenols, and carboxylic acid anhydrides are derived from petroleum resources (Figure 9.2). The number of the literatures regarding bio-based epoxy hardeners is less than that of the bio-based epoxy resins. Basically, bio-based polyamines, polyphenols, and carboxylic acid anhydrides can be used as epoxy hardeners. As a bio-based polyamine hardener, we reported the GPE or PGPE cured with ε-poly(L-lysine) (PL) [47]. The PL is produced by aerobic bacterial fermentation using *Streptomyces albulus* in a culture medium containing glucose, citric acid, and ammonium sulfate [48, 49]. The PL has been used as food preservatives [50], while has not yet been applied to the industrial polymeric materials. PL differs from usual proteins in that the amide linkage is not between the α-amino and carboxylic groups in typical of peptide bonds, but is between the ε-amino and carboxyl group. The pendant α-amino groups are expected to react with epoxy groups.

As a bio-based polyphenol hardener, we reported the GPE, PGPE, SPE, or ESO cured with tannic acid (TA) [51–53]. Commercial TA is composed of mixtures of gallotannins from sumac galls, Aleppo oak galls, or sumac leaves [54]. The chemical formula for commercial TA is often given as $C_{76}H_{52}O_{46}$ as shown in Figure 9.9. But, in fact, it contains a mixture of related compounds. Its structure is based

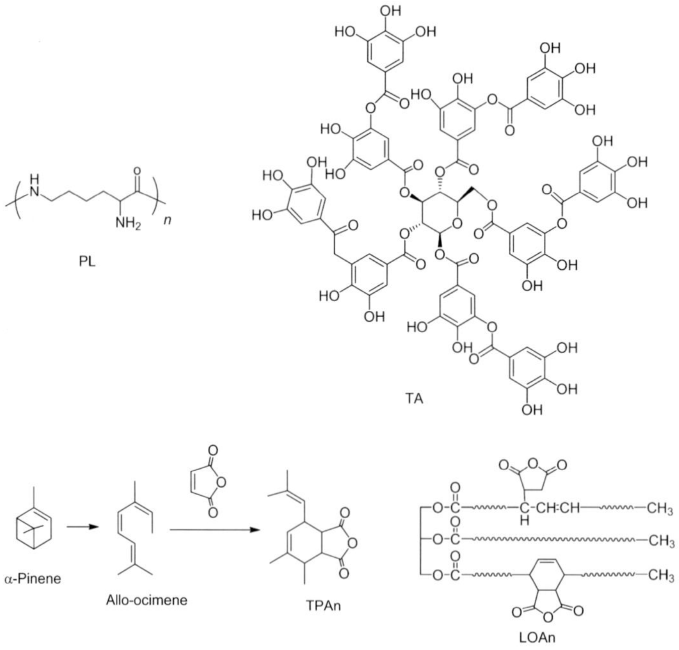

Figure 9.9 Examples of bio-based hardeners for bio-based epoxy resins.

mainly on glucose ester of gallic acid. As pyrogallol is produced by the decarboxylation of gallic acid, pyrogallol-based oligomeric compounds can be used as epoxy hardeners.

As a bio-based acid anhydride hardener, we reported the ESO cured with terpene-derived acid anhydride (TPAn) [55]. TPAn is synthesized by the Diels–Alder reaction of maleic anhydride and allo-ocimene obtained by the isomerization of α-pinene and/or β-pinene [56–58]. Also, the ESO cured with maleinated linseed oil (LOAn) was reported by Warth et al. [59].

9.2.3
Properties of Cured Bio-Based Epoxy Resins

Physical properties of the bio-based aliphatic epoxy resins investigated by our group are summarized in Table 9.1 [47, 51–53, 55]. Although bio-based polyphenol hardener, TA is industrially available from various Makers (e.g., Fuji Chemical Industry, Co., Ltd. [Wakayama, Japan]), we used the reagent grade TA of Kanto Chemical Co., Inc. (Tokyo, Japan). Bio-based polyamine hardener, PL (degree of polymerization: 25–35) was supplied from Chisso Corporation (Tokyo, Japan) as a 25% aqueous solution. Acid anhydride hardeners, TPAn (Epikure YH306, neutralization equivalent 117 g/eq.), HPAn, and LOAn (HIMALEIN LN-10, acid value 45.3 mg-KOH/g, iodine value 140 g-I_2/100 g) were supplied from Japan Epoxy Resins (Tokyo, Japan), New Japan Chemical (Osaka, Japan), and Itoh Oil Chemicals (Yokkaichi, Mie, Japan), respectively.

Thermal and mechanical properties for the GPE, PGPE, and SPE cured with PL and TA are summarized in comparison with the thermal properties of the SPE cured with petroleum-based PN (softening point 82 °C, weight per phenolic hydroxyl equivalent 104 g/eq., Sumitomo Berklite Co. Ltd. [Tokyo, Japan]) in Table 9.2. As GPE, PGPE, and PL are water-soluble, GPE/PL or PGPE/PL were mixed

Table 9.1 Physical properties of the bio-based aliphatic epoxy resins used in this study.

Epoxy resin (abbreviation)	Supplier (trade name)	Epoxy functionality	Epoxy equivalent weight (g/eq.)	Viscosity (cps, 25 °C)
Glycerol polyglycidyl ether (GPE)	Nagase ChemteX, Co. (DENACOL® EX-313)	2.0	140	150
Polyglycerol polyglycidyl ether (PGPE)	Nagase ChemteX, Co. (DENACOL® EX-512)	4.1	169	1300
Sorbitol polyglycidyl ether (SPE)	Nagase ChemteX, Co. (DENACOL® EX-614B)	3.6	172	5000
Epoxidized soybean oil (ESO)	Kao Chemical Co., Inc. (KAPOX® S-6)	4	239 (oxirane oxygen 6.7%)	

Table 9.2 Thermal and mechanical properties of the GPE, PGPE, and SPE cured with various hardeners.

Sample	Feed ratio[a]	Curing condition (°C/h)	T_g (°C)[b]	Tensile strength (MPa)	Tensile modulus (MPa)	T_d (°C)[c]
GPE-PL	1/1	110/24	52.5 (37)[d]			
	2/1		36.0 (33)[d]			
PGPE-PL	1/1	110/24	58.0 (47)[d]	3.98	5.73	298
	2/1		(35)[d]			
GPE-TA	1/1	160/3	73	36.5	2430	317
	1/1.2		79	31.2	2360	312
	1/1.4		91	32.1	2260	
PGPE-TA	1/1	160/3	77.0	63.5	2710	316
SPE-TA	1/1	160/2	95	60.6	1710	314
	1/1.2		109	25.3	1698	
	1/1.4		111	30.0	1275	
SPE-PN	1/0.8	170/3	81.0			
	1/0.9		80.6			
	1/1		78.1			346.3

a) Feed ratio means molar ratio of epoxy/amine or epoxy/hydroxy.
b) The T_g obtained from tan δ peak temperature by the dynamic mechanical analysis.
c) The T_d means 5% weight loss temperature measured by thermogravimetric analysis.
d) The values in parentheses are the T_g measured by differential scanning calorimetry.

in water in epoxy/amine molar ratio 1/1 and 2/1, and cured finally at 110 °C [45]. Figure 9.10 shows the FT-IR spectra over the wavelength range of 1200–700 cm^{-1} for the reactants and the obtained cured resins. In the spectra of GPE and PGPE, the absorption peaks characteristic to epoxy ring are observed at 900 and 845 cm^{-1}. These peaks are almost disappeared for the cured products with epoxy/amine ratio 1/1. On the other hand, although the theoretical stoichiometry of epoxy/amine ratio is 2/1, the peaks related to epoxy ring are slightly remaining in the spectra of the cured products with epoxy/amine ratio 2/1. It is considered that the secondary amino group formed by the reaction of one epoxy and one amino group is sterically hindered and is hard to react with the second epoxy group. Also, the band due to C=O stretching of the peptide bond was observed at 1630 cm^{-1} for the cured resins in a similar manner to PL, suggesting the peptide bond is retained even after the curing reaction at 110 °C. The GPE-PL (1/1) and PGPE-PL (1/1) showed higher T_g than GPE-PL (2/1) and PGPE-PL (2/1), respectively. The result that the resin cured with epoxy/amine ratio 1/1 has higher T_g than the 2/1 cured resin suggests that the former has higher cross-linking density than the latter in agreement with the result of FT-IR. When the cured resins with the same epoxy/amine ratio are compared, the PGPE-PL had higher T_g than the GPE-PL did. This result is attributed to a higher epoxy functionality of PGPE than GPE. The PGPE-PL (2/1) showed the tensile strength and modulus lower than 10 MPa, because the material

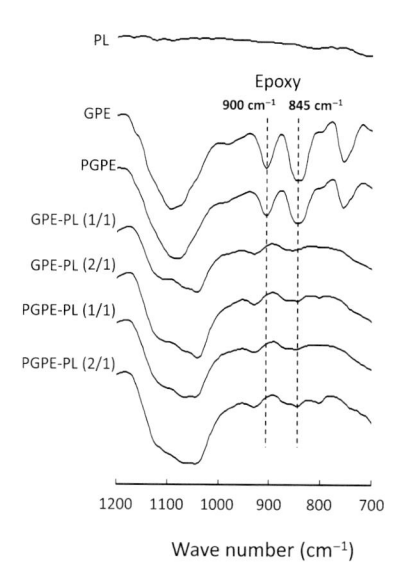

Figure 9.10 FT-IR spectra of PL, GPE, PGPE, and their cured materials. (Reproduced from [47] with permission from John Wiley & Sons, Inc.)

Table 9.3 Thermal and mechanical properties of the ESO cured with various hardeners.

Sample	Feed ratio[a]	Curing condition (°C/h)	T_g [DMA] (°C)	Tensile strength (MPa)	Tensile modulus (MPa)
ESO-TA	1/0.8	210/2	46	6.0	116
	1/1		57	12.7	409
	1/1.2		58	12.7	450
	1/1.4		58	15.1	458
ESO-TPAn	1/1	150/24	67.2	21.9	745
ESO-HPAn	1/1	150/24	59.0	4.4	24
ESO-LOAn	1/1	150/24	−41.0	1.2	25

a) Feed ratio means molar ratio of epoxy/acid anhydride.

is in rubbery state at room temperature at which tensile properties were measured. The mechanical properties were improved by the addition of nanofiller in the following section.

Next, the properties of the bio-based epoxy resins cured with TA are discussed [49–51]. Because TA is soluble in liquid GPE and SPE, GPE/TA and SPE/TA can be mixed easily. In case of PGPE/TA and ESO/TA, it is necessary to add ethanol to get homogenous solutions. Table 9.3 summarizes the thermal and mechanical properties of the ESO cured with various hardeners. When the T_gs of the materials cured at epoxy/hydroxy 1/1 are compared, the higher order of T_g is SPE-TA (95 °C) > PGPE-TA (77.0 °C) > GPE-TA (73 °C) > ESO-TA (57 °C), as is supposed from the relationship between epoxy functionality and the distance between cross-linked points. When petroleum-based PN is used as a hardener of SPE, T_gs measured by DMA for SPE-PN resins cured at epoxy hydroxy ratios 1/0.8–1/1 were 81.0–78.1 °C, which were lower than those of SPE/TA resins cured at epoxy hydroxy ratios 1/1–1/1.4(95–111 °C). Although the T_g increased with decreasing epoxy/hydroxy ratio (1/1, 1/1.2, 1/1.4) for GPE/TA and SPE/TA, the cured materials at epoxy/hydroxy ratio 1/1 showed the highest tensile strength and modulus. In case of ESO-TA, both the tensile properties and T_g increased with decreasing epoxy/hydroxy ratio. When acid anhydride hardeners are used for the curing of ESO, ESO-TPAn had the highest T_g and tensile strength and modulus, which were higher than those of ESO-TA (1/1.4) [55].

9.3
Bio-Based Epoxy Resins/Clay Nanocomposites

9.3.1
Nanocomposites Based on Polyglycerol Polyglycidyl Ether

In the previous section, the PGPE-PL (1/1) had considerably lower tensile strength, modulus, and T_g than the PGPE-TA. The layered silicate nanocomposites of PGPE/PL were investigated in order to overcome the drawbacks of thermal and mechanical properties [47]. Sodium (Na^+) montomorillonite (MMT, Kunipia F, cation exchange capacity [CEC]: 115 meq./100 g, Kunimine Industries Co. Ltd., Tokyo, Japan) was used as a layered silicate. When water-soluble PGPE and PL are used, MMT can be directly used for the preparation of nanocomposite, and it is not necessary to organically modify the MMT with a hydrophobic quaternary ammonium ion. Synthetic procedure of PGPE-PL/MMT nanocomposite is as follows: To ion-exchanged water (11.7 g) was added MMT (0.363 g), and the suspension was stirred for 24 h at room temperature. The obtained 3 wt% MMT suspension in water was added to a 10 wt% aqueous solution of PGPE (2.91 g, epoxy group 17.2 mmol), and the mixture was stirred for 24 h at room temperature, and then a 25 wt% aqueous solution (8.80 g, amino group 17.2 mmol) of PL was added and stirred for 1 h at room temperature. The obtained mixture was poured on a poly(tetrafluoroethylene) plate and cured at 80 °C for 24 h and then at 110 °C for 24 h to give PGPE-PL/MMT composite with MMT content 6.6 wt% (PGPE-PL/M7). The PGPE-PL/MMT composites with MMT content 10.8, 14.7, and 20.3 wt% (PGPE-PL/M11, PGPE-PL/M15, and PGPE-PL/M20) were also prepared in a similar manner with PGPE-PL/M7.

The X-ray diffraction (XRD) charts of the PGPE-PL/MMT composites prepared by mixing of PGPE, PL, and MMT in water and subsequent curing at finally 110 °C are shown in Figure 9.11. For PGPE-PL/MMT composites with MMT content 7–15 wt%, there appeared no XRD peak other than the peak around 2θ = 2.3° observed in PGPE-PL, suggesting the formation of exfoliated nanocomposites. On the other hand, the PGPE-PL/M20 showed a strong peak at 4.54°, which

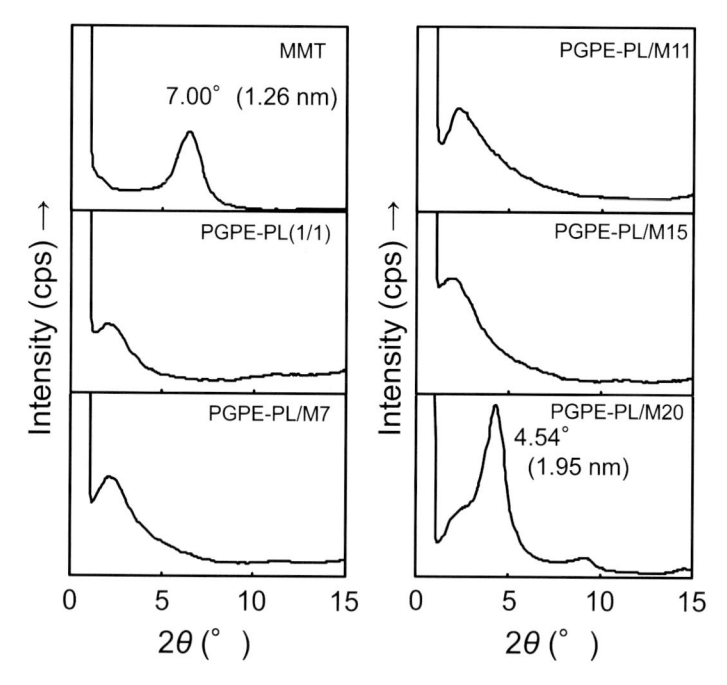

Figure 9.11 XRD charts of MMT and PGPE-PL/MMT composites. (Reproduced from [47] with permission from John Wiley & Sons, Inc.)

is corresponding to the interlayer distance (d) of 1.95 nm. The value is higher than the interlayer distance of MMT (1.26 nm), suggesting the formation of intercalated nanocomposite.

Figure 9.12 shows the transmission electron microscope (TEM) images of PGPE-PL/M7 and PGPE-PL/M20. It is obvious that exfoliated nanocomposite where MMT is homogeneously distributed is formed for PGPE-PL/M7. On the other hand, aggregated clay platelets where matrix resins are intercalated are observed for PGPE-PL/M20, indicating that MMT content is too large to be a homogeneous exfoliated nanocomposite. These results are in agreement with the XRD data.

Figure 9.13 shows DMA curves of PGPE-PL/MMT composites. Although the improvement of storage modulus (E') at −20 to 20 °C corresponding to the glassy state of the composites was not so large, the E' at around 60 to 120 °C corresponding to the rubbery state increased with increasing MMT content. The E' below 60 °C of PGPE-PL/M20 was lower than that of PGPE-PL/M15.

Table 9.4 summarizes the T_gs measured by DSC and DMA and 5% weight loss temperature (T_d) measured by TGA for the PGPE-PL/MMT composites. Both T_g and T_d increased with increasing MMT content until 14.7%, and decreased at 20.3%. The improvement of heat resistance is attributed to the restriction of

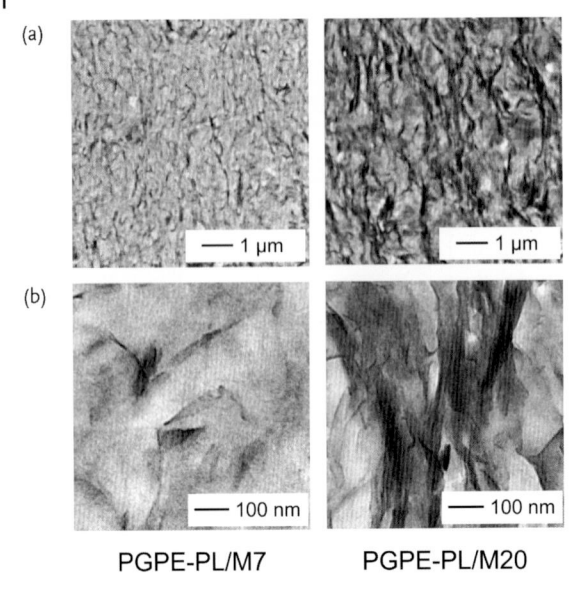

202 | *9 Bio-Based Epoxy Resin/Clay Nanocomposites*

Figure 9.12 TEM images of PGPE-PL/M7 and PGPE-PL/M20: (a) low magnification and (b) high magnification. (Reproduced from [47] with permission from John Wiley & Sons, Inc.)

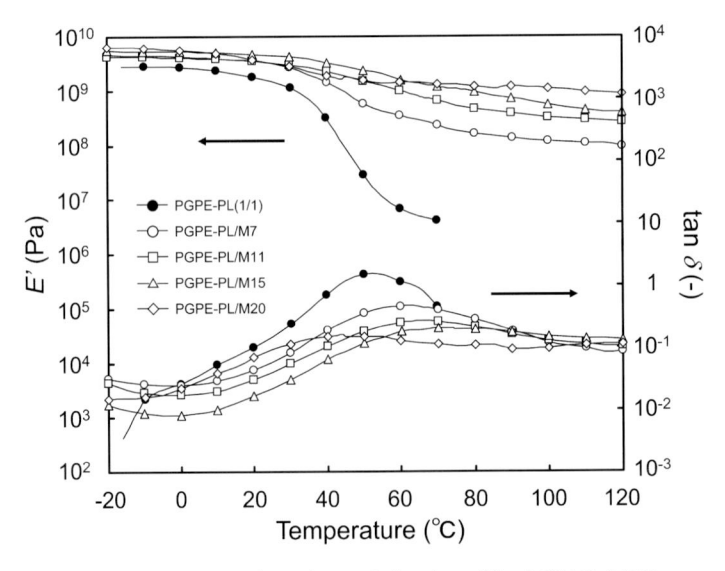

Figure 9.13 Temperature dependency of E' and tan δ for PGPE-PL/MMT composites. (Reproduced from [47] with permission from John Wiley & Sons, Inc.)

Table 9.4 Thermal properties of PGPE-PL/MMT nanocomposites.

Sample abbreviation	MMT content (wt%)	T_g [DSC] (°C)	T_g [DMA] (°C)	5% Weight loss temperature (°C)
PGPE-PL	0	47	58	298
PGPE-PL/M7	6.6	58	62	312
PGPE-PL/M11	10.8	59	66	317
PGPE-PL/M15	14.7	56	73	322
PGPE-PL/M20	20.3	34	44	316

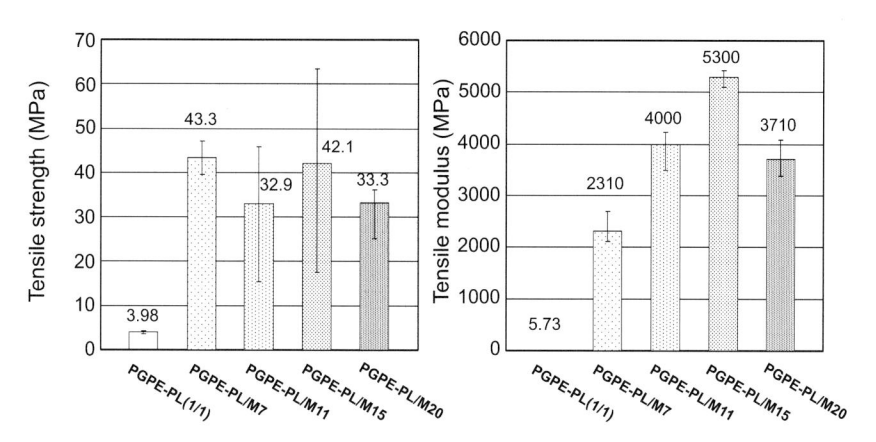

Figure 9.14 Tensile properties of PGPE-PL and PGPE-PL/MMT composites. (Reproduced from [47] with permission from John Wiley & Sons, Inc.)

molecular motion by the formation of exfoliated nanocomposite. The lowering of T_g and T_d of PGPE-PL/M20 is related to the aggregation of clay platelets and/or the hindrance of the reaction of epoxy and amine caused by the presence of excess clay.

Figure 9.14 shows the tensile properties at 25 °C for the PGPE-PL/MMT composites. In agreement with the result of DMA, the tensile modulus increased with increasing MMT content in the range of 0–14.7% and decreased at 20.3%. The prominent improvement by the addition of MMT is attributed to the reinforcement effect by the layered silicate and is also related to the transition from rubbery state to glassy state of the samples at the measurement temperature 25 °C. Regarding tensile strength, clear tendency on MMT content was not elucidated because the error bar was large. However, all the nanocomposites had considerably higher tensile strength than the PGPE-PL without MMT. For example, the tensile modulus and strength of PGPE-PL/M7 (2.3 GPa, 43 MPa) are almost comparable to those of the bisphenol A DGE cured with diethylene triamine (2.1 GPa, 69 MPa) [60].

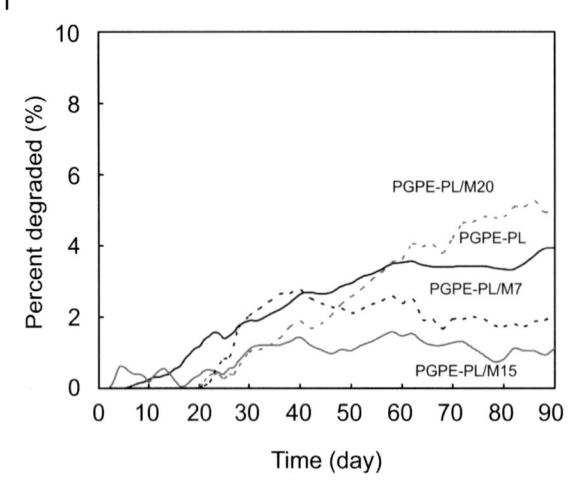

Figure 9.15 Biodegradability measured by biochemical oxygen demand in an aqueous media containing activated sludge for PGPE-PL and PGPE-PL/MMT composites. (Reproduced from [47] with permission of John Wiley & Sons, Inc.)

Figure 9.15 shows biodegradability by measuring biochemical oxygen demand in the aerobic aqueous medium containing activated sludge for the PGPE-PL and PGPE-PL/MMT nanocomposites. Biodegradability after 90 days of PGPE-PL resin was ca. 4%. The biodegradability of PGPE-PL/MMT composites decreased with increasing MMT content until 14.7 wt%, probably due to the interference of substance diffusion by the formation of exfoliated nanocomposites. On the other hand, PGPE-PL/M20 showed a little higher biodegradability than PGPE-PL. It is considered that a high water uptake in the aggregated part of intercalated clay platelets of PGPE-PL/M20 as was shown in the TEM image is related to the enhancement of biodegradation.

9.3.2
Nanocomposites Based on Epoxidized Vegetable Oil

As several studies regarding the nanocomposites of epoxidized vegetable oil and layered silicate had been previously reported, some of them are reviewed below. Uyama *et al.* reported the nanocomposites prepared by the curing of ESO using latent cationic catalyst (a benzylsulfonium hexafluoroantimonate derivative, Sun Aid SI-60L, provided by Sanshin Chemical Industry Co., Yamaguchi, Japan) in the presence of octadecylammonium-modified montmorillonite (OMM) at 150 °C [61]. The XRD and TEM analyses revealed that the nanocomposite with a clay content of 5% is an exfoliated nanocomposite, and the nanocomposites with more than 10% OMM are intercalated nanocomposites. It is reported that the T_gs measured by DMA for the nanocomposites with 5% and 15% OMM are −2 °C and 4 °C,

9.3 Bio-Based Epoxy Resins/Clay Nanocomposites

Table 9.5 Thermal and tensile properties of ESO/TETA/Cloisite® 30B nanocomposites.

Cloisite® 30B (wt%)	T_g [DMA] (°C)	Tensile strength (MPa)	Tensile modulus (MPa)	Elongation (%)
0	11.8	1.27	1.30	127
5	20.7	2.98	2.61	133
8	19.4	4.54	3.64	151
10	20.2	4.34	3.54	148

respectively, indicating that these materials are soft materials in a rubbery state at room temperature.

Liu et al. reported the nanocomposites prepared by the curing of ESO using TETA in the presence of methyl tallow bis(2-hydroxyethyl) ammonium-modified montmorillonite (Cloisite® 30B, Southern Clay Products, Gonzales, TX) at 120 °C [62]. The ratio of epoxy group to hydrogen of amino group is 1 to 1.37. The XRD and TEM data revealed that the intercalated nanocomposites are formed. The T_g measured by DMA and tensile properties are summarized in Table 9.5. The nanocomposite with 5 wt% clay content is 20.7 °C. As the clay loading increases, the resulting nanocomposites show no significant difference in the glass transition temperature. The tensile strength of the nanocomposites increased rapidly from 1.27 to 4.54 MPa with increasing clay content from 0 to 8 wt%. Similarly, the modulus increased rapidly from 1.20 to 3.64 MPa with increasing clay content from 0 to 8 wt%. Slight decrease in tensile modulus is observed when the clay content increases beyond 8–10 wt%.

Miyagawa et al. reported the nanocomposites prepared by the curing of a mixture of diglycidyl ether of bisphenol F (DGEBF) and ELO with MTHPA in the presence of Cloisite® 30B at 160 °C [63, 64]. The XRD and TEM data revealed that exfoliated nanocomposites are formed for DGEBF/ELO(80/20)/MTHPA with 5 wt% clay content. Although the storage modulus (ca. 3.2 GPa) at 30 °C for the nanocomposite is higher than that (ca. 3.9 GPa) of the corresponding cured resin without clay, the former nanocomposite had a lower T_g (ca. 113 °C) measured by DMA than the latter resin (ca. 121 °C) had. Also, the heat distortional temperature (ca. 108 °C) of the former nanocomposite was lower than that of the latter (ca. 103 °C), and the former had a little lower Izod impact strength than the latter. It should be noted that exfoliated silicate layers sometimes disturb the cross-linking reaction of epoxy resin and hardener especially for the high T_g epoxy curing system where molecular motion is restrained by the rigid structure [46].

These results indicate that methodology by the layered silicate nanocomposite is effective on the improvement of T_g for the vegetable oil-based network polymer with the T_g around room temperature, but not for the network polymer with the T_g around 100 °C, although the rigidity can be increased for both the network polymers.

9.4
Conclusion

The replacement of petroleum-based epoxy resin/hardener with bio-based epoxy resin/hardener is very important from the viewpoint of the conservation of limited petroleum resources and the protection of global environment. However, the use of aliphatic bio-based epoxy resins such as GPE, PGPE, SPE, and ESO sometimes results in the formation of the network polymers with lower T_g, strength, and rigidity than the materials using conventional petroleum-based DGEBA. Reinforcing method by the formation of exfoliated and/or intercalated layered silicate nanocomposites is effective in order to improve the thermal and mechanical properties of the network polymer produced by the curing of bio-based epoxy resin/hardener. Thermal and mechanical properties of the flexible bio-based epoxy resins with hydrophilic moiety such as PGPE, GPE, and SPE cured with flexible and hydrophilic bio-based hardeners such as PL were improved by the formation of exfoliated/intercalated layered silicate nanocomposite. Also, in case of bio-based water-soluble epoxy resin and hardener, there is a merit that sodium montmorillonite can be directly used for the formation of exfoliated/intercalated nanocomposites. On the other hand, in case of using epoxidized vegetable oils such as ESO and ELO, it is necessary to use organically modified layered silicates. It should be noted that the addition of layered silicates to the rigid aromatic epoxy resins with high T_g sometimes causes the lowering of T_g and strength.

References

1 Flich, E.W. (1994) *Epoxy Resins, Curing Agents, Compounds, and Modifiers, Second Edition: An Industrial Guide*, Noyes Publications, Park Ridge.
2 Lee, H. (1981) *Handbook of Epoxy Resins*, McGraw-Hill, New York.
3 Im, S.S., Kim, Y.H., Yoon, J.S., and Chin, I.J. (eds) (2005) Bio-based polymers: recent progress. *Macromol. Symp.*, **224** (1), 1–376.
4 Wool, R.P. and Sun, X.S. (2005) *Bio-Based Polymers and Composites*, Elsevier Academic Press, Burlington.
5 Kaplan, D.L. (1998) Introduction in biopolymers from renewable resources, in *Biopolymers from Renewable Resources* (ed. D.L. Kaplan), Springer-Verlag, Berlin, pp. 1–29.
6 Yu, L., Petinakis, S., Dean, K., Bilyk, A., and Wu, D. (2007) Green polymeric blends and composites from renewable resources. *Macromol. Symp.*, **249/250** (1), 535–539.
7 Yu, L., Dean, K., and Li, L. (2006) Polymer blends and composites from renewable resources. *Prog. Polym. Sci.*, **31** (6), 576–602.
8 Ronda, J.C., Lligadas, G., Galià, M., and Cádiz, V. (2013) A renewable approach to thermosetting resins. *React. Funct. Polym.*, **73** (2), 381–395.
9 Kim, J.R. and Sharma, S. (2012) The development and comparison of bio-thermoset plastics from epoxidized plant oils. *Ind. Crops Prod.*, **36** (1), 485–499.
10 Raquez, J.M., Deléglise, M., Lacrampe, M.F., and Krawczak, P. (2010) Thermosetting (bio)materials derived from renewable resources: a critical review. *Prog. Polym. Sci.*, **35** (4), 487–509.
11 Effendi, A., Gerhauser, H., and Bridgwater, A.V. (2008) Production of renewable phenolic resins by thermochemical conversion of biomass:

a review. *Renew. Sust. Energy Rev.*, **12** (8), 2092–2116.
12 Wool, R.P. (2005) Properties of triglyceride-based thermosets, in *Bio-Based Polymers and Composites* (eds R.P. Wool and X.S. Sun), Elsevier Academic Press, Burlington, pp. 202–255.
13 Mittal, V. (ed.) (2011) *Nanocomposites with Biodegradable Polymers: Synthesis, Properties and Future Perspectives*, Oxford University Press, Oxford, New York.
14 Das, G. and Karak, N. (2010) Thermostable and flame retardant *Mesua ferrea* L. seed oil based non-halogenated epoxy resin/clay nanocomposites. *Prog. Org. Coat.*, **69** (4), 495–503.
15 Lu, J. and Wool, R.P. (2008) Additive toughening effects on new bio-based thermosetting resins from plant oils. *Compos. Sci. Technol.*, **68** (3–4), 1025–1033.
16 Wool, R.P. (2005) 15 – nanoclay biocomposites, in *Bio-Based Polymers and Composites* (eds R.P. Wool and X.S. Sun), Elsevier Academic Press, Burlington, pp. 523–550.
17 Haas, M.J., McAloon, A.J., Yee, W.C., and Foglia, T.A. (2006) A process model to estimate biodiesel production costs. *Bioresour. Technol.*, **97** (4), 671–678.
18 Meffert, A. and Kluth, H. (1989) Process for the preparation of modified triglycerides. United States Patent 4 886 893.
19 Park, S.J., Jin, F.L., and Lee, J.R. (2004) Synthesis and thermal properties of epoxidized vegetable oil. *Macromolecular Rapid Commun.*, **25** (6), 724–727.
20 Swern, D., Billen, G.N., Findley, T.W., and Scanlan, J.T. (1945) Hydroxylation of monounsaturated fatty materials with hydrogen peroxide. *J. Am. Chem. Soc.*, **67** (10), 1786–1789.
21 Wheeler, D.H., White, J., and Mills, G. (1967) Dimer acid structures. The thermal dimer of normal linoleate, methyl 9-*cis*, 12-*cis* octadecadienoate. *J. Am. Oil Chem. Soc.*, **44** (5), 298–302.
22 Chrysanthos, M., Galy, J., and Pascault, J.P. (2011) Preparation and properties of bio-based epoxy networks derived from isosorbide diglycidyl ether. *Polymer*, **52** (16), 3611–3620.
23 Koide, T. (2012) Progress in development of epoxy resin systems based on wood biomass in Japan. *Polym. Eng. Sci.*, **52** (4), 701–717.
24 Doherty, W., Mousavioun, P., and Fellows, C.M. (2011) Value-adding to cellulosic ethanol: lignin polymers. *Ind. Crops Prod.*, **33** (2), 259–276.
25 Zhang, W., Zhang, Y., and Zha, D. (2012) Synthesis of liquefied corn barn-based epoxy resins and their thermodynamic properties. *J. Appl. Polym. Sci.*, **125** (3), 2304–2311.
26 Delmas, G.H., Benjelloun-Mlayah, B., Bigot, Y.L., and Delmas, M. (2012) Biolignin™ based epoxy resins. *J. Appl. Polym. Sci.* doi: 10.1002/app.37921
27 Tai, S., Nagata, M., Nakano, J., and Migita, N. (1967) Research on utilization of lignin (4): epoxidation of thiolignin. *Mokuzai Gakkaishi*, **13** (3), 102–107.
28 Tai, S., Nagata, M., Nakano, J., and Migita, N. (1967) Research on utilization of lignin (5): adhesive property of lignin epoxy resin. *Mokuzai Gakkaishi*, **13** (6), 257–263.
29 Tai, S., Nagata, M., Nakano, J., and Migita, N. (1968) Research on utilization of lignin (6): activation of thiolignin by phenol. *Mokuzai Gakkaishi*, **14** (1), 40–45.
30 Ito, H. and Shiraishi, N. (1987) Adhesives of epoxy resins prepared from thiolignin. *Mokuzai Gakkaishi*, **33** (5), 393–399.
31 Shiraishi, N. (1989) Recent progress in wood dissolution and adhesives from kraft lignin. *ACS Symp. Ser.*, **397** (38), 488–495.
32 Tomita, B., Kurozumi, K., Takemura, A., and Hosoya, S. (1989) Ozonized lignin–epoxy resins. *ACS Symp. Ser.*, **397** (39), 496–505.
33 Lee, H.J., Tomita, B., and Hosoya, S. (1991) Development of ozonized kraft lignin/epoxy resin adhesives. *Wood Ind.*, **46** (9), 412–417.
34 Nonaka, Y., Tomita, B., and Hatano, Y. (1996) Viscoelastic properties of lignin/epoxy resins and their adhesive strength. *Wood Ind.*, **51** (6), 250–254.
35 Nonaka, Y., Tomita, B., and Hatano, Y. (1997) Synthesis of lignin/epoxy resins in aqueous systems and their properties. *Holzforschung*, **51** (2), 183–187.

36 Ismail, T.N.M.T., Hassan, H.A., Hirose, S., Taguchi, Y., Hatakeyame, T., and Hatakeyama, H. (2010) Synthesis and thermal properties of ester-type crosslinked epoxy resins derived from lignosulfonate and glycerol. *Polym. Int.*, **59** (2), 181–186.

37 Hirosa, S., Hatakeyama, T., and Hatakeyama, H. (2003) Synthesis and thermal properties of epoxy resins from ester-carboxylic acid derivative of alcoholysis lignin. *Macromol. Symp.*, **197**, 157–169.

38 Nakamura, Y., Sawada, T., and Nakamoto, Y. (1998) Basic study on synthesis of resin from exploded lignin. *J. Netw. Polym., Jpn*, **19** (3), 148–155.

39 Nakamura, Y. (1999) Synthesis of the thermosetting resin from methanol-soluble lignin. *Cellulose Commun.*, **6** (2), 85–88.

40 Kagawa, H., Okabe, Y., Nakazawa, Y., and Enomoto, Y. (2010) Application of lignin-derived epoxy resin to electrical and electric equipment. *Mater. Stage*, **10** (7), 36–39.

41 Funaoka, M. (2003) Lignin: its functions and successive flow. *Macromol. Symp.*, **197** (1), 213–221.

42 Funaoka, M. (2006) A new epoxy resins from bioresources-based lignophenol. *J. Netw. Polym., Jpn*, **27** (3), 118–125.

43 Hasegawa, Y., Shikinaka, K., Katayama, Y., Kajita, S., Masai, E., Nakamura, M., Otsuka, Y., Ohara, S., and Shigehara, K. (2009) Tenacious epoxy adhesives prepared from lignin-derived stable metabolic intermediate. *Sen-i Gakkaishi*, **65** (12), 359–362.

44 Zhang, W., Iijima, T., Fukuda, W., and Tomoi, T. (1997) Synthesis and curing of epoxy resins containing cyclic monoterpene units in the backborn. *J. Netw. Polym., Jpn*, **18** (2), 59–65.

45 Qi, B., Zhang, Q.X., Bannister, M., and Mai, Y.-W. (2006) Investigation of the mechanical properties of DGEBA-based epoxy resin with nanoclay additives. *Compos. Struct.*, **75** (1–4), 514–519.

46 Becker, O., Varley, R., and Simon, G. (2002) Morphology, thermal relaxations and mechanical properties of layered silicate nanocomposites based upon high-functionality epoxy resins. *Polymer*, **43** (16), 4365–4373.

47 Takada, Y., Shinbo, K., Someya, Y., and Shibata, M. (2009) Preparation and properties of bio-based epoxy montomorillonite nanocomposites derived from polyglycerol polyglycidyl ether and ε-polylysine. *J. Appl. Polym. Sci.*, **113** (1), 479–484.

48 Kahar, P., Iwata, T., Hiraki, J., Park, E.Y., and Okab, E.M. (2001) Enhancement of ε-polylysine production by *Streptomyces albulus* strain 410 using pH control. *J. Biosci. Bioeng.*, **91** (2), 190–194.

49 Kahar, P., Kobayashi, K., Kojima, M., and Okabe, M. (2002) Production of ε-polylysine in an airlift bioreactor. *J. Biosci. Bioeng.*, **93** (3), 274–280.

50 Kamioka, H. (1993) Use of polylysine pharmaceuticals in processed foods. *New Food Ind.*, **35** (10), 23–31.

51 Shibata, M. and Nakai, K. (2010) Preparation and properties of biocomposites composed of bio-based epoxy resin, tannic acid, and microfibrillated cellulose. *J. Polym. Sci. [B]*, **48** (4), 425–433.

52 Shibata, M., Teramoto, N., Takada, Y., and Yoshihara, S. (2010) Preparation and properties of biocomposites composed of glycerol-based epoxy resins, tannic acid, and wood flour. *J. Appl. Polym. Sci.*, **118** (5), 2998–3004.

53 Shibata, M., Teramoto, N., and Makino, K. (2011) Preparation and properties of biocomposites composed of epoxidized soybean oil, tannic acid, and microfibrillated cellulose. *J. Appl. Polym. Sci.*, **120** (1), 273–278.

54 Salunkhe, D.K., Chavan, J.K., and Kadam, S.S. (1989) Tannins, in *Dietary Tannins: Consequences and Remedies*, CRC Press, Boca Raton, pp. 10–17.

55 Takahashi, T., Hirayama, K., Teramoto, N., and Shibata, M. (2008) Biocomposites composed of epoxiduzed soybean oil cured with terpene-based acid anhydride and cellulose fibers. *J. Appl. Polym. Sci.*, **108** (3), 1596–1602.

56 Goldblatt, L.A. and Palkin, S. (1941) Vapor phase thermal isomerization of α- and β-pinene. *J. Am. Chem. Soc.*, **63** (12), 3517–3522.

57 Gustorf, E.K. and Letich, J. (1968) Cis–trans isomerization of alloocimene with dienophiles and other π-acids. *Tetrahedron Lett.*, **9** (45), 4689–4692.

58 Milks, J.E. and Lancaster, J.E. (1965) Reaction products of alloocimene and maleic anhydride. *J. Org. Chem.*, **30** (3), 888–891.

59 Warth, H., Mülhaupt, R., Hoffmann, B., and Lawson, S. (1997) Polyester networks based upon epoxidized and maleinated natural oils. *Die Angew. Makromol. Chem.*, **249** (1), 79–92.

60 Asp, L.E., Berglund, L.A., and Gudmundson, P. (1995) Effects of a composite-like stress state on the fracture of epoxide. *Compos. Sci. Technol.*, **53** (1), 27–37.

61 Uyama, H., Kuwabara, M., Tsujimoto, T., Nakano, M., Usuki, A., and Kobayashi, S. (2003) Green nanocomposites from renewable resources: plant oil–clay hybrid materials. *Chem. Mater.*, **15** (13), 2492–2494.

62 Liu, Z., Erhan, S., and Xu, J. (2005) Preparation, characterization and mechanical properties of epoxidized soybean oil/clay nanocomposites. *Polymer*, **46** (23), 10119–10127.

63 Miyagawa, H., Misra, M., Drzal, L.T., and Mohanty, A.K. (2005) Novel biobased nanocomposites from functionalized vegetable oil and organically-modified layered silicate clay. *Polymer*, **46** (12), 445–453.

64 Miyagawa, H., Misra, M., Drzal, L.T., and Mohanty, A.K. (2005) Fracture toughness and impact strength of anhydride-cured biobased epoxy. *Polym. Eng. and Sci.*, **45** (4), 487–495.

10
Electrical Properties and Electromagnetic Interference Shielding Response of Electrically Conducting Thermosetting Nanocomposites

Parveen Saini

10.1
Introduction

Electromagnetic interference (EMI) is a serious but unavoidable offshoot of explosive growth of electronics and telecommunications. In precise scientific sense, it is an undesired electromagnetic (EM) induction in the electronic circuitry of an instrument due to external emissive sources [1–5]. The mutual interference among electronic items and appliances can lead to degradation of device performance and may also adversely affect the human health [4–9]. Therefore, extensive efforts have been made in the past to develop suitable countermeasures for suppression (or even elimination) of EMI using wide range of materials including metals, carbonaceous substances (e.g., carbon black, graphite, carbon fibers, carbon nanotubes (CNTs), and graphene), intrinsically conducting polymers (ICPs), and their combinations (e.g., alloys, blends, composites, and multilayered/sandwiched structures) [3–20]. In particular, electrically conducting thermosetting nanocomposites represent a special class of hybrid materials that possess unique combination of electrical, thermal, and mechanical properties and potential applications in area of EMI shielding/microwave absorption. However, designing a composite for EMI shield with a certain level of attenuation, meeting a set of physical criteria (e.g., high conductivity, low density, corrosion resistance, thermal/environmental stability, and adequate mechanical properties) and controlling the dominant shielding mechanism is not a straightforward task and involves complex interplay of intrinsic properties of shield material (e.g., electrical conductivity (σ), permittivity (ε), and magnetic permeability (μ)) and logical selection of extrinsic parameters (e.g., thickness or geometry) [2–4, 9, 14, 19]. This encourages the need of basic understanding of shielding theory, governing theoretical equations and relevant measurement techniques a prime prerequisite.

10.2
EMI Shield and Shielding Effectiveness

In EM terminology, shield is defined as a barrier to regulate the transmission of EM waves. For example, EM shield for an electronic instrument refers to an enclosure that completely encloses the product and prevents the EM signals from an outside source to deteriorate its performance. Conversely, EM shield also prevents the emissions of the instrument from disturbing the outside electronic items or from exerting any harmful effect on the health of its users. The efficiency of EM shield is expressed in terms of a quantity called "shielding effectiveness (SE)" which is measured in terms of reduction in the magnitude of incident power/field upon transition across the shield. Mathematically, total shielding effectiveness (SE_T) can be expressed in logarithmic scale as [1, 3, 9, 11, 21]:

$$SE_T(dB) = 10\log_{10}\left(\frac{P_T}{P_I}\right) = 20\log_{10}\left(\frac{E_T}{E_I}\right) = 20\log_{10}\left(\frac{H_T}{H_I}\right) \quad (10.1)$$

where P_I (E_I or H_I) and P_T (E_T or H_T) are the power (electric or magnetic field intensity) of incident and transmitted EM waves, respectively. The most common unit to express SE is decibel (dB) such that −20, −30, and −40 dB corresponds to attenuation of 99% (P_T/P_I = 0.01), 99.9% (P_T/P_I = 0.001), and 99.999% (P_T/P_I = 0.0001), respectively.

10.2.1
Shielding Effectiveness: Theoretical Expressions

As schematically represented in Figure 10.1, three distinct mechanisms namely reflection (R), absorption (A), and multiple internal reflections (MIRs) contribute toward overall attenuation with SE_R, SE_A, and SE_M as corresponding SE components, that is, [9, 11]:

$$SE_T(dB) = SE_R(dB) + SE_A(dB) + SE_M(dB) \quad (10.2)$$

These mechanisms are related to intrinsic and extrinsic parameters of shielding materials through a set of theoretical expressions that may serve as a guide in designing a material for a particular application [2, 9].

10.2.1.1 Shielding Due to Reflection
The reflection loss (SE_R) occurs due to impedance (Z) mismatch between free space ($Z_o \sim 377\,\Omega$) and shield material ($Z < Z_o$). The magnitude of SE_R under plane wave (far-field conditions) can be expressed as [9, 11]

$$SE_R(dB) = -10\log_{10}\left(\frac{\sigma_T}{16\omega\varepsilon_r\varepsilon_o\mu_r}\right) \quad (10.3)$$

where σ_T is the total conductivity, ω is the angular frequency, ε_r is the relative permittivity, and μ_r is the relative permeability referred to free space (ε_o). There-

10.2 EMI Shield and Shielding Effectiveness

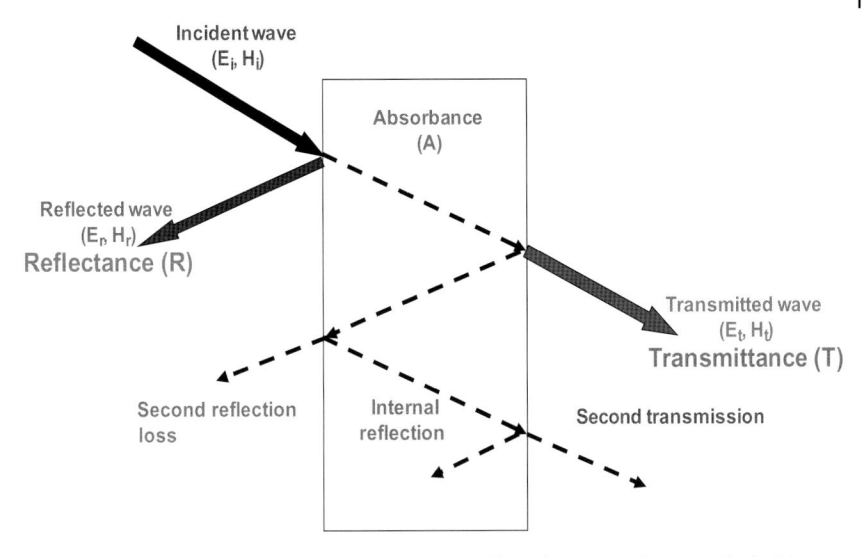

Figure 10.1 Schematic representation of interaction of incident EM radiation with shield material.

fore, selection of a material with high σ_T and low ε_r or μ_r helps in achieving high SE_R and vice versa.

10.2.1.2 Shielding Due to Absorption

The absorption loss (SE_A) occurs due to dynamic loss processes leading to molecular friction as well as induced current in the medium, which produce ohmic losses and heating of the material. The theoretical SE_A (dB) is related to skin depth (δ) and thickness (t) of shield material as [3, 9, 11]:

$$SE_A(dB) = -20\frac{t}{\delta}\log_{10} e = -8.68\left(\frac{t}{\delta}\right) \tag{10.4}$$

Further, skin depth that is defined as the distance required by the wave to be attenuated to $1/e$ or 37% of its original strength can be expressed as

$$\delta = \left(\frac{2}{\sigma_T \omega \mu}\right)^{1/2}$$

Therefore,

$$SE_A(dB) = -8.68t\left(\frac{\sigma_T \omega \mu_r}{2}\right)^{1/2} \tag{10.5}$$

The above expression revealed that a good absorbing material should possess high σ_T or μ_r and sufficient thickness to achieve the required number of skin depths even at the lowest frequency of concern.

10.2.1.3 Shielding Due to MIRs

The part of the incident wave that traversed through the thickness of the shield to reach the other face gets reflected back toward first face where it is partially rereflected again towards second boundary so that the process is continued again, as depicted in Figure 10.1. The shielding contribution due to these MIRs, that is, SE_M, can be expressed as [2, 9]

$$SE_M(dB) = 20\log_{10}\left|(1 - 10^{-SE_A/10})\right| \qquad (10.6)$$

It can be seen that SE_M is closely related to absorption loss (SE_A) and can be neglected (SEM ≈ 0) in all cases where SE_A is ≥10 dB [2, 3, 9, 22] at very high frequencies (~GHz or high).

10.2.2
Shielding Effectiveness: Experimental Details

Experimentally, shielding is measured using instruments called network analyzers (Figure 10.2). Both scalar and vector network analyzers (SNA or VNA) can be used, though SNA measures only the amplitude of signals compared to VNA which measures magnitude as well as phases of the incident/transmitted signals.

The incident and transmitted waves in a two-port VNA can be expressed in terms of complex scattering parameters (or S parameters), that is, S_{11} (or S_{22}) and S_{12} (or S_{21}), respectively. The first digit of the subscript number of S parameter represents the port from where the incident signal is coming, whereas the second digit denotes the port where the signal was measured. For example, S_{11} represents reflected signal (incident from port 1 and measured at port 1) and S_{21} represents transmitted signal (incident from port 2 and measured at port 1). The reflectance (R) and transmittance (T) can be expressed as: $T = |E_T/E_I|^2 = |S_{12}|^2 = |S_{21}|^2$, $R = |E_R/E_I|^2 = |S_{11}|^2 = |S_{22}|^2$, giving absorbance ($A$) as: $A = (1 - R - T)$. The intensity

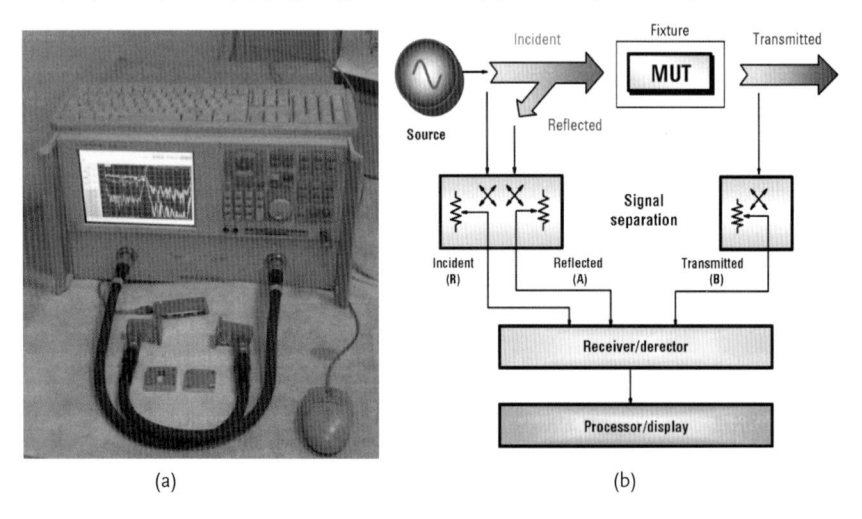

Figure 10.2 A two-port VNA (a) and its internal block diagram (b).

of the EM wave inside the shield after primary reflection is based on quantity (1 − R), so that effective absorbance $\{A_{\text{eff}} = [(1 − R − T)/(1 − R)]\}$ can be defined and reflection and absorption losses can be expressed as [3, 9, 14, 22]

$$SE_R(dB) = 10\log_{10}(1 − R) \tag{10.7}$$

$$SE_A(dB) = 10\log_{10}(1 − A_{\text{eff}}) = 10\log_{10}\left[\frac{T}{(1−R)}\right] \tag{10.8}$$

Therefore, from the measured S parameters, VNA automatically computes "R" and "T" that give attenuation due to reflection and absorption, that is, SE_R and SE_A, respectively.

10.2.3
Materials for EMI Shielding: Polymer-Based Nanocomposites

The careful analysis of theoretical shielding expressions revealed that in order to meet design requirements and for extending efficient shielding action, shield should possess a balanced combination of electrical conductivity (σ), dielectric permittivity (ε), and magnetic permeability (μ) [1, 3, 4, 6, 7, 9, 11, 23–26]. Further, the primary mechanism of EMI shielding is reflection from the front face of the shield, for which the shield must possess free charge carriers to cause ohmic (heating) losses in the shield. The secondary EMI shielding mechanism is absorption, for which shield should possess electric and/or magnetic dipoles that can interact with the electric (E) and magnetic (H) field components of the incident EM radiation [7, 9].

The earliest known materials for EMI shielding are metals [1, 2, 4–9, 11, 21, 22] mainly due to their high electrical/thermal conductivity and good mechanical properties. However, metal-based compositions are suffered from problems [1, 3, 4, 9, 16, 27, 28] such as high reflectivity, poor corrosion resistance, weight penalty, processing difficulties, and low values of specific properties (e.g., specific strength/modulus, thermal/electrical conductivity). In this direction, electrically conducting polymer composites represent a special class of materials that possess unique combination of electrical, thermal, dielectric, magnetic, and/or mechanical properties for realizing required electromagnetic shielding performance [23, 29–50]. The simplest example of natural polymer composite is wood which is a combination of reinforcing cellulose fibers (dispersed phase or filler) and lignin (continuous phase or matrix) binder. Similarly, carbon fiber reinforced epoxy is an example of man-made composite. Nanocomposites represent a special case of composites for which dimensions of filled inclusions are of nanometer (1 nm = 10^{-9} m; e.g., diameter of human hair~10^4 nm) range. These nanofillers facilitate very high interfacial area per volume to host polymer matrix leading to remarkable changes in the properties (of nanocomposites) compared to their bulk counterparts, for example, better specific strength/modulus, electrical/thermal properties, and SE [9, 15, 29, 31, 32, 34, 41–43, 51–53]. Polymer is a versatile choice as a matrix material due to advantages like low density, mechanical flexibility, facile processing, and corrosion resistance. The properties of polymer-based nanocomposites are governed by the nature and concentration of their constituents (i.e., filler and matrix) and level of interaction between them [2, 5, 41, 54]. Interestingly, most polymeric matrices possess poor electrical, dielectric, or magnetic properties and are transparent to electromagnetic radiations [9, 54]. Therefore, most of the electrical and electromagnetic properties of the conventional nanocomposites are mainly contributed by the incorporated fillers (nature and concentration), and matrix simply plays the role of holding the filler particles [41, 54]. The selection of a particular filler depends on the desired properties, for example, when electrical conduction is important, electrically conductive fillers are used, whereas for dielectric/magnetic properties, fillers with magnetic/electric dipoles are preferred. Here, the utilization of X-linkable, thermally/dimensionally stable, and mechanically strong yet moderate density thermosetting host matrix can offer an attractive solution over conventional thermoplastic polymer-based matrices [5, 18, 32, 44, 50, 55–57] that suffer from disadvantages like inferior mechanical properties and poor thermal stability. However, the handling of nanofillers and their incorporation (maintaining uniform dispersion as well as checking agglomeration) within thermosets is an extremely challenging task, mainly because of agglomeration tendency of nanofillers, difficult viscosity control of thermoset matrix, and trade-off between filler–matrix interaction and filler's intrinsic properties. These often resulted in failure to efficiently translate the intrinsic properties of nanofillers into the bulk properties of resultant nanocomposites. In fact, handling and efficient dispersion of nanofiller are the biggest challenges for thermoset nanocomposite technology. Consequently, the development of state-of-the-art commercially viable EMI shielding materials requires coherent strategies and dedicated efforts to achieve strict control over above-mentioned electromagnetic attributes. Therefore, in the past, a wide range of nanomaterials with good EM properties has been employed as fillers, for example, conducting polymers, carbon black, graphite and CNT, graphene, and dielectric (titanates) or magnetic (ferrites) nanoparticles [4, 5, 12, 18, 20, 34, 41–46, 55, 58–63]. The following section focuses on the EMI shielding properties of thermosetting nanocomposites with special reference to above fillers.

10.2.4
Synthesis of Thermosetting Nanocomposites

The generalized scheme for the fabrication of electrically conducting thermosetting nanocomposite is shown in Figure 10.3.

Both filler and thermosetting polymer/resin are separately dissolved/dispersed in a common liquid medium (under continuous stirring/sonication) and mixed thereafter followed by shaping (casting/molding) and stabilization (X-linking/setting/curing). In particular, X-linking can be achieved by incorporating an additional chemical called curing agent/hardener or by simply exposing the sample to high-energy radiations. Depending upon the nature of thermosetting resin, presence and nature of curing agent, and X-linking mechanism, external parameters

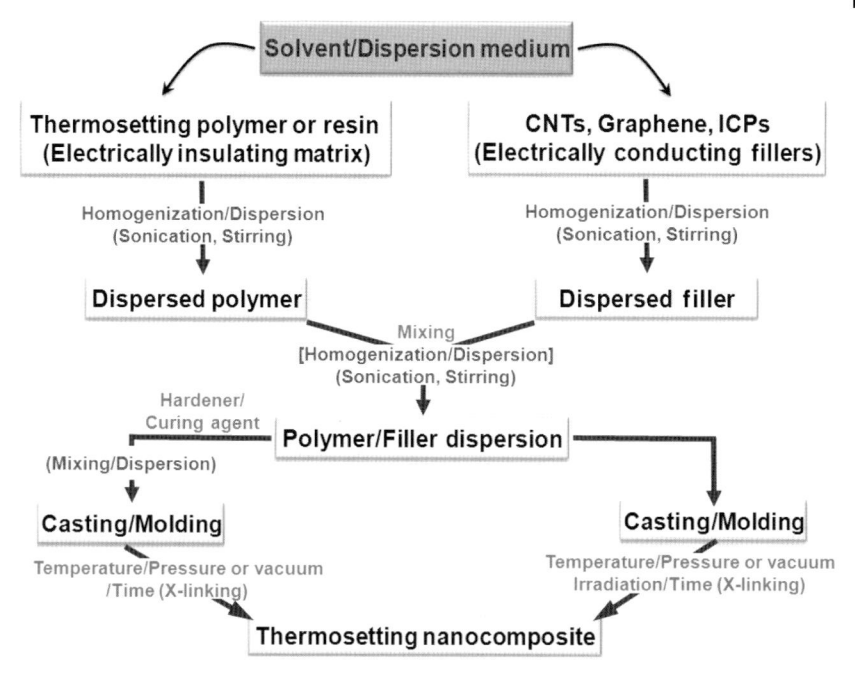

Figure 10.3 Schematic representation of steps involved in the formation of electrically conducting thermosetting nanocomposites.

(such as temperature and pressure/vacuum) and reaction/setting time have to be strictly controlled. It is interesting to note that during the X-linking reaction, the molecular weight increases to a point so that the melting point is higher than the surrounding ambient temperature and the material is transformed into an infusible/insoluble solid mass. As the decomposition temperature is higher than the melting point; therefore, once X-linked, a thermoset material cannot be melted or reshaped. Therefore, thermosetting materials (with 3D cross-linked network of bonds) are generally much stronger and thermally stable than thermoplastic materials. Consequently, they are more suitable for high-performance applications (engineering/aerospace/defense) with stringent requirements like high strength/modulus, good dimensional stability, and stable performance at elevated temperatures (up to the decomposition temperature) [35, 41, 52, 59, 62].

10.2.5
Electrical Properties of Thermosetting Nanocomposites

The thermosetting nanocomposites where the fillers are electrically conducting in nature (e.g., conducting polymers, graphite, carbon black, CNTs, and graphene) lead to onset of electrical conductivity within otherwise insulating matrix [4, 5, 12, 15, 18–20, 24, 33, 41, 44, 51, 53, 55, 56, 64, 65]. This can be attributed to the

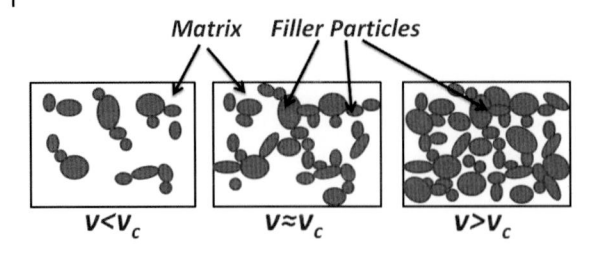

Figure 10.4 Schematic representation of interparticle contact and formation of conducting filler networks as a function of filler loading (v), before percolation ($v < v_c$), at percolation ($v \approx v_c$), and beyond percolation ($v > v_c$).

formation of 3D electrically conductive networks (schematic representation in Figure 10.4) within host thermoset matrix so that electrons can easily hop/tunnel between dispersed filler particles. The percolation threshold (the minimum filler loading where the first continuous networks of filler particles are formed within matrix polymer) for a given matrix–filler combination can be quantitatively estimated by plotting the electrical conductivity as a function of the reduced volume fraction of filler [5, 9, 32, 41, 53, 55, 56] and performing data fitting using a power law function [9, 55]:

$$\sigma = \sigma_o (v - v_c)^t \tag{10.9}$$

where σ is the electrical conductivity of the composite, σ_o is the characteristic conductivity, v is the volume fraction of filler, v_c is the volume fraction at the percolation threshold, and t is the critical exponent. In practical situations, where densities of polymer matrix and filled inclusion are same (e.g., for organic fillers like ICPs, CNTs, or graphene), the mass fraction (p) and volume fraction (v) of the filler can be assumed same. It is important to point out that the above-mentioned establishment of electrical conductivity leads to proportionate enhancement of EMI SE [5, 18, 20, 41, 55, 56]. Therefore, the next section is devoted to electrical properties and EMI shielding response of thermosetting nanocomposites, especially those based on ICPs, CNTs, and graphene as functional fillers.

10.2.5.1 ICP-Based Nanocomposites

ICPs especially those with metallic conductivity (also known as synthetic metals) represent important alternatives for metals particularly for electrical and EMI shielding applications [1, 3, 8, 9, 11, 14, 18, 22, 24, 28, 37, 66–72]. These materials possess distinguished advantages compared to metals, for example, low density, corrosion resistance, facile processing (via solution or melt blending routes), and good compatibility with various polymeric matrices. Furthermore, the regulation of their electrical conductivity by controlling parameters such as oxidation state, doping level, morphology, and chemical structure makes ICPs a powerful candidate for various techno-commercial applications [1, 9, 14, 73–88]. Figure 10.5 shows the structure of some of the well-known conducting polymers in their respective undoped or base forms along with comparison of electrical conductivity

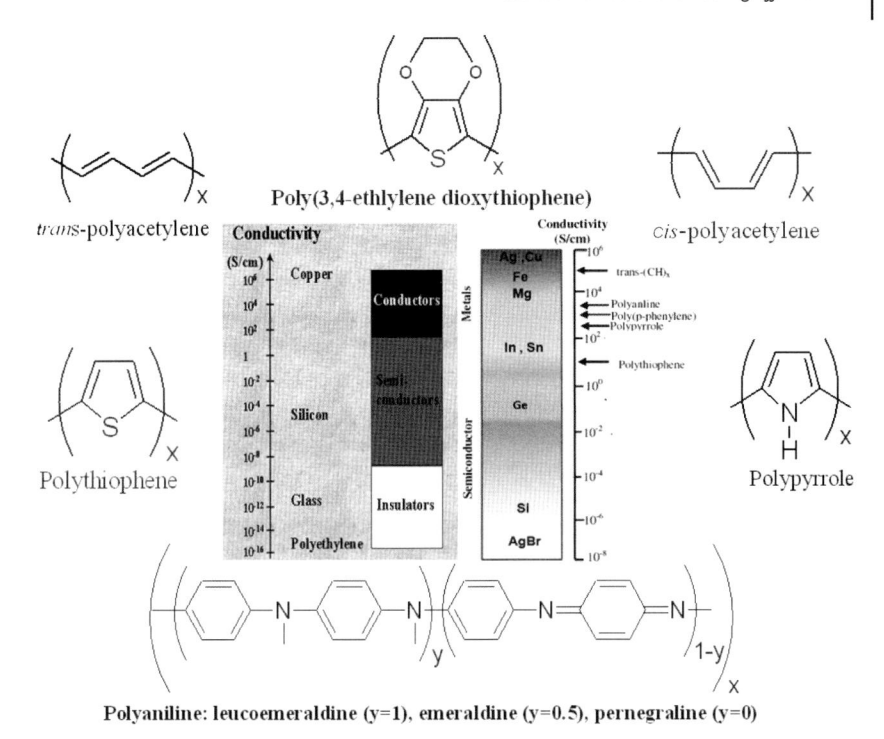

Figure 10.5 Chemical structures of repeat units of some conjugated polymers in respective undoped/base forms along with comparative conductivity details of their highly doped forms relative to common metals, semiconductors, and insulators.

value of their highly doped forms compared to various metals, conventional semiconductors, and common insulators.

The frequency dependence of intrinsic conductivity of conjugated polymers especially in the field of microwave absorption range (100 MHz–20 GHz) has inspired many ideas to use them for microwave and EMI shielding applications [1, 4, 9, 22, 24, 25, 69, 82, 89–91]. The unique properties like tunable conductivity and controllable EM attributes (permittivity/permeability) further strengthen their candidature as futuristic shielding material. In addition, they display unique shielding mechanism of reflection plus absorption compared to dominant reflection for metals or carbon-based compositions. However, despite above advantages, pure ICPs possess poor mechanical properties and have to be blended with thermoplastic or thermosetting matrices so as to combine the electronic properties of ICPs with the superior mechanical properties of the matrix polymer [4, 9–12, 16, 18, 44, 56, 66].

It has been observed that, for a given ICP–matrix system, the electrical properties of the resultant composites critically depend on loading level of the filler and the involved processing technique [18, 33, 37, 56], as shown in Figure 10.6. The higher filler content leads to the enhancement of number of conducting links and

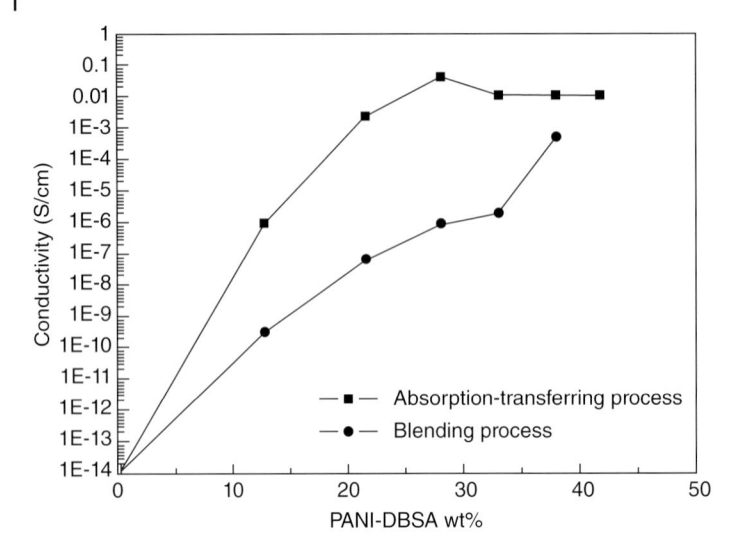

Figure 10.6 Electrical conductivity of the epoxy composites plotted with respect to the PANI-DBSA concentration. Reprinted with permission from [56] Copyright (2008) American Chemical Society.

improvement of interparticle charge transport (by tunneling or hopping phenomenon) resulting in improvement of electrical conductivity. Similarly, selection of proper processing technique helps in the improvement of dispersion of ICPs particles so that continuous conducting network can be achieved at relatively lower loading level. Conversely, at same loading level, a well-dispersed conducting filler (e.g., by absorption-transferring process in Figure 10.6) leads to higher conductivity compared to that less-dispersed one (e.g., by blending process in Figure 10.6).

It is also worth pointing out that incorporation of ICP in polymeric matrices not only leads to establishment and improvement of electrical conductivity but also contributes towards improvement of dielectric properties (e.g., real/imaginary permittivity, Figure 10.7) [8, 9, 12, 18, 24, 56, 92].

Such polarization and related relaxation phenomenons contribute towards energy storage and losses, respectively. Nevertheless, the establishment and enhancement of electrical conductivity is of paramount importance because it leads to improvement of both reflection and absorption loss components (according to Eqs. (10.3–10.5)) leading to enhancement (Figure 10.8) of overall shielding effectiveness.

Recently, it has also been shown that well-dispersed ICPs (e.g., PANI nanoparticles within epoxy matrix) not only provide a continuous conducting network but also facilitate better charge delocalization leading to huge negative permittivity (Figure 10.7) which is a characteristic signature of left-handed materials [18, 56]. Interestingly, though EMI shielding is closely related to electrical conductivity, the factors affecting EMI are more complex because they involve not only conduction (leading to ohmic losses) but also polarization phenomenon (due to the presence

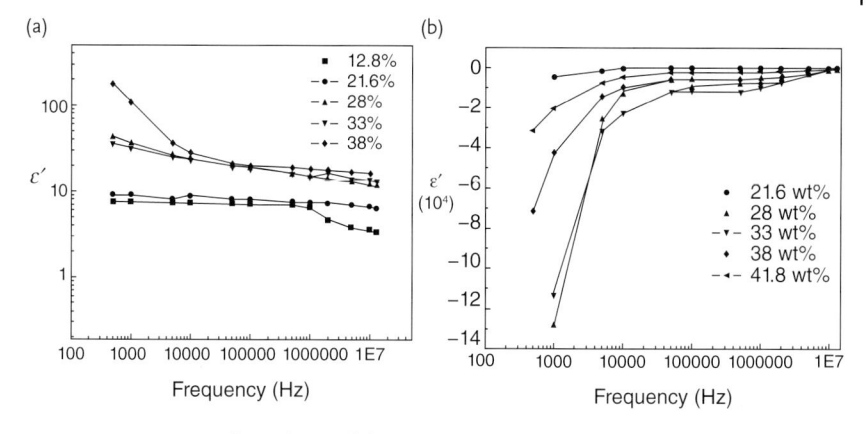

Figure 10.7 Frequency dependence of the dielectric constant (ε') for PANI-DBSA/epoxy hybrids prepared using the (a) blending and (b) absorption-transferring processes. Reprinted with permission from [56] Copyright (2008) American Chemical Society.

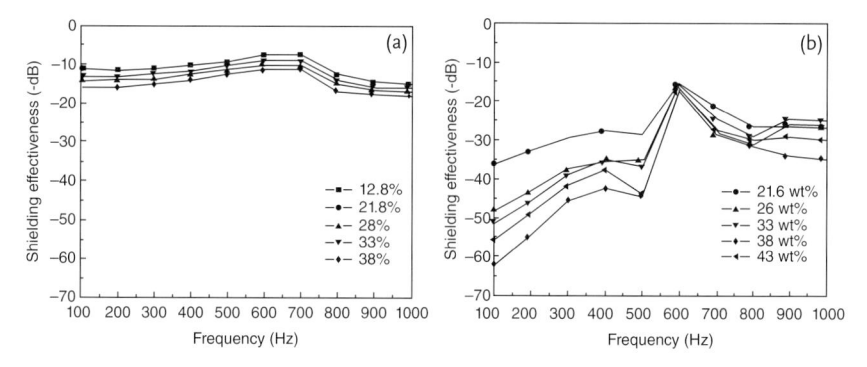

Figure 10.8 EMI shielding effectiveness toward electric fields plotted with respect to frequency (100 MHz–1 GHz) for PANI-DBSA/epoxy hybrids prepared using the (a) blending and (b) absorption-transferring processes. Reprinted with permission from [56] Copyright (2008) American Chemical Society.

of polarons/bipolarons, dipoles, and filler–matrix interfacial polarization). For examples, when polyaniline doped with dodecylbenzene sulfonic acid (PANI-DBSA) nanoparticles (filler) were added in the epoxy (matrix) [56], both the conductivity and dipole density increased (Figures 10.6 and 10.7) due to the formation of conductive networks. But any excess addition of PANI-DBSA (i.e., beyond the percolation threshold of 28 wt% PANI-DBSA) did not benefit the electrical conduction because of nanoparticle aggregation. However, it did result in more dipolar and interfacial polarization in the hybrid material. Therefore, 38 wt% PANI-DBSA/epoxy improved the EMI shielding efficiency (Figure 10.8) to −30 to −60 dB (in 100–1000 MHz range) without much enhancement of conductivity.

When ICPs are combined with other conducting fillers, significant reduction in percolation threshold, higher conductivity, and better shielding performance are observed as compared to pristine ICPs [91–93]. This may be attributed to bridging of metallic islands of ICPs (granular metals) as well as better dispersion of ICP-coated fillers within various host matrices. For example, PANI-grafted-MWCNT/epoxy composites display improved electrical and mechanical properties (e.g., tensile strength, Young's modulus, flexural strength, and flexural modulus) compared with neat epoxy or MWCNT/epoxy systems. Similarly, PANI-coated MWNT epoxy composites [92, 93] display improved microwave absorption response [91]. For many applications, for example, radar absorbers shield should contain electric and/or magnetic dipoles that can interact with the orthogonally pulsating electric and magnetic fields of the incident EM radiation [2, 4, 6, 7, 9, 10, 27, 66, 67, 94]. Combination of ICPs with various inorganic fillers with high permittivity or permeability (e.g., ferrites, titanates or other oxides) can give unique combination of properties like moderate electrical conductivity and good dielectric/magnetic properties so that superior shielding performance can be realized. Therefore, numerous attempts have also been made to introduce dielectric or fillers along with ICP [4, 9, 10, 12, 26, 50, 58, 66, 95–100] to obtain enhanced shielding response arising from balanced combination of ohmic, dielectric, and magnetic losses. The electric field loss is caused by the dielectric relaxation effect associated with permanent and induced molecular dipoles. As the frequency of incident EM wave increases (especially in the microwave region), dipole present in the system fails to maintain in-phase movement with rapidly pulsating electric vector. Such out-of-phase movement of dipoles leads to molecular friction resulting in energy dissipation in the form of heat. In contrast, magnetic losses are related to permeability of the material and occur due to phenomenon such as hysteresis, eddy currents, domain wall movement, or ferromagnetic resonance. Therefore, microwave absorption is a combined effect of dielectric and magnetic losses along with finite conductivity and matching thickness. For example, dielectric PANI/epoxy composites display only dielectric losses (Figure 10.9) compared to hybrid PANI/Fe_3O_4/epoxy nanocomposites [50], which display additional magnetic losses (Figure 10.10).

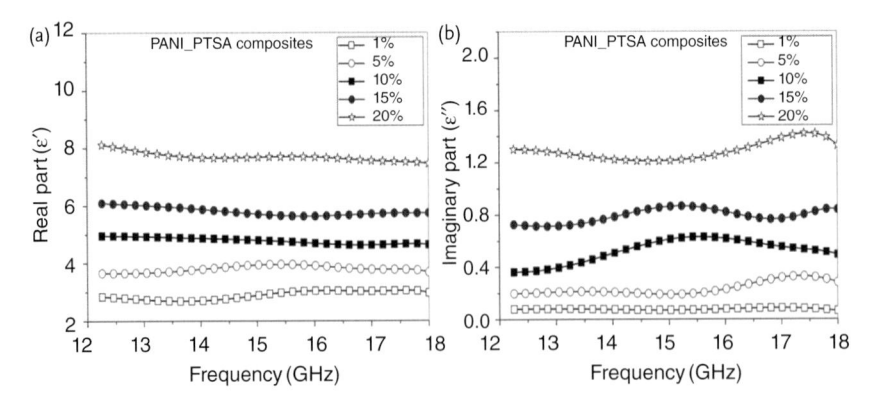

Figure 10.9 The PANI/epoxy resin composites frequency dependence of (a) real part of complex permittivity and (b) imaginary part of complex permittivity. Reprinted from [50], Copyright (2012), with permission from Elsevier.

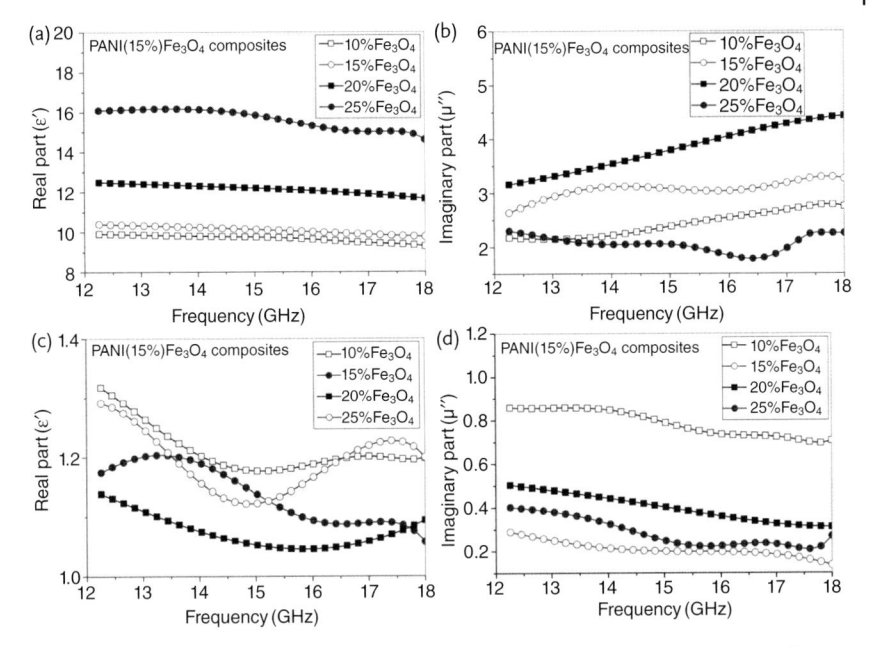

Figure 10.10 The PANI–PTSA (15%)/Fe$_3$O$_4$/epoxy resin hybrid composites frequency dependence of (a) real part of complex permittivity, (b) imaginary part of complex permittivity, (c) real part of complex permeability, and (d) imaginary part of complex permeability. Reprinted from [50], Copyright (2012), with permission from Elsevier.

It is worth mentioning that due to underlying dielectric and magnetic losses, materials with high permittivity and/or permeability are expected to display good microwave absorption response [1, 8–10, 12, 24, 27, 50, 61, 66, 101], for example, epoxy composite (Figure 10.11) containing 15% of PANI and 10% of Fe$_3$O$_4$ ($\varepsilon' = 10$) gives return loss of −42 dB (at 16.3 GHz), whereas composite of 15% of PANI and 25% of Fe$_3$O$_4$ ($\varepsilon' = 17$) gives return loss of −37.4 dB (at 14.85 GHz). In comparison, a purely dielectric composite with 20% of PANI ($\varepsilon' = 8.5$) shows a minimum reflection coefficient of only −11 dB (at 18 GHz).

10.2.5.2 CNT-Based Nanocomposites

The discovery of CNTs by Iijima [102] has revolutionized the nanotechnology because of their potential to replace metals especially for filled nanocomposites [5–7, 15, 29, 30, 39, 47, 53, 59, 63, 103]. CNTs consist of graphene sheets rolled up in the form of single or multiple concentric cylinders (Figure 10.12) constituting single-walled carbon nanotubes (SWCNTs) or multiwalled carbon nanotubes (MWCNTs), respectively.

In the recent past, CNTs has emerged as a promising filler due to outstanding properties such as metallic electrical conductivity, excellent corrosion resistance, and low density along with ultrahigh strength/modulus and thermal conductivity

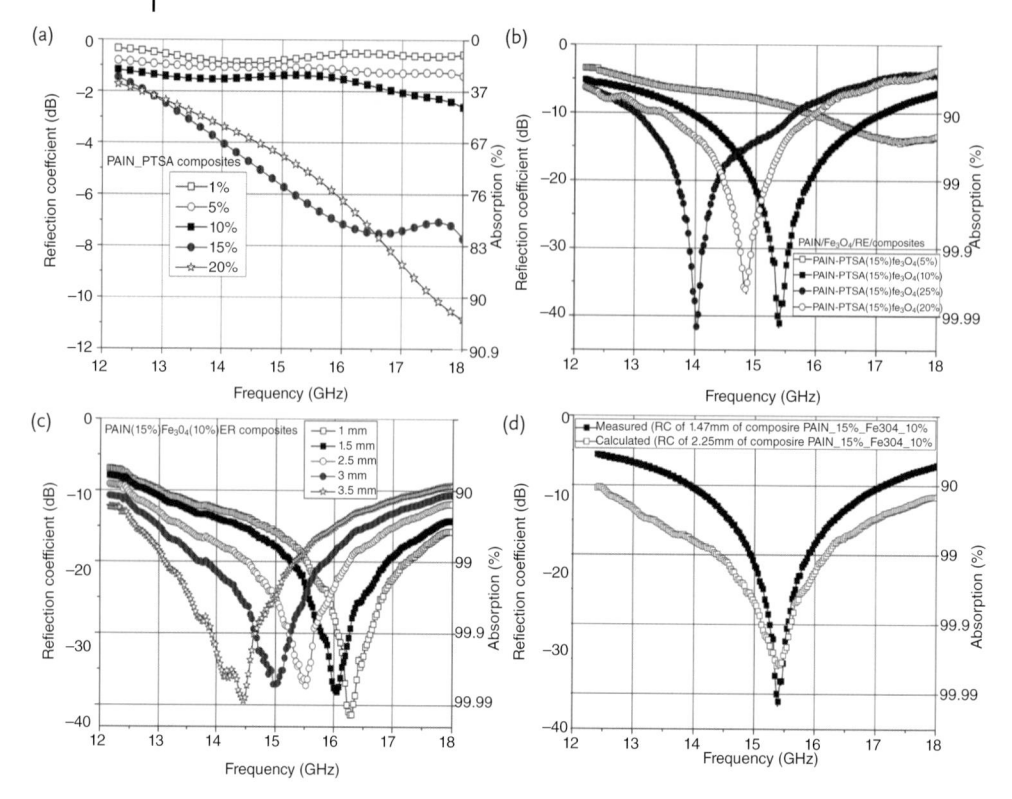

Figure 10.11 The composites frequency dependence of reflection coefficient at different filler's amount: PANI–PTSA/epoxy resin composites (a), PANI–PTSA/Fe₃O₄/epoxy resin hybrid composites (b), PANI-PTSA (15%)/Fe₃O₄ (10%)/epoxy resin hybrid composites at different thickness (c), and the comparison of measured and calculated reflection coefficient for PANI–PTSA (15%)/Fe₃O₄ (10%)/epoxy resin hybrid composite (d). Reprinted from [50], Copyright (2012), with permission from Elsevier.

[3, 9, 15, 29, 30, 32, 104]. In particular, higher intrinsic conductivity (10^4–10^6 S/cm) and aspect ratio (length/diameter ratio) of CNTs (100–5000) compared to other carbon-based fillers (e.g., carbon black, graphite, carbon fiber) permit the realization of conductivity (in nanocomposites) at much lower loadings (very low percolation threshold) [5, 9, 29, 30, 32, 53, 55]. As in case of ICPs, formation of continuous networks of CNTs within matrix (SEM image Figure 10.13) also leads to improvement in electrical, mechanical, and EM properties of resultant nanocomposites [5, 53, 55].

For example, Figure 10.14 shows the dc conductivity (σ) of SWNTs/epoxy composites as a function of SWNTs mass fraction (p) revealing that below 0.6 wt% loading, the conductivity displays a dramatic increase of 10 orders of magnitude indicating the formation of percolating network. Further, as shown in inset of Figure 10.11 for the log(σ) versus log($(p - p_c)/p_c$) plot, the conductivity of SWNT/epoxy composite agrees very well with the percolation p_c ~0.062% and t ~ 2.68

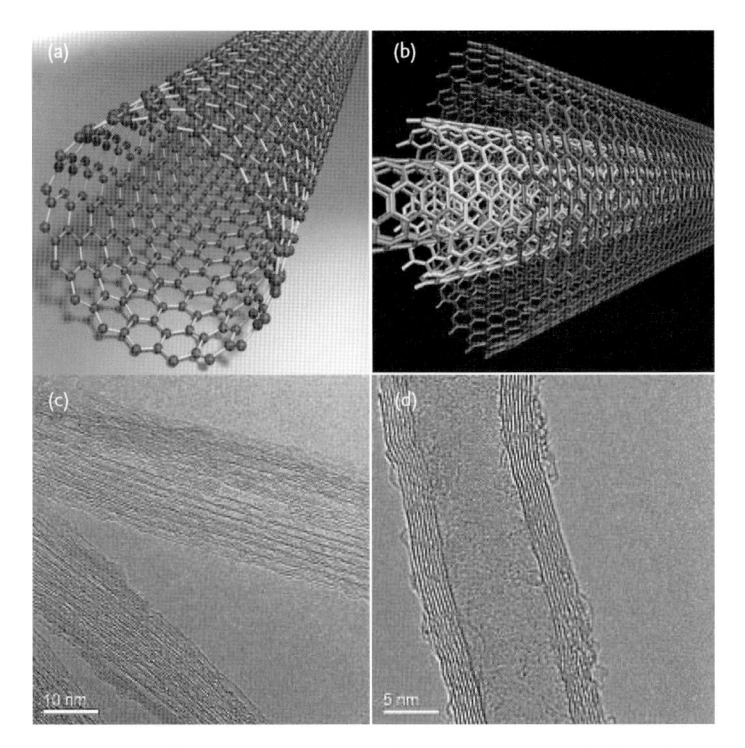

Figure 10.12 Schematic representation of (a) single-walled carbon nanotubes (SWCNTs) and (b) multiwalled carbon nanotubes (MWCNTs) along with their representative high-resolution TEM images showing (c) SWCNT bundles and (d) individual MWCNT (multilayered wall plus inner cavity).

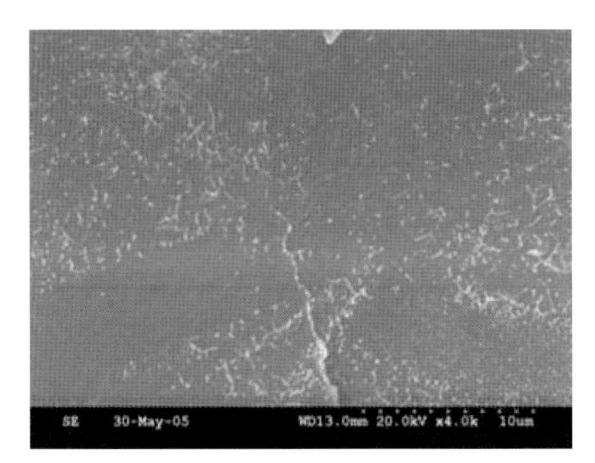

Figure 10.13 SEM image of the cross section of SWNTs-long/epoxy composites with 10 wt% loading. Reprinted with permission from [5] Copyright (2006) American Chemical Society.

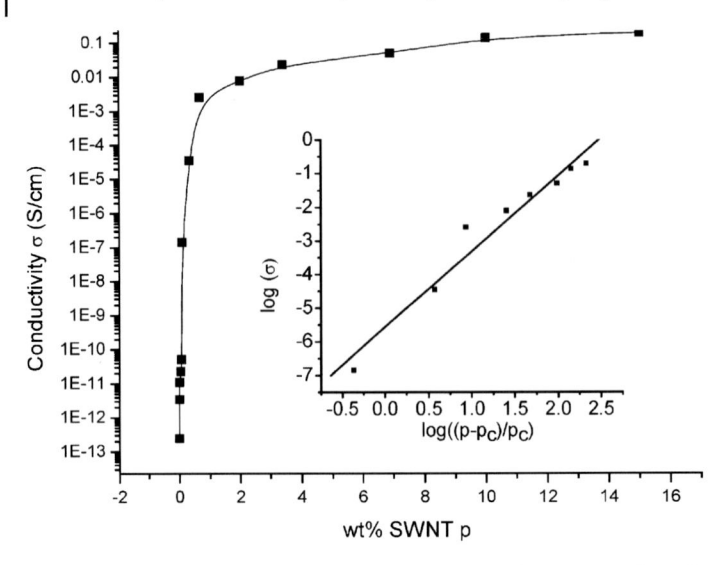

Figure 10.14 log DC conductivity (σ) versus mass fraction (p) of SWNTs-long composites measured at room temperature. Inset: log–log plot for σ versus $[(p - p_c)/p_c]$ for the same composites. The straight line in the inset is least-squares fit to the data using Eq. (10.9) returning the best fit values $p_c \sim 0.062\%$ and $t \sim 2.68$. Reprinted with permission from [5] Copyright (2006) American Chemical Society.

indicating a very efficient dispersion of SWNTs into the epoxy matrix [5]. Interestingly, these SWCNT–polymer composites possess high real permittivity (polarization, ε') as well as imaginary permittivity (adsorption or electric loss, ε'') [55], indicating that such composites could be used as electromagnetic wave absorbers, for example, for cell phone electronic protection [5]. Figure 10.15 shows the complex permittivity spectra of the composites containing 0.01–15% of long SWCNTs [55] that shows that the real (ε') and imaginary (ε'') permittivity increase dramatically as the concentration of the SWCNTs increases from 0.01 to 15 wt%. The highest values of the real and imaginary permittivity parts for the composite with 15% SWCNT loading reach 67 and 76, respectively. Overall, the real and imaginary parts of permittivity for the composites (Figure 10.15) with 15 wt% of SWCNTs range from 67–42 and 76–60, respectively, in the X-band (frequency range of 8.2–12.4 GHz).

The low percolation threshold extends an added advantage because low filler loading seldom leads to any significant disturbance of the physical and mechanical properties of the receptor thermoset matrix. In practice, as the loading level of the filler (hence conductivity) increases, both electrical conductivity (Figure 10.14) as well as dielectric properties (Figure 10.15) increase leading to proportionate enhancement of SE (Figure 10.16) [5, 55].

It is worth pointing that electrical conductivity of CNT-based polymer nanocomposites depends on many factors including type of CNTs, intrinsic conductivity, aspect ratio, surface functionalization, processing method, loading level, and

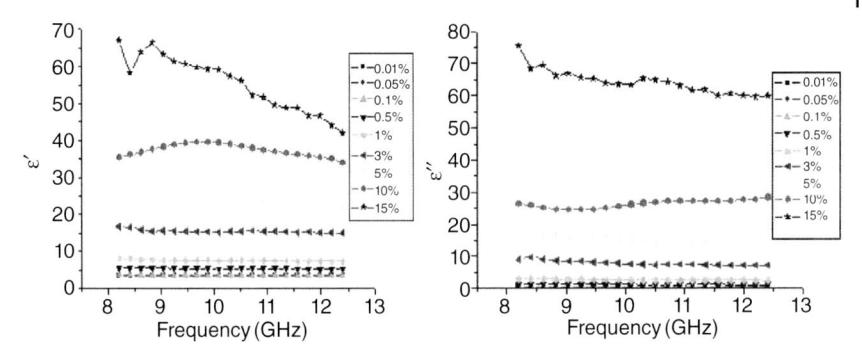

Figure 10.15 Complex permittivity spectra of the composites using "long-SWCNTs" with loading from 0.01 to 15 wt%. Reprinted from [55], Copyright (2007), with permission from Elsevier.

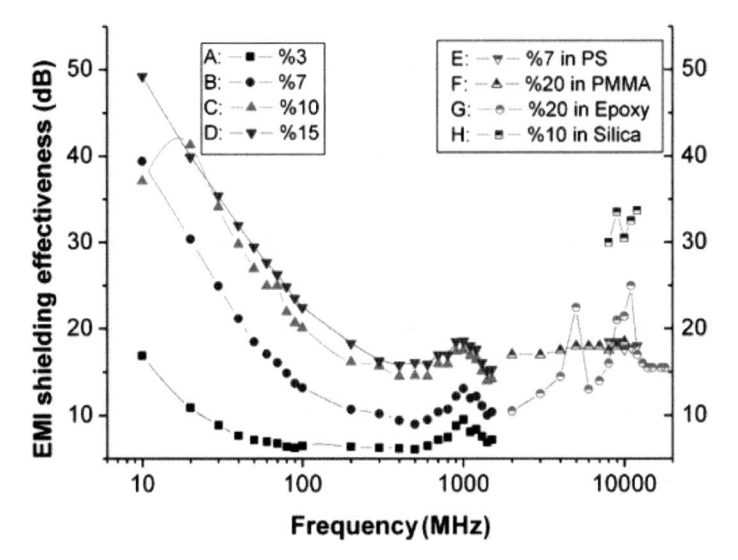

Figure 10.16 EMI shielding effectiveness (plots labeled A–D) for SWNT-polymer materials (3–15 wt%) studied in this work (10 MHz–1.5 GHz). Plots labeled E–H are higher frequency data for MWNT-based material presented for comparison: (e) MWNTs in PS; F, MWNTs in PMMA MWNTs in epoxy resin and the value of the y-axis for G is the reflection loss, and H MWNTs in silica. Reprinted with permission from [5] Copyright (2006) American Chemical Society.

nature of host polymer matrix [5, 29, 30, 32, 55, 104]. In practice, besides intrinsic conductivity and aspect ratio of the filler, the electrical and electromagnetic properties of the CNT/polymer nanocomposites [5, 55] also depend on the actual length of CNTs as well as pretreatment such as annealing (Figure 10.17).

From the simplified percolation and EMI theories for the isotropic dispersion of a random rod network, the CNT composites with higher bundle aspect ratios

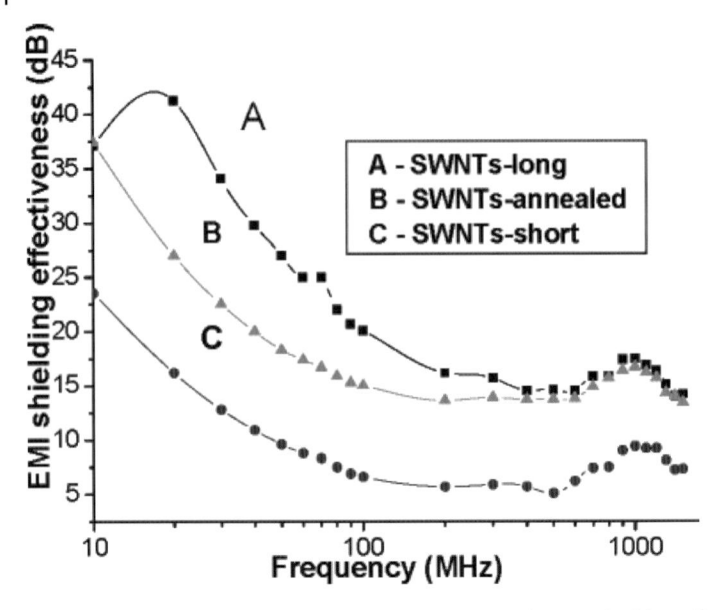

Figure 10.17 Impact of wall integrity and aspect ratio on the EMI shielding effectiveness of the composites containing 10 wt% SWNTs. Reprinted with permission from [5] Copyright (2006) American Chemical Society.

are predicted to have lower percolation threshold concentrations, higher conductivities, and better EMI shielding performances under the same weight percent loading. It is well known that high-temperature annealing of SWNTs in inert gas or vacuum can remove wall defects. This annealing treatment is therefore expected to improve the dc conductivity and thus the EMI SE. It is important to note that, though CNTs impart conductivity to nanocomposites, mechanical properties remained underachieved due to poor dispersion and agglomeration of CNTs especially at high loading levels [15, 92, 93, 105, 106]. Such problems get reflected in terms of inferior mechanical properties due to lack of interfacial adhesion between CNT and matrix polymer leading to poor load transfer characteristics. Further, any attempts to improve the CNT dispersion by chemical functionalization leads to deterioration of intrinsic properties (electrical/thermal conductivity, tensile strength/modulus) of CNTs due to damage to tube's outer wall causing partial loss of conjugation. To counter such problems, several attempts have also been made to enwrap (surface coating involving physical interaction) [9, 92] the CNTs with polymeric materials leading to better CNT dispersion along with improved dielectric properties and enhanced microwave absorption response. Hybrid approach where CNT-decorated fillers such as carbon fibers/cloth are used as multiscale conducting reinforcement has also been adopted to fabricate composites with improved electrical, thermal, and electromagnetic response [106]. Besides, high CNT content (>50 wt%) yet mechanically strong (tensile strength~2088 MPa, modulus~169 GPa) CNT nanocomposites have also been prepared [107] by adopt-

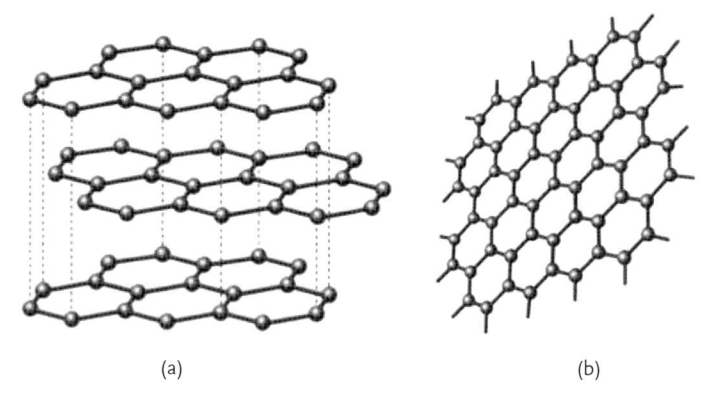

Figure 10.18 Schematic representation of graphite (a) with stacked graphene layers and individual graphene sheet (b).

ing simple mechanical stretching and prepregging (preresin impregnation) processes on initially randomly dispersed, commercially available sheets of millimeter-long MWNTs that lead to substantial alignment enhancement, good dispersion, and unprecedentedly high electrical conductivity (5500 S/cm). These nanocomposites with unique integration of superior electrical, thermal, and mechanical properties have opened the door for developing structurally strong EMI shielding materials.

10.2.5.3 Graphene-Based Nanocomposites

Graphene is a two-dimensional conducting nanofiller that can be considered as ultimate carbon allotrope made up of atomic sheet of sp^2 carbon atoms (Figure 10.18) arranged in a hexagonal lattice [13, 41, 108].

In fact graphene is known as the thinnest (one-carbon-atom thick) yet strongest material (on the basis of specific strength) compared to other carbon allotropes (graphite, carbon fibers, fullerenes, CNTs, etc.) or conventional metals. In addition, graphene also possesses exceptional electrical and thermal properties making it promising candidate for electronics and EMI shielding applications [13, 20, 36, 41, 108–112].

Recently, ability to synthesize and isolate individual graphene sheets has attracted enormous attention due to novel properties [13, 41, 109, 113] and number of related applications. Again, the promising EMI shielding response of graphene (as in case of ICPs and CNTs) can be attributed to outstanding electrical conductivity and high aspect ratio. The dispersion of these nanosheets within polymeric matrices leads to onset of electrical conductivity (at very low percolation threshold, Figure 10.19) which scales with loading level [20, 41]. It is also found that for a given matrix and loading level, percolation threshold is much lower for graphene than other carbon-based fillers [41].

However, though much progress has been made in the science and technology of graphene-based thermoplastic nanocomposites [36, 111, 112, 114, 115],

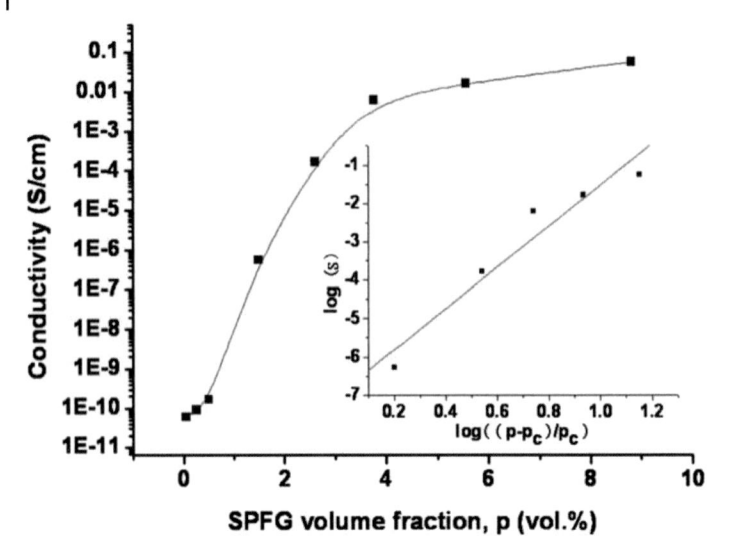

Figure 10.19 $\log_{10}$ DC conductivity (s) versus volume fraction (p) of solution processable functionalized graphene (SPFG)/epoxy composites measured at room temperature. Inset: log–log plot for s versus $[(p - p_c)/p_c]$ for the same composites along with straight line fit of data by least squares technique. The best fit gave values $p_c \sim 0.52$ vol% and $t \sim 5.37$ with a correlation factor of 0.97. Reprinted from [20], Copyright (2009), with permission from Elsevier.

graphene-filled thermosetting nanocomposites for EMI shielding applications are still at an early stages of development [20]. This may be primarily attributed to the processing difficulties, for example, agglomeration tendency of graphene sheets/particles (due to extremely high-specific surface area), poor compatibility with relatively polar thermosetting matrices (leading to nonuniform dispersion), and viscosity built-up of thermoset resin (upon nanofiller incorporation). Interestingly, the controlled functionalization of graphene leads to the formation of electrically conducting nanocomposites with very low percolation thresholds. For example, composites based on graphene sheets made by incorporating solution-processable functionalized graphene (SPFG) into an epoxy matrix [20] display percolation (Figure 10.19) at only 0.52 vol% SPGF loading. This is attributed to high aspect ratio of the graphene sheets and their homogeneous dispersion in the thermoset epoxy matrix.

Here, it is worth mentioning that, even at percolation, the conductivity is not sufficient to extend any significant EMI shielding (Figure 10.20) action (e.g., SE <−5 dB up to 1% loading). This can be attributed to the presence of EM radiation transparent patches (regions lacking graphene links) of nonconducting epoxy matrix. However, as the SPGF content increases, beyond 1.0 wt% loading, the conducting link density increases rapidly, which ultimately gets reflected in the improved EMI shielding performance (e.g., attenuation of −21 dB at 15 wt% or 8.8 vol% loading) in the X band (8.2–12.4 GHz frequency range). Nevertheless, the

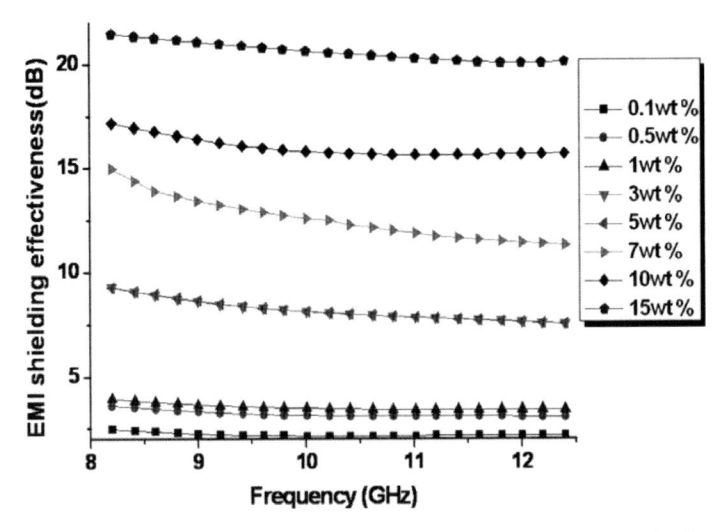

Figure 10.20 EMI SE of graphene/epoxy composites with various SPFG loadings as a function of frequency in the X-band. Reprinted from [20], Copyright (2009), with permission from Elsevier.

graphene-based thermosetting nanocomposites research is still at very early stage of evolution, and issues like processing difficulties and detailed understanding of their EMI shielding/microwave absorption properties require further attention so as to utilize the full technological potential of these materials.

10.3
Conclusions

Based on the exhaustive review of electrically conducting thermosetting nanocomposites, it is clear that conducting fillers like ICPs, CNTs, and graphene present an attractive solution for realizing structurally strong EMI shielding materials. However, despite of facile processing, ability to accommodate dielectric/magnetic fillers, and good compatibility of ICPs with various matrices, their low intrinsic conductivity (compared to metals/carbonaceous materials), high percolation threshold (lower aspect ratios), and lack of mechanical reinforcing ability are still a matter of great concern. Similarly, regardless of their outstanding properties and high aspect ratios, CNTs/graphene-based thermosetting nanocomposites have still not been able to touch the theoretically predicted/ultimate limits. In particular, problems involving uniform dispersion, prevention of agglomeration, and improved interfacial interaction with host matrices must be addressed to realize their full technical potential for development of advanced nanocomposites for structural and EMI shielding applications. Nevertheless, to overcome the limitations of existing materials for realization of state-of-the-art EMI shielding materials, complex methodologies are suggested including strategic combination of materials (conducting polymers, carbon-based materials, and dielectric/magnetic nanofillers) as well as coherent designs including multilayered structures, multiscale materials, and meta-nanocomposites.

Acknowledgments

Author is thankful to Director NPL for according permission of publication. Special thanks to Dr. Manju Arora and Dr. B.P. Singh for useful discussions and suggestions. I am also thankful to all those who pushed me beyond my limits.

References

1 Joo, J. and Epstein, A.J. (1994) Electromagnetic radiation shielding by intrinsically conducting polymers. *Appl. Phys. Lett.*, **65** (18), 2278–2280.
2 Ott, H.W. (2009) *Electromagnetic Compatibility Engineering*, John Wiley & Sons, Inc., New Jersey.
3 Saini, P., Choudhary, V., Singh, B.P., Mathur, R.B., and Dhawan, S.K. (2009) Polyaniline–MWCNT nanocomposites for microwave absorption and EMI shielding. *Mater. Chem. Phys.*, **113** (2–3), 919–926.
4 Olmedo, L., Hourquebie, P., and Jousse, F. (1997) *Handbook of Organic Conductive Molecules and Polymers*, vol. 2, John Wiley & Sons, Ltd, Chichester, UK.
5 Li, N., Huang, Y., Du, F., He, X., Lin, X., Gao, H., Ma, Y., Li, F., Chen, Y., and Eklund, P.C. (2006) Electromagnetic interference (EMI) shielding of single-walled carbon nanotube epoxy composites. *Nano Lett.*, **6** (6), 1141–1145.
6 Chung, D.D.L. (2000) Materials for electromagnetic interference shielding. *J. Mater. Eng. Perform.*, **9** (3), 350–354.
7 Chung, D.D.L. (2001) Electromagnetic interference shielding effectiveness of carbon Materials. *Carbon*, **39** (2), 279–285.
8 Lakshmi, K., John, H., Mathew, K.T., Joseph, R., and George, K.E. (2009) Microwave absorption, reflection and EMI shielding of PU-PANI composite. *Acta Mater.*, **57**, 371–375.
9 Saini, P., Choudhary, V., Singh, B.P., Mathur, R.B., and Dhawan, S.K. (2011) Enhanced microwave absorption behavior of polyaniline–CNT/polystyrene blend in 12.4–18.0 GHz range. *Synth. Met.*, **161** (15–16), 1522–1526.
10 Abbas, S.M., Dixit, A.K., Chatterjee, R., and Goel, T.C. (2005) Complex permittivity and microwave absorption properties of $BaTiO_3$–polyaniline composite. *Mater. Sci. Eng. B*, **123** (2), 167–171.
11 Colaneri, N.F. and Shacklette, L.W. (1992) EMI shielding measurements of conductive polymer blends. *IEEE Trans. Instrum. Meas.*, **41**, 291–297.
12 Das, C.K. and Mandal, A. (2012) Microwave absorbing properties of DBSA-doped polyaniline/$BaTiO_3$–$Ni_{0.5}Zn_{0.5}Fe_2O_4$ nanocomposites. *J. Mater. Sci. Res.*, **1** (1), 45–53.
13 Geim, A.K. and Novoselov, K.S. (2007) The rise of graphene. *Nat. Mater.*, **6**, 183–191.
14 Joo, J., Song, H.G., Jang, K.S., and Oh, E.J. (1999) Electromagnetic interference shielding efficiency of polyaniline mixtures and multilayer films. *Synth. Met.*, **102** (1–3), 1346–1349.
15 Ajayan, P.M., Schadler, L.S., Giannaris, C., and Rubio, A. (2000) Single-walled carbon nanotube–polymer composites: strength and weakness. *Adv. Mater.*, **12** (10), 750–753.
16 Shacklette, L.W., Colaneri, N.F., Kulkarni, V.G., and Wessling, B. (1992)

EMI shielding of intinsically conductive polymers. *J. Vinyl Technol.*, **14** (2), 118–122.

17 Huang, C.-Y. and Wu, C.-C. (2000) The EMI shielding effectiveness of PC/ABS/nickel-coated-carbon-fiber composites. *Eur. Polym. J.*, **36** (12), 2729–2737.

18 Hsieh, C.-H., Lee, A.-H., Liu, C.-D., Han, J.-L., Hsieh, K.-H., and Lee, S.-N. (2012) Polyaniline nano-composites with large negative dielectric permittivity. *AIP Adv.*, **2** (1), 012127 (8 pages).

19 Qin, F. and Brosseau, C. (2012) A review and analysis of microwave absorption in polymer composites filled with carbonaceous particles. *J. Appl. Phys.*, **111** (6), 061301 (24 pages).

20 Liang, J., Wang, Y., Huang, Y., Ma, Y., Liu, Z., Cai, J., Zhang, C., Gao, H., and Chen, Y. (2009) Electromagnetic interference shielding of graphene/epoxy composites. *Carbon*, **47** (3), 92–925.

21 Schulz, R.B., Plantz, V.C., and Brush, D.R. (1988) Shielding theory and practice. *IEEE Trans.*, **30** (3), 187–201.

22 Saini, P. and Choudhary, V. (2013) Enhanced electromagnetic interference shielding effectiveness of polyaniline functionalized carbon nanotubes filled polystyrene composites. *J. Nanopart. Res.*, **15**, 1415. doi: 10.1007/s11051-012-1415-2

23 Pomposo, J.A., Rodriguez, J., and Grande, H. (1999) Polypyrrole-based conducting hot melt adhesives for EMI shielding applications. *Synth. Met.*, **104** (2), 107–111.

24 Saini, P. and Choudhary, V. (2013) Structural details, electrical properties, and electromagnetic interference shielding response of processable copolymers of aniline. *J. Mater. Sci.*, **48** (2), 797–804.

25 Olmedo, L., Hourquebie, P., and Jousse, F. (1995) Microwave properties of conductive polymers. *Synth. Met.*, **69** (1–3), 205–208.

26 Kurlyandskaya, G.V., Cunanan, J., Bhagat, S.M., Aphesteguy, J.C., and Jacobo, S.E. (2007) Field-induced microwave absorption in Fe_3O_4 nanoparticles and Fe_3O_4/polyaniline composites synthesized by different methods. *J. Phys. Chem. Solids*, **68**, 1527–1532.

27 Abbas, S.M., Chatterjee, R., Dixit, A.K., Kumar, A.V.R., and Goel, T.C. (2007) Electromagnetic and microwave absorption properties of (Co^{2+}–Si^{4+}) substituted barium hexaferrites and its polymer composite. *J. Appl. Phys.*, **101** (7), 074105 (6 pages).

28 Saini, P., Choudhary, V., Sood, K.N., and Dhawan, S.K. (2009) Electromagnetic interference shielding behavior of polyaniline/graphite composites prepared by *in situ* emulsion pathway. *J. Appl. Polym. Sci.*, **113** (5), 3146–3155.

29 Ajayan, P.M., Stephan, O., Colliex, C., and Trauth, D. (1994) Aligned carbon nanotube arrays formed by cutting a polymer resin–nanotube composite. *Science*, **265** (5176), 1212–1214.

30 Baughman, R.H., Zakhidov, A.A., and de Heer, W.A. (2002) Carbon nanotubes the route toward applications. *Science*, **297** (5582), 787–792.

31 Mathur, R.B., Singh, B.P., and Pandey, S. (2010) *Polymer Nanotubes Nanocomposites, Synthesis Properties and Applications* (ed. V. Mittal), Wiley Scrivener, Beverly, MA.

32 Moniruzzaman, M. and Winey, K.I. (2006) Polymer nanocomposites containing carbon nanotubes. *Macromolecules*, **39** (16), 5194–5205.

33 Pud, A., Ogurtsov, N., Korzhenko, A., and Shapoval, G. (2003) Some aspects of preparation methods and properties of polyaniline blends and composites with organic polymers. *Prog. Polym. Sci.*, **28** (12), 1701–1753.

34 Thostenson, E.T., Li, C., and Chou, T.-W. (2005) Nanocomposites in context. *Compos. Sci. Technol.*, **65** (3–4), 491–516.

35 Tong, X.C. (2009) *Advanced Mater. and Design For Electromagnetic Interference Shielding*, CRC Press Taylor & Francis, London, NY.

36 Varrla, E., Venkataraman, S., and Sundara, R. (2011) Functionalized Graphene–PVDF foam composites for EMI shielding. *Macromol. Mater. Eng.*, **296** (10), 894–898.

37 Wessling, B. (1999) Polyaniline on the metallic side of the insulator-to-metal

transition due to dispersion: the basis for successful nano-technology and industrial applications of organic metals. *Synth. Met.*, **102** (1–3), 1396–1399.
38 Wojkiewicz, J.L., Fauveaux, S., and Miane, J.L. (2003) Electromagnetic shielding properties of polyaniline composites. *Synth. Met.*, **135–136**, 127–128.
39 Yang, Y., Gupta, M.C., Dudley, K.L., and Lawrence, R.W. (2005) Novel carbon nanotube–polystyrene foam composites for electromagnetic interference shielding. *Nano Lett.*, **5** (4), 2131–2134.
40 Yang, Y., Gupta, M.C., Dudley, K.L., and Lawrence, R.W. (2005) Conductive carbon nanofiber-polymer foam structures. *Adv. Mater.*, **17** (16), 1999–2003.
41 Kuilla, T., Bhadra, S., Yao, D., Kim, N.H., Bose, S., and Lee, J.H. (2010) Recent advances in graphene based polymer composites. *Prog. Polym. Sci.*, **35**, 1350–1375.
42 Hussain, F., Hojjati, M., Okamoto, M., and Gorga, R.E. (2006) Polymer–matrix nanocomposites, processing, manufacturing, and application: an overview. *J. Compos. Mater.*, **40** (17), 1511–1575.
43 Kumar, R., Dhakate, S.R., Saini, P., and Mathur, R.B. (2013) *RSC Advances*. doi: 10.1039/c3ra00121k
44 Belaabed, B., Wojkiewicz, J.L., Lamouri, S., El Kamchi, N., and Redon, N. (2012) Thermomechanical behaviors and dielectric properties of polyaniline-doped para-toluene sulfonic acid/epoxy resin composites. *Polym. Adv. Technol.*, **23** (8), 1194–1201.
45 Du, J. and Cheng, H.-M. (2012) The fabrication, properties, and uses of graphene/polymer composites. *Macromol. Chem. Phys.*, **213** (10–11), 1060.
46 Verdejo, R., Bernal, M.M., Romasanta, L.J., and Manchado, M.A.-L. (2011) Graphene filled polymer nanocomposites. *J. Mater. Chem.*, **21** (10), 3301–3310.
47 Liu, L., Kong, L.B., Yin, W.Y., Chen, Y., and Matitsine, S. (2010) Microwave dielectric properties of carbon nanotube composites. *Carbon Nanotubes*, InTech,

ISBN 978-953-307-054-4. doi: 10.5772/39420
48 Jou, W.-S., Cheng, H.Z., and Hsu, C.-F. (2006) A carbon nanotube polymer-based composite with high electromagnetic shielding. *J. Electron. Mater.*, **35** (3), 462–470.
49 Ma, P.-C., Siddiqui, N.A., Marom, G., and Kim, J.-K. (2010) Dispersion and functionalization of carbon nanotubes for polymer-based nanocomposites: a review. *Compos. A Appl. Sci. Manuf.*, **41** (10), 1345–1367.
50 Belkacem, B., Jean, L., Wojkiewicz, S.L., Noureddine, E.K., and Tuami, L. (2012) Synthesis and characterization of hybrid conducting composites based on polyaniline/magnetite fillers with improved microwave absorption properties. *J. Alloys Compd.*, **527**, 137–144.
51 Choudhary, V. and Gupta, A. (2011) *Carbon Nanotubes-Polymer Nanocomposites* (ed. S. Yellampalli), InTech.
52 Alexandre, M. and Dubois, P. (2000) Polymer-layered silicate nanocomposites: preparation, properties and uses of a new class of materials. *Mater. Sci. Eng. R*, **28** (1–2), 1–63.
53 Ramasubramaniam, R., Chen, J., and Liu, H. (2003) Homogeneous carbon nanotube/polymer composites for electrical applications. *Appl. Phys. Lett.*, **83** (14), 2928 (3 pages).
54 Riande, E. and Diaz-Calleja, R. (2004) *Electrical Properties of Polymers*, CRC Press, Boca Raton, FL, ISBN 978-1-4200-3047-1.
55 Huang, Y., Li, N., Ma, Y., Du, F., Li, F., He, X., Lin, X., Gao, H., and Chen, Y. (2007) The influence of single-walled carbon nanotube structure on the electromagnetic interference shielding efficiency of its epoxy composites. *Carbon*, **45** (8), 1614–1621.
56 Liu, C.-D., Lee, S.-N., Ho, C.-H., Han, J.-L., and Hsieh, K.-H. (2008) Electrical properties of well-dispersed nanopolyaniline/epoxy hybrids prepared using an absorption-transferring process. *J. Phys. Chem. C*, **112** (41), 15956–15960.
57 Nam, I.W., Lee, H.K., and Jang, J.H. (2011) Electromagnetic interference

shielding/absorbing characteristics of CNT embedded epoxy composites. *Compos. A Appl. Sci. Manuf.*, **42**, 1110–1118.

58 Kamchi, N.E., Belaabed, B., Wojkiewicz, J.-L., Lamouri, S., and Lasri, T. (2012) Hybrid polyaniline/nanomagnetic particles composites: high performance materials for EMI shielding. *J. Appl. Polym. Sci.*, (7 pages). doi: 10.1002/APP.38036

59 Che, R.-C., Peng, L.-M., Duan, X.-F., Chen, Q., and Liang, X.-L. (2004) Microwave absorption enhancement and complex permittivity and permeability of Fe encapsulated within carbon nanotubes. *Adv. Mater.*, **16** (5), 401–405.

60 Lu, J., Moon, K.-S., Kim, B.-K., and Wong, C.P. (2007) High dielectric constant polyaniline/epoxy composites via *in situ* polymerization for embedded capacitor applications. *Polymer*, **48** (6), 1510–1516.

61 Azadmanjiri, J., Hojati-Talemi, P., Simon, G.P., Suzuki, K., and Selomulya, C. (2011) Synthesis and electromagnetic interference shielding properties of iron oxide/polypyrrole nanocomposites. *Polym. Eng. Sci.*, **51** (2), 247–253.

62 Mehdipour, A., Rosca, I.D., Trueman, C.W., Sebak, A.-R., and Van Hoa, S. (2012) Multiwall carbon nanotube–epoxy composites with high shielding effectiveness for aeronautic applications. *IEEE Trans. Electromagn. Compat.*, **54** (1), 28–36.

63 Sun, X.-G., Gao, M., Li, C., and Wu, Y. (2011) Microwave absorption characteristics of carbon nanotubes. *Carbon Nanotubes – Synthesis, Characterization, Applications*, ISBN 978-953-307-497-9.

64 Liu, L. and Grunlan, J.C. (2007) Clay assisted dispersion of carbon nanotubes in conductive epoxy nanocomposites. *Adv. Funct. Mater.*, **17** (14), 2343–2348.

65 Liu, Z., Bai, G., Huang, Y., Ma, Y., Du, F., Li, F., Guo, T., and Chen, Y. (2007) Reflection and absorption contributions to the electromagnetic interference shielding of single-walled carbon nanotube/polyurethane composites. *Carbon*, **45** (4), 821–827.

66 Abbas, S.M., Chandra, M., Verma, A., Chatterjee, R., and Goel, T.C. (2006) Complex permittivity and microwave absorption properties of a composite dielectric absorber. *Compos. A Appl. Sci. Manuf.*, **37** (11), 2148–2154.

67 Chandrasekhar, P. and Naishadham, K. (1999) Broadband microwave absorption and shielding properties of a poly(aniline). *Synth. Met.*, **105** (2), 115–120.

68 Gairola, S.P., Verma, V., Kumar, L., Dar, M.A., Annapoorni, S., and Kotnala, R.K. (2010) Enhanced microwave absorption properties in polyaniline and nano-ferrite composite in X-band. *Synth. Met.*, **160** (21–22), 2315–2318.

69 Naishadham, K. and Kadaba, P.K. (1991) Measurement of the microwave conductivity of a polymeric material with potential applications in absorbers and shielding. *IEEE Trans. Microw. Theory Technol.*, **39** (7), 1158–1164.

70 Saini, P., Arora, M., Gupta, G., Gupta, B.K., Singh, V.N., and Choudhary, V. (2013) High permittivity polyaniline/barium titanate nanocomposites with excellent electromagnetic interference shielding response. *Nanoscale*. doi: 10.1039/C3NR00634D

71 Saini, P., Choudhary, V., and Dhawan, S.K. (2009) Electrical properties and EMI shielding behavior of highly thermally stable polyaniline/colloidal graphite composites. *Polym. Adv. Technol.*, **20** (4), 355–361.

72 Taka, T. (1991) EMI shielding measurements on poly(3-octyl thiophene) blends. *Synth. Met.*, **41** (3), 1177–1180.

73 Saini, P. and Arora, M. (2012) Microwave absorption and EMI shielding behavior of nanocomposites based on intrinsically conducting polymers, graphene and carbon nanotubes, in *New Polymers for Special Applications* (ed. A.D. Gomes), Intech, Croatia. doi: 10.5772/48779 http://www.intechopen.com/download/pdf/38964

74 Chiang, C.K., Fincher, C.R.J., Park, Y.W., Jr., Heeger, A.J., Shirakawa, H., Louis, E.J., Gau, S.C., and MacDiarmid, A.G. (1977) Electrical conductivity in doped polyacetylene. *Phys. Rev. Lett.*, **39** (17), 1098–1101.

75 Chiang, C.K., Gau, S.C., Fincher, C.R.J., Park, Y.W., MacDiarmid, A.G., and

Heeger, A.J. (1978) Polyacetylene, (CH)$_x$: n-type and p-type doping and compensation. *Appl. Phys. Lett.*, **33** (1), 18 (3 pages).

76 Chiang, C.K., Druy, M.A., Gau, S.C., Heeger, A.J., Louis, E.J., MacDiarmid, A.G., Park, Y.W., and Shirakawa, H. (1978) Synthesis of highly conducting films of derivatives of polyacetylene, (CH)$_x$. *J. Am. Chem. Soc.*, **100**, 1013–1015.

77 Ellis, J.R. (1986) *Handbook of Conducting Polymers*, 1st edn, T.A. Skotheim, Marcel Dekker, New York.

78 Heeger, A.J. (2001) Semiconducting and metallic polymers: the fourth generation of polymeric materials (Nobel Lecture). *Angew. Chem. Int. Ed*, **40**, 2591–2611.

79 Heeger, A.J. (2001) The fourth generation of polymeric materials Semiconducting and metallic polymers: Nobel Lecture. *Rev. Mod. Phys.*, **73** (3), 681–700.

80 Kathirgamanathan, P. (1993) Novel cable shielding materials based on the impregnation of microporous membranes with inherently conducting polymers. *Adv. Mater.*, **5** (4), 281–283.

81 Mattosso, L.H.C., Faria, R.M., Bulhoes, L.O.S., MacDiarmid, A.G., and Epstein, A.J. (1994) Synthesis, doping, and processing of high molecular weight poly(o-methoxyaniline). *J. Polym. Sci. Part A Polym. Chem.*, **32** (11), 2147–2153.

82 Nalwa, H.S. (1997) *Handbook of Organic Conductive Molecules and Polymers (Four Volumes)*, John Wiley & Sons, Inc., New York.

83 Saini, P., Choudhary, V., and Dhawan, S.K. (2007) LiSIPA doped polyaniline colloidal graphite composites: synthesis and characterization. *Indian J. Eng. Mater. Sci.*, **14** (6), 436–442.

84 Saini, P., Jalan, R., and Dhawan, S.K. (2008) Synthesis and characterization of processable polyaniline doped with novel dopant NaSIPA. *J. Appl. Polym. Sci.*, **108** (3), 1437–1446.

85 Shirakawa, H. (2001) The discovery of polyacetylene film: the dawning of an era of conducting polymers (Nobel Lecture). *Angew. Chem. Int. Ed.*, **40** (14), 2575–2580.

86 Shirakawa, H., Louis, E.J., MacDiarmid, A.G., Chiang, C.K., and Heeger, A.J. (1977) Synthesis of electrically conducting organic polymers: halogen derivatives of polyacetylene, (CH)$_x$. *Chem. Commum.*, 578–580.

87 Stafstrom, S., Bredas, J.L., Epstein, A.J., Woo, H.S., Tanner, D.B., Huang, W.S., and MacDiarmid, A.G. (1987) Polaron lattice in highly conducting polyaniline: theoretical and optical studies. *Phys. Rev. Lett.*, **59** (13), 1464–1467.

88 Wang, Z.H., Li, C., Scherr, E.M., MacDiarmid, A.G., and Epstein, A.J. (1991) Three dimensionality of "metallic" states in conducting polymers: polyaniline. *Phys. Rev. Lett.*, **66** (13), 1745–1748.

89 Javadi, H.H.S., Cromack, K.R., MacDiarmid, A.G., and Epstein, A.J. (1989) Microwave transport in the emeraldine form of polyaniline. *Phys. Rev. B*, **39** (6), 3579–3584.

90 Joo, J., Oblakowski, Z., Du, G., Pouget, J.P., Oh, E.J., Weisinger, J.M., Min, Y., MacDiarmid, A.G., and Epstein, A.J. (1994) Microwave dielectric response of mesoscopic metallic regions and the intrinsic metallic state of polyaniline. *Phys. Rev. B*, **49** (4), 2977–2980.

91 Trivedi, D.C. (1997) *Handbook of Organic Conductive Molecules and Polymers*, John Wiley & Sons, Ltd, Chichester, UK, p. 2.

92 Ting, T.H., Jau, Y.N., and Yu, R.P. (2012) Microwave absorbing properties of polyaniline/multi-walled carbon nanotube composites with various polyaniline contents. *Appl. Surf. Sci.*, **258**, 3184–3190.

93 Xu, J., Yao, P., Jiang, Z., Liu, H., Li, X., Liu, L., Li, M., and Zheng, Y. (2012) Preparation, morphology, and properties of conducting polyaniline-grafted multiwalled carbon nanotubes/epoxy composites. *J. Appl. Polym. Sci.*, **125** (S1), E334–E341.

94 Knott, E.F., Schaeffer, J.F., and Radar, M.T. (1993) *Cross Section Handbook*, Artech House, New York.

95 Pant, H.C., Patra, M.K., Verma, A., Vadera, S.R., and Kumar, N. (2006) Study of the dielectric properties of barium titanate–polymer composites. *Acta Mater.*, **54** (12), 3163–3169.

96 Xu, P., Han, X.J., Jiang, J.J., Wang, X.H., Li, X.D., and Wen, A.H. (2007) Synthesis and characterization of novel coralloid polyaniline/BaFe$_{12}$O$_{19}$ nanocomposites. *J. Phys. Chem. C*, **111** (34), 12603–12608.

97 Yang, C., Du, J., Peng, Q., Qiao, R., Chen, W., Xu, C., Shuai, Z., and Gao, M. (2009) Polyaniline/Fe$_3$O$_4$ nanoparticle composite: synthesis and reaction mechanism. *J. Phys. Chem. B*, **113** (15), 5052–5058.

98 Yang, C.C., Gung, Y.J., Hung, W.C., Ting, T.H., and Wub, K.H. (2010) Infrared and microwave absorbing properties of BaTiO$_3$/polyaniline and BaFe$_{12}$O$_{19}$/polyaniline composites. *Compos. Sci. Technol.*, **70** (3), 466–471.

99 Yang, C.C., Gung, Y.J., Shih, C.C., Hung, W.C., and Wub, K.H. (2011) Synthesis, infrared and microwave absorbing properties of (BaFe$_{12}$O$_{19}$ + BaTiO$_3$)/polyaniline composite. *J. Magn. Magn. Mater.*, **323** (7), 933–938.

100 Aphestegu, J.C., Damiani, A., DiGiovanni, D., and Jacobo, S.E. (2012) Microwave absorption behavior of polyaniline magnetic composite in the X-band. *Physica B*, **407** (16), 3168–3171.

101 Cho, H.S. and Kim, S.S. (1999) M-hexaferrites with planar magnetic anisotropy and their application to high-frequency microwave absorbers. *IEEE Trans. Magn.*, **35** (5), 3151–3153.

102 Iijima, S. (1991) Helical microtubules of graphitic carbon. *Nature*, **354**, 56–58.

103 Gupta, A. and Choudhary, V. (2011) Electromagnetic interference shielding behavior of poly(trimethylene terephthalate)/multi-walled carbon nanotube composites. *Compos. Sci. Technol.*, **71**, 1563–1568.

104 Bal, S. and Samal, S.S. (2007) Carbon nanotube reinforced polymer composites – a state of the art. *Bull. Mater. Sci.*, **30** (4), 379–386.

105 Singh, B.P., Prabha, Saini, P., Gupta, T., Garg, P., Kumar, G., Pandey, I., Pandey, S., Seth, R.K., Dhawan, S.K., and Mathur, R.B. (2011) *J. Nanopart. Res.*, **70** (12), 7065–7074.

106 Singh, B.P., Choudhary, V., Saini, P., and Mathur, R.B. (2012) Designing of epoxy composites reinforced with carbon nanotubes grown carbon fiber fabric for improved electromagnetic interference shielding. *AIP Adv.*, **2**, 6.

107 Cheng, Q., Bao, J., Park, J., Liang, Z., Zhang, C., and Wang, B. (2009) High mechanical performance composite conductor: multi-walled carbon nanotube sheet/bismaleimide nanocomposites. *Adv. Funct. Mater.*, **19**, 3219–3225.

108 Al-Hartomy, O.A., Al-Ghamdi, A., Dishovsky, N., Shtarkova, R., Iliev, V., Mutlay, I., and El-Tantawy, F. (2012) Dielectric and microwave properties of natural rubber based nanocomposites containing graphene. *Mater. Sci. Appl.*, **3** (7), 453–459.

109 Geim, A.K. (2009) Graphene: status and prospects. *Science*, **324** (5934), 1530–1534.

110 Stankovich, S., Dmitriy, A.D., Geoffrey, H.B.D., Kevin, M.K., Zimney, E.J., Stach, E.A., Piner, R.D., Nguyen, S.T., and Ruoff, R.S. (2006) Graphene-based composite materials. *Nature*, **442**, 282–286.

111 Stankovich, S., Dikin, D.A., Piner, R.D., Kohlhaas, K.M., Kleinhammes, A., Jia, Y., Wu, Y., SonBinh, T.N., and Rodney, S.R. (2007) Synthesis of graphene-based nanosheets via chemical reduction of exfoliated graphite oxide. *Carbon*, **45** (7), 1558–1565.

112 Zhang, H.-B., Yan, Q., Zheng, W.-G., He, Z., and Yu, Z.-Z.T. (2011) Tough graphene-polymer microcellular foams for electromagnetic interference shielding. *ACS Appl. Mater. Interfaces*, **3** (3), 918–924.

113 Meyer, J.C., Geim, A.K., Katsnelson, M.I., Novoselov, K.S., Booth, T.J., and Roth, S. (2007) The structure of suspended graphene sheets. *Nature*, **446** (7131), 60–63.

114 In, K.M., Junghyun, L., Ruoff, R.S., and Lee, H. (2010) *Nat. Commum.*, **1**, 73–79. doi: 10.1038/ncomms1067

115 Ramanathan, T., Abdala, A.A., Stankovich, S., Dikin, D.A., Herrera-Alonso, M., Piner, R.D., Adamson, D.H., Schniepp, H.C., Chen, X., Ruoff, R.S., Nguyen, S.T., Aksay, I.A., Prud'Homme, R.K., and Brinson, L.C. (2008) *Nat. Nanotechnol.*, **3**, 31.

Index

a

absorbance 215
absorbers, electromagnetic wave 226
absorption loss 213
absorption-transferring process 220, 221
acidity, Lewis 80
acids
– acid-catalyzed curing 168
– layered silicic 132
– tannic 196
acrylic matrix 183
acrylic rubber 119
activated monomer (AM) mechanism 22
activation energy 80, 82, 83
acyphosphine oxide 50
adsorbents, HPBs 157–161
adsorption capacity, hydrogels 173
adsorption isotherms 160, 161
agglomeration 80
alcoholysis 193
alignment, clay platelets 120
aliphatic diamine 116
aliphatic epoxy monomer 20
aliphatic epoxy resins 197
aliphatic HPBs 148
aliphatic polyester (APES) 144
alkoxydes 32, 33
alkoxysilane 18
alkyl ammonium ions, primary 110
alkyl groups, hydroxy 51, 52
aluminosilicates, MMT, see montmorillonite
amine catalysts, functional 58
3-aminopropyltrimethoxysilane (APTMS) 88, 92, 94
ammonium cations, quaternary 58
ammonium ions 9
– primary alkyl 110
anhydride curing 168
anhydride matrix, epoxy/ 123

antimicrobial coatings 179
antimicrobial surfaces 149–157
AOETMA 50
APDMMS 50
Arrhenius relationship 72
aspect ratio
– SWCNTs 228
– vermiculites 6, 7

b

Bacillus subtilis 156, 157, 179
bacteria 155, 179
basal plane spacing 11
benzoxazine, quaternized monomers 170
bilayer, local 9
bio-based epoxy/clay nanocomposites 189–209
bio-based polyphenol hardener 196
bio-derived clay–UP nanocomposites 144
biochemical oxygen demand 204
biocidal agents 155
biodegradability 204
– nanocomposites 144
bisphenol, limonene 195
bisphenol A diglycidyl ether 2, 117, 123, 189
blending 220, 221
– melt 166
boehmite 29
bottom-up UV-cured epoxies 32–35
branched polyol 53
branching, HPBs 147
bridge interaction 172
bulk-initiated method 175
butadiene rubbers 181
butanediol 46
BzC16 3, 8–11
BzC18OH 8–11

c

C18-Swy MMT 44, 45
carbon nanotubes (CNTs) 223–229
– MWCNTs, see multiwalled carbon nanotubes
– SWCNTs, see single-walled carbon nanotubes
catalysts
– functional amine 58
– Grubbs 174, 175
– petroleum-based 190
cation exchange capacity (CEC) 4, 6–8
cationic disinfectants 155
cationic polymerization 176
– ring-opening 21, 168
cations
– exchangeable 45
– organoammonium 43
– quaternary ammonium 58
– soft 43
– , see also ions
CE (3,4-epoxycyclohexylmethyl-3′,4′-epoxycyclohexane carboxylate) 23
cell phones 226
cellulose
– decomposition 92–95
– nanocomposites 76–88
ceramic fillers 19
chain extenders 41
chain transfer agent 168
char yield 136–139
chemical-controlled curing 84
chemiluminescence (CL) emission 154, 155
chemistry
– colloidal 17
– PUs 41
– thiol-ene 180
chitosan 177
clay minerals 131–135
clay nanocomposites 28
– bio-based 144, 189–209
– epoxy– 117
– ESO 113
– HPBs 147–163
– mechanical performance 109–128
– phenol–formaldehyde/ 169
– polybenzoxazine/ 169, 170
– polysulfone/ 174
– preparation 166–176
– PU– 39–67
– SMC 113
– UPR hybrid 129–146
clays
– dispersions 39–67
– excess 203
– exchangeable cations 45
– intragallery space 56
– Lewis acidity 80
– organo-, see organoclays
– organophilic character 153
– platelet alignment 120
– polar 120
– SMC 113
– smectite 132
– surface-modified 133–135, 147–163
– tensile modulus 56, 57
click reaction, cycloaddition 182
Cloisite 30B 26, 27, 46
coatings
– antimicrobial 179
– epoxy 34
– epoxy–vermiculite nanocomposites 2
coercivity 159
colloidal chemistry 17
complex permittivity 222, 226
composite films 5, 6
condensation
– ceramic fillers 19
– sol–gel process 32
conducting nanocomposites 211–237
conductivity, electrical 215
conjugated polymers 219
contact-specific surface area 115
conventional thermogravimetry (CTG) 90–95
creep-resistant materials 1
critical load 27
critical strain, relative 119, 120
critical stress, relative 122
cross-link density 130, 136–139
cross-linking
– conducting nanocomposites 216
– epoxide groups 1
– polymerization 174
Cr(VI) water treatment 157–161
CTG (conventional thermogravimetry) 90–95
cure kinetics 72–74, 76–88
cured epoxies
– bio-based 197–200
– UV- 17–37
curing
– acid-catalyzed 168
– diffusion-controlled 84
– layer-by-layer 24
cycloaddition click reaction 182
cycloaliphatic epoxy monomers 20

d

deagglomeration 3
decomposition, cellulose 92–95
decomposition peak 90, 91
deformation mechanism 116
degradation kinetics 72–74, 88–98
degree of conversion 72, 83
delamination 3, 4
– time 8
dendritic polymers 147
density, cross-link 130, 136–139
derivative TG (DTG) 90–93
dialkoxycarbenium ions 23
diamine, aliphatic 116
dicyandiamide (DICY) 189, 190
dicycloaliphatic epoxides 30
dielectric constant 221
dielectric permittivity 215
differential scanning calorimetry (DSC) 70
– temperature-modulated 74, 75, 84–88
diffusion-controlled curing 84
diglycidyl ester of dimer aid (DGEDA) 191, 192
diglycidyl ether of bisphenol A (DGEBA) 2, 117, 123, 189
diglycidyl ether of isosorbide 192
diisocyanates 41, 46
– PMDI 48
diol-functionalized MMT, PU chains 171
4,4'-diphenyl methane diisocyanate (MDI) 41, 53–56
direct photoinitiation method 28
disinfectants, cationic 155
disordered morphology 115
dispersed phase, dimensions 17
dispersions
– clay 39–67
– PU 43
distance
– interlayer 44
– , see also spacing
DMF 11
double bond conversion 152
d-spacing 11
– HPBs 150
DTG (derivative TG) 90–93
dynamic loss processes 213
dynamic mechanical analysis (DMA) 98–100
dynamic vitrification temperature 86, 87
dynamical mechanical thermal analysis (DMTA) 111–114

e

EGPs (exfoliated graphite platelets) 31
elastic modulus 114, 136–139
elastomers, thermoplastic polyurethane 42, 43
electrical conductivity 215
electrical properties 211–237
electromagnetic interference (EMI) shielding 211–237
electromagnetic wave absorbers 226
electron transfer and polymerization process 180
elongation
– nanocomposites 100, 101
– at failure 136–139
end groups, vinylic 148
epichlorohydrin (ECH) 189
epoxides
– dicycloaliphatic 30
– photopolymerization 20–23
epoxidized lignocresol 194
epoxidized linseed oil (ELO) 191, 192
epoxidized soybean oil (ESO) 191, 204, 205
– clay nanocomposites 113
epoxidized vegetable oil 204, 205
epoxies
– (cyclo)aliphatic monomers 20
– coatings 34
– epoxy–clay nanocomposites 117
– epoxy–vermiculite nanocomposites 1–16
– epoxy/anhydride matrix 123
– hardeners 190–200
– UV-cured 17–37
epoxy functions 20
epoxy group, dialkoxycarbenium ions 23
epoxy resins 2, 167
– aliphatic 197
– bio-based 189–209
– graphene/epoxy composites 230, 231
– PANI/ 220–222
3,4-epoxycyclohexylmethyl-3',4'-epoxycyclohexane carboxylate (CE) 23
esters
– DGEDA 191, 192
– vinyl 143
ethers
– bisphenol A diglycidyl 2
– DGEBA 2, 117, 123, 189
– diglycidyl ether of isosorbide 192
– glycidyl 20
– hexanedioldiglycidyl 34
– PGPE 198–204
excess clay 203
exchangeable cations 45

exfoliated graphite platelets (EGPs) 31
exfoliated morphology 26, 134
exfoliated platelets 7
exfoliated structure 54, 167
exothermic peak temperatures 76, 77

f

far-field conditions 212
feed ratio 198, 199
fillers 5, 6
– ceramic 19
– conducting network 218
– micro- 109
– modified 10
– nano-, *see* nanofillers
– PU-nanocomposites 43
films, composite 5, 6
fire growth rate index (FIGRA) 141
flame retardance 141–144
flexural modulus 136–139
foams, PU 43
formaldehyde-based nanocomposites 69–108
– phenol/clay 169
formaldehyde-based thermosetting polymers 69
fracture toughness 120–123, 136–139
free radical polymerization 176
Freundlich process 161
friction, molecular 213
FT-IR analysis, MWCNTs 88–90
functional amine catalysts 58
functionalities, polar 47
functionalized graphene, solution processable 230, 231
functionalized graphene sheets (FGS) 30, 31
functionalized nanoparticles 30

g

gallery onium ions 118
gel-like structure 135
gelation 78
glass-transition temperature 112, 114, 136–139
– HPBs 152
glassy matrix 112, 116, 117
glassy state 86
glucose oxidase 183, 184
glycidyl ether 20
gold nanocomposites 181–184
Gram-negative bacteria 155, 179
Gram-positive bacteria 179

graphene 71
– FGS 30, 31
– nanocomposites 229–231
graphite 229
graphite platelets, EGPs 31
green nanocomposites 168
groups
– epoxy 23
– ester carbonyl 23
– hydroxy alkyl 51, 52
– hydroxy ethyl 47, 48
– hydroxyl methyl 89
– methylol 54
– vinylic end 148
Grubbs catalyst 174, 175

h

hard matrix 112
hardeners, epoxy 190–200
heat capacity 85, 86
heat flow 72, 75
– nonreversing 85, 86
– reversing 85, 86
heat of reaction 78
HEMA 151–157
hexanedioldiglycidyl ether (HDGE) 34
H_2O_2 treatment 3
hybrid materials
– conducting nanocomposites 211
– organic–inorganic 18
hybrid nanocomposites
– toughened UP resin 142
– UPR/clay 129–146
hydrogels
– adsorption capacity 173
– thermoresponsive 182
hydrolysis
– ceramic fillers 19
– sol–gel process 32
hydroperoxides 154
hydrophilicity 159
hydroxy alkyl groups 51, 52
hydroxy ethyl groups 47, 48
hydroxyl methyl groups 89
hyperbranched polymers (HPBs) 147–163
– structure 150

i

immersion, monomer 167
impact strength 136–139
in situ polymerization 19, 166
– intercalative 131
in situ template synthesis 131

inorganic volume fraction 4
integrity 228
intercalated morphology 25, 134
intercalated structure 54, 167
intercalation 170
interfaces, organic inorganic 111
interference, electromagnetic 211–237
intergallery surface-initiated method 175
interlamellar spacing 117, 153
Interlayer distance 44
internal reflections, multiple 212–214
interparticle contact 218
intragallery space 56
intrinsically conducting polymers (ICPs) 211, 218–223
iodonium ions 190
ion exchange 4, 5
ions
– ammonium, *see* ammonium ions
– dialkoxycarbenium 23
– gallery onium 118
– iodonium 190
– sulfonium 190
– , *see* also cations
iron oxide nanoparticles 30
isobutylene–isoprene rubbers 181
isoconversional analysis, CTG 94, 95
isoconversional equation 74
isoconversional method 83, 84
isocyanates 44
isodimensional nanoparticles 39
isosorbide, diglycidyl ether 192

k
kayak 24
kinetics
– cure 72–74, 76–88
– degradation 72–74, 88–98
– sol–gel process 33
– thermal 72–100
Kissinger method 74, 80–83
kraft lignin 193

l
lamellar silicates 25
lamination, "wet edge" 25
laws and equations
– Arrhenius relationship 72
– degree of conversion 72
– isoconversional equation 74
– total shielding effectiveness 212
layer-by-layer curing 24
layered silicic acids 132

layered structure 131, 132
layers, graphene 229
Lewis acidity, clays 80
light-induced synthesis 178, 179
lignocresol, epoxidized 194
limonene bisphenol (LBP) 195
limonene diepoxide (LMDE) 191
linear heating rate 75
linseed oil, epoxidized 191, 192
liquid composite molding (LCM) 141
liquid-rubbery state 86
load, critical 27
local bilayer 9
localized reactions 24
loop interaction 172
loss, absorption 213

m
magnetic permeability 215
magnetic sorbents 157
maleinization 27
MAOETMA 50
MAPDMMS 50
materials 39
– EMI shielding 215, 216
– thermosetting polymeric 1
matrices
– acrylic 183
– epoxy/anhydride 123
– glassy 112, 116, 117
– hard 112
– interactions with nanoclays 172
– particle/matrix interaction 115
– polymeric 17
– rubbery 112, 116, 117
– soft 112
– UP 110
MDI (4,4'-diphenyl methane diisocyanate) 41, 53–56
mean particle size 5, 6
mechanical analysis
– DMA 98–100
– DMTA 111–114
mechanical properties 100–103, 109–128, 135–141
melamine-formaldehyde (MF) resins 69, 70
melt blending 166
melt intercalation 131
metal nanocomposites
– polysulfone/ 184, 185
– preparation 176
– UV-cured 35
metal salts 178

methacrylates 49
methods and techniques
– AM mechanism 22
– direct photoinitiation method 28
– isoconversional method 83, 84
– Kissinger method 74, 80–83
– LCM 141
– model-free kinetics method 83
– multiheating rate methods 73, 82
– Ozawa method 73, 80–83
– sequential/simultaneous mixing 173
– steam explosion method 194
methyl modifiers 48
methylol groups 54
MF resin/clay/cellulose nanocomposites, cure kinetics 76–80
microfillers 109
microhardness 157
microstructure, fillers 5, 6
milling 3, 7
minerals, clay 131–135
MIRs (multiple internal reflections) 212–214
mixing, sequential/simultaneous 173
modelfree kinetics method 83
modifications, organic 39–67
modified fillers 10
modifiers, (non)reactive 43, 44
modulated TG (MTG) 75, 76, 95–98
modulus
– elastic 114, 136–139
– flexural 136–139
– relative Young's 117
– storage 98, 99, 112, 113, 136–139, 202
– tensile 56, 57, 103, 198, 199
– UTM 136–139
– Young's 114–118, 136–139
molding, liquid composite 141
molecular friction 213
monolayer adsorption 161
monomers
– immersion 167
– multifunctional 147
– quaternized 170
montmorillonite (MMT) 40
– C18-Swy 44, 45
– diol-functionalized 171
– epoxy–vermiculite nanocomposites 2
– Na–MMT 200, 201
– OMM 204
– OMMT 134, 135
– organic-MMT content 121
– polydicyclopentadiene/MMT nanocomposites 175

– sodium 27, 28, 133, 171
– UP–MMT nanocomposites 140
morphology 165
– disordered 115
– exfoliated 26, 134
– intercalated 25, 134
MUF (melamine-urea-formaldehyde) resins 69, 70
multifunctional monomers 147
multiheating rate methods 73, 82
multiple internal reflections (MIRs) 212–214
multistep reactions 96
multiwalled carbon nanotubes (MWCNTs) 30, 31, 71, 72, 223–225, 227
– PF resin/MWCNT/cellulose nanocomposites 88–100
– surface-modified 88–90

n
nanocomposites
– biodegradable 144
– clay, *see* clay nanocomposites
– CNT-based 223–229
– conducting 211–237
– epoxy–vermiculite 1–16
– formaldehyde-based 69–108, 169
– graphene-based 229–231
– green 168
– hyperbranched polymers 147–163
– ICP-based 218–223
– magnetic behavior 159
– metal 176
– MF resin/clay/cellulose 76–80
– PF resin/clay/cellulose 80–88
– PGPE 198–204
– polydicyclopentadiene/MMT 175
– polymer, *see* polymer nanocomposites
– polysulfone/metal/ 184, 185
– precursors 40
– preparation 4, 5, 165–188
– PU–clay 39–67
– rubber/silver 181
– thermal stability 91
– UP–MMT 140
– UV-cured epoxies 17–37
– UV-cured metal 35
– VEO 143
nanofillers
– EMI shielding 215, 217, 218
– HPBs 148
– reinforcing 109

nanoparticles
– functionalized 30
– iron oxide 30
– isodimensional 39
– , see also particles
nanosheets 17, 39, 40
nanotubes 17, 39, 40
– CNTs 223–229
– MWCNTs, see multiwalled carbon nanotubes
– SWCNTs, see single-walled carbon nanotubes
nanowhiskers 17, 39, 40
network
– conducting 218
– percolating 224
network analyzers, SNA/VNA 214
nonreactive modifiers 43, 44
nonreversing heat flow 75, 85, 86

o
octadecylammonium-modified montmorillonite (OMM) 204
octahedral sheet 131, 133
one-tail modifiers 48
onium ions, gallery 118
organic–inorganic hybrids 18
organic inorganic interface 111
organic-MMT content 121
organic modifications 39–67
organically modified MMT (OMMT) 134, 135
organoammonium cations 43
organoclays 44
– modifier structures 55
organophilic boehmite 29
organophilic character, treated clays 153
organovermiculites 10
oscillatory temperature forcing function 76
OX-MWCNTs 92, 97
Ozawa method 73, 80–83

p
PANI–DBSA 220, 221
PANI–PTSA 223
particles
– mean size 5, 6
– particle/matrix interaction 115
– , see also nanoparticles
PBS (poly(butylene succinate)) 53
peak, exothermic 76, 77
peak heat release rate (PHRR) 141
PEGDMA 151–157
PEI (polyethyleneimine) 158–161

percolating network 224
permeability 215
– relative 212
permittivity
– complex 222, 226
– dielectric 215
petroleum-based catalysts 190
PF resin/clay nanocomposites 169
PF resin/clay/cellulose nanocomposites, cure kinetics 80–88
PF resin/MWCNT/cellulose nanocomposites 88–100
phenol-formaldehyde (PF) resins 69
phenolic resins 168, 169
photoinitiation 28
photopolymerization 149
– epoxides 20–23
photosensitizers 21
plasticizing effect 118
plastics, thermosetting 130
platelets
– alignment 120
– EGPs 31
– exfoliated 7
– thickness 132
PNCs (polymer nanocomposites) 39–41
– UV-cured epoxies 17–37
polar clays 120
polar functionalities 47
polyamines 189
polyaniline 219
polybenzoxazine/clay nanocomposites 170
poly(butylene succiate) (PBS) 53
polydicyclopentadiene/MMT nanocomposites 175
polyester
– aliphatic 144
– resins 129
polyethyleneimine (PEI) 158–161
polyglycerol polyglycidyl ether (PGPE) 198–204
polymer–gold nanocomposites 181–184
polymer-layered silicate nanocomposites 109
polymer nanocomposites (PNCs) 39–41
– UV-cured epoxies 17–37
polymer–silver nanocomposites 177–181
polymeric 4,4′-diphenyl methane diisocyanate (PMDI) 48
polymeric materials, thermosetting 1
polymeric matrices 17
polymerization
– and electron transfer 180
– cationic 176

– cationic ring-opening 21, 168
– cross-linking 174
– free radical 176
– *in situ* 19, 166
– intercalative 131
– radical-initiated 22
– RAFT 182, 183
polymers
– dendritic 147
– formaldehyde-based thermosetting 69
– hyperbranched 147–163
– ICPs 211, 218–223
– thermoset 165
polyols 41
– branched 53
polyphenol hardener 196
polysulfone/clay nanocomposites 174
polysulfone/metal nanocomposites 184, 185
poly(tetramethylene glycol) 53, 54
polyurethanes (PUs)
– clay nanocomposites 39–67
– dispersions 43
– foams 43
– interaction with MMT 171, 172
– thermoplastic elastomers 42, 43
– thermoset 42, 43
precursors, nanocomposite 40
preparation
– clay nanocomposites 166–176
– metal nanocomposites 176
– nanocomposites 4, 5
– , *see also* synthesis
primary alkyl ammonium ions 110
Pseudomonas aeruginosa 156, 157, 179
PUs, *see* polyurethanes
2-pyrone-4,6-dicarboxylic acid 195

q

quaternary ammonium cations 58
quasi-Newtonian behavior 29
quaternized benzoxazine 170

r

radical-initiated polymerization 22
RAFT polymerization 182, 183
reaction rate 72
reactions
– alcoholysis 193
– condensation 19, 32
– cycloaddition 182
– hydrolysis 19, 32
– localized 24
– maleinization 27

– multistep 96
– polymerization, *see* polymerization
reactive modifiers 43, 44
reflectance 213
reflection coefficient 224
reflection shielding 212, 213
reinforced UP nanocomposites 130, 131
reinforcing nanofillers 109
relative critical strain 119, 120
relative critical stress 122
relative permeability 212
relative Young's modulus 117
resins
– aliphatic epoxy 197
– epoxy, *see* epoxy resins
– MF 69, 76–80
– PANI/epoxy resin 220–222
– PF 69, 70, 80–100
– phenolic 168, 169
– polyester 129
– toughened UP 142
– UP 129–146
resole PF resins 84–86
– tensile strength 102, 103
– viscosity 88
resorcinol-formaldehyde (RF) resins 69
reversing heat capacity 85, 86
reversing heat flow 85, 86
rigidity 114–118
rubber
– acrylic 119
– rubbery matrices 112, 116, 117
– silicon 181
– silver nanocomposites 181

s

salts, metal 178
scalar network analyzers (SNA) 214
scratch resistance test 27, 28
sequential mixing 173
sheets
– FGS 30, 31
– graphene 229
– nano- 17, 39, 40
– octahedral 131, 133
– tetrahedral 131, 133
shielding, EMI 211–237
shielding effectiveness, total 212
silane-modified clay (SMC) 113
silane treatment 121
silicate nanocomposites, polymer-layered 109
silicates, lamellar 25
silicic acids, layered 132

silicon rubbers 181
silver nanocomposites 177–181
simultaneous mixing 173
single-walled carbon nanotubes (SWCNTs) 71, 223, 225–228
sinusoidal modulation 75
skin depth 213
smectite clays 132
sodium borohydride 177
sodium dodecyl sulfate (SDS) 90
sodium montmorillonite (Na–MMT) 27, 28
– PGPE 200, 201
– PU chains 171
– structure 133
soft cations 43
soft matrix 112
sol–gel process 32, 33
solution intercalation 131
solution processable functionalized graphene (SPFG) 230, 231
sonication 4
sorbents, magnetic 157
sorbitol polyglycidyl ether (SPE) 191
sorption behavior, UP nanocomposites 142
soybean oil, epoxidized 113, 191, 204, 205
spacing
– basal plane 11
– interlamellar 117, 153
– , see also distance
d-spacing 11
– HPBs 150
sporeforming bacteria 179
stability, thermal 91
Staphylococcus aureus 179
steam explosion method 194
storage modulus 98, 99, 112, 113, 136–139
– temperature dependency 202
strain
– at break 118–120
– relative critical 119, 120
stress
– at break 120–123
– relative critical 122
– yield 135
structural control 147
structure
– bio-based epoxy resins 191
– CE 23
– conducting polymers 219
– epoxy functions 20
– exfoliated 54
– gel-like 135
– HPBs 150

– intercalated 54
– intercalated/exfoliated 167
– layered 131, 132
– modifiers 48, 49
– Na–MMT 133
– structure–property relationship 39
– surfactants 51
sulfonium ions 190
surface area, contact-specific 115
surface-modified clays 133–135
– hyperbranched polymers 147–163
surface-modified MWCNTs 88–90
surface-modified vermiculites 10
surfaces, antimicrobial 149–157
surfactants
– SDS 90
– structure 51
surfmers 149
suspensions, fillers 11
SWCNTs (single-walled carbon nanotubes) 71, 223, 225–228
swelling 142
– organoclays 44
synthesis
– conducting nanocomposites 216, 217
– light-induced 178, 179

t

tactoids 26
tail interaction 172
tannic acid (TA) 196
temperature
– dynamic vitrification 86, 87
– glass transition 112, 114, 136–139
temperature-modulated DSC (TMDSC) 74, 75, 84–88
in situ template synthesis 131
tenacity 123
tensile modulus 103, 198, 199
– clays 56, 57
– ultimate 136–139
tensile strength 101–103, 198, 199
– UTS 136–139
TETA (triethylenetetramine) 189, 205
tetraethoxy-orthosilicate (TEOS) 33, 34
tetraethylenepentamine (TEPA) 2, 9
tetrahedral sheet 131, 133
theories and models
– deformation mechanism 116
– EMI shielding 212–214
– TMDSC 74, 75
thermal kinetic analysis 95–98
thermal properties

– clay–UP nanocomposites   135–141
– formaldehyde-based nanocomposites   69–108
thermal stability   91
thermogravimetric analysis (TGA)
– epoxy–vermiculite nanocomposites   5, 9, 10
– HPBs   150, 151
thermogravimetry
– conventional   90–95
– DTG   90–93
– MTG   75, 76, 95–98
thermoplastic polyurethane elastomers (TPUs)   42, 43
thermoresponsive hydrogels   182
thermoset formaldehyde-based nanocomposites   69–108
thermoset polymers   1, 165
– formaldehyde-based   69
thermoset PUs   42, 43
thermosetting plastics   130
thiol-ene chemistry   180
top-down UV-cured epoxies   24–31
total reaction heat   72
total shielding effectiveness   212
toughened UP resin   142
toughness, fracture   120–123
transmittance   213
triethylenetetramine (TETA)   189, 205
trimethylol propane (TMP)   41, 54
two-port VNA   214
two-tail modifiers   48

u

ultimate tensile modulus (UTM)   136–139
ultimate tensile strength (UTS)   136–139
unsaturated polyester resin (UPR), hybrid clay nanocomposites   129–146
unsaturated polyester (UP) matrix   110
urea-formaldehyde (UF) resins   69
UV-cured epoxies   17–37
UV-cured metal nanocomposites   35

v

vector network analyzers (VNA)   214
vegetable oil, epoxidized   204, 205
vermiculites
– epoxy–vermiculite nanocomposites   1–16
– surface-modified   10
vinyl ester oligomer (VEO) nanocomposites   143
vinylic end groups   148
viscoelasticity analysis   111–114
viscosity
– aliphatic epoxy resins   197
– resole PF resins   88
vitrification   78, 86

w

water treatment, Cr(VI)   157–161
WAXRD
– epoxy–vermiculite nanocomposites   5, 11–13
– organoclay swelling   44
"wet edge lamination"   25
whiskers, nano-   17, 39, 40
wood chip   193

x

X-linking, *see* cross-linking
X-ray diffractometry (XRD)
– Cloisite 30B   26, 27
– epoxy–vermiculite nanocomposites   5, 11–13

y

yield stress   135
Young's modulus   114–118, 136–139
– relative   117